The role of legumes in conservation tillage systems

J. F. Power, editor

The proceedings of a national conference
University of Georgia, Athens
April 27-29, 1987

SOIL
CONSERVATION
SOCIETY
OF AMERICA

Published by the

Soil Conservation Society of America

with the support of the
Agricultural Research Service/U.S. Department of Agriculture
National Fertilizer Development Center/Tennessee Valley Authority

The role of legumes in conservation tillage systems

$12.00

ISBN 0-935734-15-5

Contents

INSECTS AND DISEASES

CROPPING PRACTICES

WEED CONTROL

EROSION AND PRODUCTIVITY

ECONOMICS

Preface

The energy crisis of the late 1970s raised the question of the wisdom of depending upon commercial fertilizers as the primary source of nitrogen for crop production—a source almost entirely dependent upon the volatile supply and demand status of the world market for natural gas. This question encouraged many scientists to rethink the N-supply picture and to initiate new research to evaluate legumes as alternative N sources.

Legumes also have other benefits, including potentially less soil erosion; improved surface water and groundwater quality; increased soil organic matter status; and, presumably, enhanced infiltration, surface stability, and soil N availability. To encourage new research and to summarize existing information, the U.S. Department of Agriculture's Agricultural Research Service held a conference on legumes in conservation tillage systems at Lincoln, Nebraska, in 1983. Scientists from throughout the United States, from both ARS and the state experiment stations, summarized a number of aspects of the use of legumes in conservation tillage systems. It was obvious at that meeting that appreciable research on the subject had been completed at only a few locations in recent years. However, many scientists were or had recently initiated new research. One recommendation of that workshop was that a more comprehensive symposium be held at a time when results from much of the new research could be presented.

Thus was born planning for this comprehensive symposium, held at Athens, Georgia, in April 1987 to coincide with the 50th anniversary of the USDA-ARS Southern Piedmont Conservation Research Station at Watkinsville, a center with a long record of research on legumes in conservation tillage systems. We are grateful that the Soil Conservation Society of America, its leaders, and members believed the subject matter of the conference was worthy of its sponsorship. Other cosponsors included the American Society of Agronomy, the Conservation Tillage Information Center of the National Association of Conservation Districts, the National Fertilizer Development Center of the Tennessee Valley Authority, the University of Georgia Center for Continuing Education, and ARS. We particularly wish to thank the staff and members of the Soil Conservation Society of America for publicizing the symposium, working with the Georgia Center in organizing and arranging the symposium, and for the publication of the papers presented in this volume.

ARS, the University of Georgia, and the National Fertilizer Development Center made substantial financial contributions to support this symposium. We also thank our various commercial exhibitors, some of whom have made financial contributions for various purposes. We wish to thank as well the invited and volunteer contributors of papers to this symposium. Their cooperation was outstanding.

Members of the Organizing Committee:
J. F. Power, USDA-ARS, Lincoln, Nebraska (chairman)
F. C. Boswell, University of Georgia, Experiment
R. R. Bruce, USDA-ARS, Watkinsville, Georgia
L. D. Davis, SCSA, Ankeny, Iowa
M. W. Frere, USDA-ARS, Watkinsville, Georgia
W. W. Frye, University of Kentucky, Lexington
W. L. Hargrove, University of Georgia, Experiment
Arnold King, USDA-SCS, Ft. Worth, Texas
G. W. Langdale, USDA-ARS, Watkinsville, Georgia
Max Schnepf, SCSA, Ankeny, Iowa
T. N. Shiflet, USDA-SCS, Lincoln, Nebraska

Introduction

J. F. Power

Throughout most of the history of agriculture, legumes and animal manure have been the primary source of N in crop production systems. This situation has changed drastically in the last several decades with the widespread use of commercial N fertilizer. With increased N fertilizer usage in the United States came a corresponding decrease in forage legume seed production. In 1960 legume seed production was more than 160,000 tons/year and N fertilizer usage was just over 2 million tons/year. By 1980 N fertilizer usage had risen to nearly 9.5 million tons/year while legume seed production had dropped to less than 70,000 tons/year. Fertilizer N has largely replaced biologically fixed N as a source of N for production of grain crops.

In recent years fertilizer N usage has exceeded the total quantity of N removed in crops harvested annually in the United States. Nitrogen input into the crop production system from all sources—fertilizer; crop residue; manure; and symbiotic fixation, including that by soybeans—may be more than two times greater than N removal in the harvested crops. It is of no surprise that Americans are concerned about the environmental effects of excess N, including the accumulation of nitrates in surface water and groundwater and the depletion of ozone in the atmosphere by reaction with nitrous oxides. These questions of controlling N in the environment on a national scale are of recent origin and parallel the decreased use of legume-fixed N as the primary N source for crop production.

Substitution of fertilizer N for biologically fixed N is a major contributor to the trend toward more extensive use of monocultures and simple rotations. This trend favors production of grain crops, and without special protection the soil is more vulnerable to erosion than it would be in legume-based cropping systems. The increased soil erosion potential resulting from widespread use of monocultures has led to the development of conservation tillage techniques. Farmers have used conservation tillage increasingly within the last decade because it controls soil erosion better and because it reduces production costs. Use of reduced tillage or no-till techniques, however, creates problems for operators who wish to use legumes in cropping systems.

The sustainability of crop production systems has been another concern expressed in the last decade. Almost all fertilizer N is derived from natural gas, which is a finite resource subject to widely fluctuating prices. What, then, is the wisdom of being so dependent on fertilizer N? Could new methods of incorporating legumes into crop production systems lessen this dependence on a highly volatile fertilizer N market while also maintaining an economically acceptable level of crop production?

These are some of the new questions that have arisen during the last several decades—questions for which scientists find few answers in old data bases of research on the more conventional legume-based crop rotations. Technology now permits widespread use of reduced tillage, no-till, double or multiple cropping, cover crops, and other developments not available to preceding generations of farmers. Can farmers economically integrate legumes into this new technology to effectively reduce risks from N in the environment, soil erosion, and unstable world supplies and prices of fertilizer N? These are some of the questions addressed by authors in this volume.

J. F. Power is the symposium chairman and research leader, Agricultural Research Service, U.S. Department of Agriculture, Agronomy Department, University of Nebraska, Lincoln.

THE NEED

The need for legume cover crops in conservation tillage production

W. L. Hargrove and W. W. Frye

Using legumes in crop rotations to enhance soil fertility and crop production is among the oldest of agricultural management practices. Writers in ancient Greece and Rome documented the use of such legumes as faba bean (*Vicia faba*), vetch (*Vicia* species), and lupines (*Lupinus* species) as important cover crops or rotation crops with grains. Planting legumes in rotations, along with manuring and marling, formed the basis of soil fertility maintenance for several centuries in Europe.

Colonial Americans, however, seldom used legumes, liming, or crop rotations to improve the soil because the newly cleared land was generally fertile. After the land's fertility was depleted, settlers moved on to clear and bring into production new land. But between 1880 and World War II, legumes became an important component of crop rotations, especially in the South. Research on the evaluation and management of legumes was a major thrust of several of the southern agricultural experiment stations in the early 1900s (*1, 4*). After World War II, inexpensive and abundant fertilizer-N diminished the role of N-fixing plants in cropping systems. But researchers continued to study the role of N-fixing plants in soil and water conservation during the 1950s and 1960s (*9, 17, 29*). Until recently, there had been little research since 1960 on winter legumes in terms of their contribution to soil and crop management.

W. L. Hargrove is an associate professor in the Agronomy Department, Georgia Agricultural Experiment Station, Experiment, Georgia 30212. W. W. Frye is a professor in the Agronomy Department, Kentucky Agricultural Experiment Station, Lexington, Kentucky. This paper is a contribution from the Agronomy Department, Georgia Agricultural Experiment Station, Experiment, Georgia 30312, and the Agronomy Department, Kentucky Agricultural Experiment Station, Lexington, Kentucky 40546. Supported by state and Hatch funds allocated to the Georgia Agricultural Experiment Station and the Kentucky Agricultural Experiment Station.

Escalating costs of fertilizer-N and concern over soil erosion have rekindled interest in legumes and their role in cropping systems. What role, then, can legumes play in modern conservation tillage production systems?

Soil conservation

Soil erosion continues to be a serious problem on U.S. cropland. According to the U.S. Department of Agriculture's recent National Resources Inventory, 44 percent of U.S. cropland is losing topsoil in excess of its soil loss tolerance level (5 tons/acre/year). Excessive soil erosion has serious on-site and off-site consequences. The primary on-site consequence is reduced productivity (*25*), while the primary off-site consequence is reduced water quality.

In the past, farmers have used a combination of practices to control soil erosion: terraces, diversion ditches, grassed waterways, and cropping practices. However, with the development of new equipment and herbicides in the past 10 years, farmers increasingly are adopting conservation tillage practices, which contribute to improved soil erosion control (Table 1). In 1984 about one-third of all harvested cropland was managed with conservation tillage practices. USDA officials believe that by the year 2000 more than two-thirds of all harvested cropland will be under conservation tillage management (*35*).

Winter legume cover crops can play significant roles in reducing erosion in conservation tillage systems. Recent data from Georgia (*15, 22*) illustrate the benefit of a winter legume cover crop in a conservation tillage production system. Winter legumes may be more effective than small grains in partitioning rainfall to the soil profile, as opposed to runoff, and in controlling soil erosion.

Similar results have been obtained with a living mulch in the Northeast. Hall and associates (*12*) reported that soil loss was 86 percent and 99 percent less in no-till corn (*Zea mays* L.) with a living mulch of birdsfoot trefoil (*Lotus corniculatus*) or crownvetch (*Coronilla varia* L.), respectively, compared with conventionally tilled corn. Clearly, a legume cover crop or intercrop can play a significant role in soil erosion control.

Water conservation

The value of a mulch in water conservation has been well documented (*5, 8, 32*). Mulches generally increase water infiltration and/or decrease evaporation from the soil surface.

Cover crops also may be of benefit on wet, somewhat poorly drained soils. Water use by a cover crop may dry the soil and allow earlier planting of the subsequent crop. Such water use, of course, would be a disadvantage on soils that retain a low amount of plant-available water, such as very sandy soils or thin soils. In particular, sand and loamy sand textural classes are subject to rapid depletion of plant-available water by actively growing cover crops. To circumvent this problem, the cover crop can be killed 10 to 14 days prior to planting the subsequent crop. This eliminates continued water use by the cover crop and enhances the probability of soil surface recharge by planting time.

In terms of specific results for legume cover crops, essentially all runoff studies during the 1940s and 1950s showed a 50 percent to 90 percent reduction in annual runoff when a legume was included in a multiple cropping system (*21*). These earlier studies suggested that infiltration rates improved during spring months when a perennial legume was monocropped or included in a grass-legume mixture. More recently, Langdale and Leonard (*22*) and Hargrove and associates (*15*) demonstrated that a winter legume cover crop can effectively reduce water runoff, especially during winter and early spring months (Table 2). During three years of a crimson clover/sorghum (*Trifolium incarnatum* L./*Sorghum bicolor* L. Moench) cropping system, runoff averaged only 0.2 inch/year, and the average number of runoff events was only 3/year. Under conditions similar to these studies, a winter legume potentially may be more effective than winter small grains in terms of water conservation due to the earlier and more complete canopy cover afforded by the legume compared to small grains. However, the consequences of water use by the cover crops and detailed studies of soil-plant-water relations in these systems are lacking.

Energy conservation

Energy conservation is implicit in conservation tillage because less fuel is used for tillage operations. However,

Table 1. Conservation tillage in the United States, 1972 to 1984.

Year	Acreage in Conservation Tillage (million acres)	Fraction of Total Harvested Cropland in Conservation Tillage (%)
1972	29.7	10.0
1975	56.2	16.7
1978	74.8	22.2
1981	99.0	27.5
1984	107.5	32.6

Source: *No-Till Farmer* magazine.

fuel savings from conservation tillage represent only a small fraction of the fossil fuel energy used for row-crop production (*13*). The greatest energy requirement for no-till corn production is for N fertilizer, representing about two-thirds of the total fossil fuel energy required. Escalation of energy prices and, thus, of fertilizer-N prices has subsided temporarily. However, it is predicted that prices will again soar. Because a winter legume cover crop may provide significant quantities of biologically fixed N, it serves a very important role in energy conservation. When energy prices again escalate, energy conservation, through lessened dependence on fertilizer-N, may represent the greatest need that winter legumes can fulfill.

Recently published studies of legume cover crops focus on the N contribution of the legume to subsequent crops. Results from Virginia (*27*), Delaware (*26*), Kentucky (*11*), Alabama (*31*), and Georgia (*14*) have shown that winter legume cover crops can replace a significant amount of fertilizer-N, ranging from 50 to 120 pounds of N/acre, for subsequent crops of corn, grain sorghum, or cotton (*Gossypium hirstum* L.). Table 3 shows representative results of the contribution of winter legumes to the N requirement of no-till corn, grain sorghum, and cotton. In these studies, hairy vetch (*Vicia villosa* L. Roth) replaced 85, 81, and 61 pounds of fertilizer-N/acre for corn, grain sorghum, and cotton, respectively. That would be all the recommended fertilizer-N for grain sorghum and cotton and more than half the recommended amount for corn. Reddy and associates (*30*) found that cereal crops following tropical summer legumes yielded more than those following summer fallow.

Winter legume cover crops may not be possible in more temperate climates because of winter-kill. However, some recent studies (*2, 3, 7, 12, 18, 19, 32*) show that perennial legumes, such as alfalfa (*Medicago sativa* L.) followed by corn or crownvetch grown as an intercrop, may replace significant quantities of fertilizer-N, ranging from 63 to 150 pounds of N/acre (Table 4), a significant energy savings.

Clearly, growing a legume prior to or with a nonlegume can reduce the fertilizer-N requirement of the nonlegume

Table 2. Hydrologic data from a 6- to 7-acre watershed comparing various cropping/tillage systems (*15, 22*).

Time Period	Cropping System*	Tillage	Annual Rainfall (inches)	Annual Runoff (inches)	Annual Soil Loss (tons/acre)
1972-1974	Fallow/soybeans	conventional	54.0	8.7	11.6
1974-1976	Barley/grain sorghum	no-till	52.0	3.5	0.2
1976-1979	Barley/soybeans	non-till + in-row chisel	46.5	0.8	0.06
1979-1983	Crimson clover/grain sorghum	no-till + in-row chisel	43.7	0.2	0002

*Soybeans = *Glycine max* L. Merr.; barley = *Hordeum vulgare* L.

Table 3. Estimates of the nitrogen contribution of winter legumes to the nitrogen requirement of no-till corn, grain sorghum, and cotton.

Location	Crop	Cover Crop	Fertilizer-N (pounds/acre)	Reference
Kentucky	Corn	Hairy vetch	85	(11)
		Big flower vetch	45	
Georgia	Grain sorghum	Crimson clover	75	(14)
		Hairy vetch	81	
		Common vetch	53	
		Subterranean clover	51	
Alabama	Cotton	Hairy vetch	61	(31)
		Crimson clover	61	

Table 4. Estimates of the nitrogen contribution of perennial legumes to the nitrogen requirement of no-till corn.

Location	Perennial Legume Treatment	Fertilizer-N Replaced (pounds/acre)	Reference
Ohio	Followed 2 yrs. of alfalfa	100 to 150*	(32)
New York	Followed 2 yrs. of alfalfa	120	(2)
Wisconsin	Followed 3 yrs. of alfalfa	63	(3)
Minnesota	Followed 1 yr. of alfalfa	65	(18)

*Estimated from the data presented

substantially. Because fertilizer-N represents a sizeable portion of the fossil fuel energy required for nonleguminous row-crop production, this represents a significant energy savings, enhancing the conservation value of a no-till production system.

Soil improvement

One of the benefits of continuous no-till production is greater soil organic matter concentrations, especially near the soil surface (6, 10,16, 20). Increased soil organic matter can improve soil aggregation, enchance structural stability, improve water infiltration and storage, and increase cation exchange capacity and nutrient retention. The influence of these factors on overall productivity can be dramatic, especially on Ultisols and Oxisols, which tend to have poor physical and/or chemical properties and inherently low fertility.

The exact role legume cover crops or legume sod crops play in rotation with nonlegumes in soil organic matter accumulation is uncertain. Hargrove (13, 14) found that soil organic matter accumulation was greater with a crimson clover/grain sorghum no-till cropping system compared to a wheat/soybean (*Triticum aestivum* L. em Thell,/*Glycine max*

L. Merr.) no-till cropping system (Table 5). However, the total amount of residue returned to the soil with the clover/sorghum system was greater than with the wheat/soybean system. Thus, at equal amounts of residue, the legume cover crop may not increase soil organic matter to a greater extent than nonleguminous residue. In fact, Larson and associates (23) found that organic matter accumulation was comparable for equal quantities of legumes, wheat straw, or even sawdust added to soil. Therefore, the benefit of a legume cover crop or sod crop is that a sizeable quantity of plant material can be returned to the soil. Dry matter accumulation for well-adapted legumes can range as high as 4 tons/acre (14), but more commonly ranges from 2 to 3 tons/acre (11, 14, 31).

As mentioned, increased organic matter can improve overall soil physical condition. There is little recent data on this subject, but research from the 1940s to 1950s documented improved soil tilth, infiltration and permeability, and macroporosity with legume cover crops (24, 28, 34, 36). In these studies, however, legumes generally were plowed under because no-till practices were not yet common. More recently, Touchton and colleagues (31) found increased infiltration rates with hairy vetch or crimson clover planted as a cover crop compared to no cover crop for no-till cotton production. Additional research is needed on this subject, especially with respect to poorly structured soils that tend to compact easily.

Increased soil organic matter also increases nutrient retention and may enhance the soil fertility status. More specifically, Hargrove (14) found that the effects of legume cover crops on soil fertility status included a lower pH, a redistribution of K^+ to the soil surface from deeper in the soil profile, and a lower C:N ratio in soil organic matter. Furthermore, concentrations of nutrients, especially the relatively immobile nutrients, would be expected to be greater near the soil surface where residues are concentrated and where more chemically active organic matter may retain some nutrients.

In summary, though not well documented with definite data, legumes may play a significant role in soil improvement through organic matter accretion. The results may be dramatic on Ultisols and Oxisols of low inherent fertility; consequently, legumes are a needed component for successfully managing these soils.

Crop rotation benefits

The benefits of crop rotation have long been recognized. But the exact mechanisms by which rotation improves crop

Table 5. Influence of 5 years of various cropping sequences and tillage on soil organic carbon and nitrogen concentrations in the surface 7.5 cm of soil (13).

Cropping Sequence	Tillage Treatment	Fertilizer-N Rate (pounds/acre/year)	Organic C	Organic N	C:N Ratio
			%	%	
Wheat/soybean	Conventional	70	1.4	0.12	11.7
Wheat/soybean	No-till	70	1.6	0.15	10.7
Clover/sorghum	No-till	0	2.2	0.17	13.0
Clover/sorghum	No-till	120	2.4	0.19	12.6

growth and yield remain somewhat nebulous. This is also true of the role of legumes in crop rotations. The benefits have long been recognized but poorly understood. In the Midwest and Northeast, corn yields following alfalfa often are greater than what can be accounted for by N effects, and researchers have attempted to quantify the rotation effect (*3, 18*). Researchers also have found rotation effects in Kentucky where corn yields following hairy vetch were superior to those following a rye (*Secale cereale* L.) cover crop, even when sufficient fertilizer-N was added to remove differences in N availability. On the other hand, Hargrove (*14*) observed no rotation effect and explained crop responses, for the most part, by the N contribution of the legume. This reemphasizes the nebulous nature of the so-called rotation effect. The only thing that can be said is that if it exists legumes probably contribute to it.

Long-term implications

The influence of soil erosion on productivity and the strong dependence of nonleguminous crop production on energy-intensive fertilizer suggests that agricultural productivity may not be sustainable over the long-term. It seems necessary to reduce soil erosion rates and to decrease the dependence on fertilizer-N to achieve a sustainable agriculture. The use of legumes in concert with conservation tillage can clearly play an important role in meeting these requirements.

Along with sustaining productivity, these systems may play a role in restoring productivity to eroded or otherwise degraded soils. Current, unpublished research by Langdale and associates at Watkinsville, Georgia, indicates that productivity can be restored at least partially to severely eroded soils in a relatively short time period (2 to 3 years) using no-till practices and winter legumes. The keys to restoration of productivity appear to be halting erosion and promoting the accumulation of organic matter. The extent to which productivity of the ravaged Ultisols of the old Cotton Belt can be restored remains to be seen.

Conclusions

Clearly, legumes incorporated into cropping systems potentially can meet several current needs in agricultural production. These include soil, water, and energy conservation; soil improvement; and enhanced overall productivity. The use of legumes to maintain soil fertility and improve crop productivity is one of the oldest management practices in plant agriculture, dating back to ancient Greek and Roman civilizations. It is ironic that the need exists to return to this tenet of the ancients to sustain modern, technologically advanced agriculture.

REFERENCES

1. Bailey, R. Y., J. T. Williamson, and J. F. Dugger. 1930. *Experiments with legumes in Alabama.* Bull. 232. Ala. Agr. Exp. Sta., Auburn.
2. Baldock, J. O., and R. B. Musgrave. 1980. *Manure and mineral fertilizer effects in continuous and rotational crop sequences in central New York.* Agron. J. 72: 511-518.
3. Baldock, J. O., R. L. Higgs, W. H. Paulson, J. A. Jakobs, and W. D. Shrader. 1981. *Legume and mineral N effects on crop yields in several crop sequences in the upper Mississippi Valley.* Agron. J. 73: 885-890.
4. Bledsoe, R. P., and S. J. Hadden. 1936. *Rates of seeding small grains and winter legumes for hay.* Bull. 194. Ga. Agr. Exp. Sta., Athens.
5. Blevins, R. L., D. Cook, S. H. Phillips, and R. E. Phillips. 1971. *Influence of no-tillage on soil moisture.* Agron. J. 63: 593-596.
6. Blevins, R. L., G. W. Thomas, M. S. Smith, W. W. Frye, and P. L. Cornelius. 1983 *Changes in soil properties after 10 years continuous non-tilled and conventionally-tilled corn.* Soil and Tillage Res. 3: 135-136.
7. Bolton, E. F., V. A. Dirks, and J. W. Aylesworth. 1976. *Some effects of alfalfa, fertilizer, and lime on corn yields in rotations on clay soil during a range of seasonal moisture conditions.* Can. J. Soil Sci. 56: 21-25.
8. Bond, J. J., and W. O. Willis. 1969. *Soil water evaporation: Surface residue rate and placement effects.* Soil Sci. Soc. Am. Proc. 33: 445-448.
9. Carreker, J. R., S. W. Wilkinson, and J. E. Box. 1977. *Soil and water management systems for sloping land.* ARS-S-160. Agr. Res. Serv., U.S. Dept. Agr., Washington, D.C.
10. Dick, W. A. 1983. *Organic carbon, nitrogen, and phosphorus concentrations and pH in soil profiles as affected by tillage intensity.* Soil Sci. Soc. Am. J. 47: 102-107.
11. Ebelhar, S. A., W. W. Frye, and R. L. Blevins. 1984. *Nitrogen from legume cover crops for no-tillage corn.* Agron. J. 76: 51-55.
12. Hall, J. K., N. L. Hartwig, and L. D. Hoffman. 1984. *Cyanazine losses in runoff from no-tillage corn in "living" and dead mulches vs. unmulched conventional tillage.* J. Environ. Qual. 13: 105-110.
13. Hargrove, W. L., ed. 1982. *Proceedings of the minisymposium on legume cover crops for conservation tillage production systems.* Spec. Publ. No. 19. Univ. Ga., Athens.
14. Hargrove, W. L. 1986. *Winter legumes as a nitrogen source for no-till grain sorghum.* Agron. J. 78: 70-74.
15. Hargrove, W. L., G. W. Langdale, and A. W. Thomas. 1984. *Role of legume cover crops in conservation tillage production systems.* Paper 84-2038. Am. Soc. Agr. Eng., St. Joseph, Mich.
16. Hargrove, W. L., J. T. Reid, J. T. Touchton, and R. N. Gallaher. 1982. *Influence of tillage practices on the fertility status of an acid soil doublecropped to wheat and soybeans.* Agron. J. 74: 684-687.
17. Hendrickson, B. H., A. P. Barnett, and O. W. Beale. 1963. *Conservation methods for soils of the southern Piedmont.* Agr. Inf. Bull. No. 269. U.S. Dept. of Agr., Washington, D.C.
18. Hesterman, O. B., C. C. Sheaffer, D. K. Barnes, W. E. Lueschen, and J. H. Ford. 1986. *Alfalfa dry matter and nitrogen production, and fertilizer nitrogen response in legume-corn rotations.* Agron. J. 78: 19-23
19. Hesterman, O. B., C. C. Sheaffer, and E. I. Fuller. 1986. *Economic comparisons of crop rotations including alfalfa, soybean, and corn.* Agron. J. 78: 24-28.
20. Lal, R., D. DeVleeschauwer, and R. Malafa Nanje. 1980. *Changes in properties of a newly cleared tropical alfisol as affected by mulching.* Soil Sci. Soc. Am. J. 44: 827-833.
21. Langdale, G. W. 1983. *Legumes in cropping systems-water conservation and use in the Southeast.* Rpt., Work-Planning Conf. on Legumes in Cons. Tillage Systems. Agr. Res. Serv., U.S. Dept. Agr., Lincoln, Nebr.
22. Langdale, G. W., and R. A. Leonard. 1983. *Nutrient and sediment losses associated with conventional and reduced tillage agricultural practices.* In R. R. Lowrence, et al [eds.]. *Nutrient cycling in agricultural ecosystems.* Coll. Agr. Spec. Publ. No. 23. Univ. Ga., Athens.
23. Larson, W. E., C. E. Clapp, W. H. Pierre, and Y. B. Morachan. 1972. *Effects of increasing amounts of organic*

residues on continuous corn: II. Organic carbon, nitrogen, phosphorus, and sulfur. Agron. J. 64: 204-208.

24. Lutz, J. F. 1954. *Influence of cover crops on certain physical properties of soils.* In *Winter Cover Crops in North Carolina.* N. Car. Agron. Res. Rpt. 12. N. Car. State Univ., Raleigh. pp. 6-9.
25. McCool, D. K., ed. 1984. *Erosion and soil productivity.* Am. Soc. Agr. Eng., St. Joseph, Mich. 289 pp.
26. Mitchell, W. H., and M. R. Teel. 1977. *Winter annual cover crops for no-tillage corn production.* Agron. J. 69: 569-573.
27. Moschler, W. W., G. M. Shear, D. L. Hallock, R. D. Sears, and G. D. Jones. 1967. *Winter cover crops for sod-planted corn: Their selection and management.* Agron. J. 59: 547-551.
28. Obenshain, S. S., and P. T. Gish. 1941. *The effect of green manure crops on certain properties of Berks silt loam.* Bull. 73. Va. Agr. Exp. Sta., Blacksburg.
29. Pieters, A. J., P. R. Henson, W. E. Adams, and A. P. Barnett. 1950. *Sericea and other perennial lespedezas for forage and soil conservation.* Circ. 863. U.S. Dept. Agr., Washington, D.C.
30. Reddy, K. C., A. R. Soffes, and G. M. Prine. 1986. *Tropical legumes for green manure. I. Nitrogen production and the effects of succeeding crops.* Agron. J. 78: 1-4.
31. Touchton, J. T., D. H. Rickerl, R. H. Walker, and C. E. Snipes. 1984. *Winter legumes as a nitrogen source for no-tillage cotton.* Soil and Tillage Res. 4: 391-401.
32. Triplett, G. B., Jr., D. M. Van Doren, and B. L. Schmidt. 1968. *Effect of corn stover mulch on no-tillage corn yield and water infiltration.* Agron. J. 60: 236-239.
33. Triplett, G. B., Jr., F. Haghiri, and D. M. Van Doren, Jr. 1979. *Plowing effect of corn yield response to N following alfalfa.* Agron. J. 71: 801-803.
34. Uhland, R. E. 1949. *Physical properties of soils as modified by crops and management.* Soil Sci. Soc. Am. Proc. 14: 361-366.
35. U.S. Department of Agriculture, Office of Planning and Evaluation. 1975. *Minimum tillage: A preliminary technology assessment.* Rpt. for the Comm. on Agr. and For., U.S. Senate, Part II. Publ. No. 57-398. U.S. Govt. Printing Office, Washington, D.C.
36. Welch, C. D., W. L. Nelson, and B. A. Krantz. 1950. *Effects of winter cover crops on soil properties and yields in a cotton-corn and in a cotton-peanut rotation.* Soil Sci. Soc. Am. Proc. 15: 229-234.

Effect of legume cover crops and tillage on soil water, temperature, and organic matter

M. Utomo, R.L. Blevins, and W.W. Frye

Use of winter annual legume cover crops in no-till corn (*Zea mays* L.) production is beneficial because legumes fix substantial amounts of atmospheric N_2, add organic matter, increase the supply of soil water and plant nutrients, improve certain physical soil properties, and provide erosion control (*3*).

The mulch of the killed cover crop, however, often may result in cooler and wetter soils, which can delay planting time. Also, the growing cover crop may deplete stored soil water before it is killed, causing more drought stress during dry periods.

We sought to determine the effect of legume cover crops and tillage on soil water content and organic matter and to compare soil temperatures measured in the row slits of the planter and between rows of no-till and conventionally tilled corn.

Study and methods

We conducted the experiment in 1984 and 1985 at Lexington, Kentucky, on a Maury soil (fine-silty, mixed, mesic, Typic Paleudalfs). We established the plots in 1976; they were maintained through 1983 under continous no-till corn with annual legumes as cover crops (*1*). In 1984 we split the plots into conventional tillage and no-till treatments. The winter cover treatments were overseeded into the corn in mid-September each year. Winter cover crops included hairy vetch (*Vicia villosa* Roth), bigflower vetch (*Vicia grandiflora* W. Koch), rye (*Secale cereale* L.), and corn residue alone as a check treatment. Seeding rates were 36, 36, and 170 pounds/acre for hairy vetch, bigflower vetch, and rye, respectively. Following planting, the plots were sprayed with 1.0 pint/acre paraquat mixed with 2.0 pounds/acre alachlor and 3.0 pounds/acre cyanazine. We took soil samples weekly at 15-cm increments throughout the growing season for gravimetric water determination. We measured daily maximum and minimum soil temperatures at a depth of 2 inches in the row slits made by the planter and between rows for the first 30 days after corn planting. Soil samples from the 0- to 2-inch depth were dried, sieved, and analyzed for organic C using a Leco-CR-12 C determinator.

Results and discussion

Soil water content at the 0- to 6-inch depth prior to corn planting in 1985 was significantly lower ($P < 0.05$) under rye and hairy vetch than under corn residue. Soil water con-

M. Utomo is a former graduate assistant and R. L. Blevins and W. W. Frye are professors in the Department of Agronomy, University of Kentucky, Lexington, 40546-0091.

tent in surface soil was similar for all cover treatments prior to planting in 1984 due to plentiful spring rainfall. Soil water under conventional tillage was lower for hairy vetch (20%) compared to rye (22%) and cornstalk residues (24%). Two weeks after cover crops were killed, the soil water in the no-till treatment was significantly higher than that under conventional tillage. The effect of mulch in conserving soil water extended through the remainder of the growing season.

The maximum soil temperature in the no-till corn, averaged over all cover crop treatments and measurement locations, was significantly lower than the temperature for conventional tillage during the first month of the corn growing season. The average maximum soil temperature with no-till averaged 4.3 °F and 2.9 °F higher under hairy vetch and corn residue, respectively, when measured in the row slits than when measured between rows (4). Soil temperature was only 2.9 °F and 2.4 °F lower than in the row slits of conventional tillage with hairy vetch and rye cover treatments, respectively. The average minimum soil temperature under no-till was not significantly different from conventional tillage.

Total soil organic matter in the top 3 inches of the no-till treatment was significantly higher than that of conventional tillage on a given sampling date. After corn harvest in both 1984 and 1985, hairy vetch with no-till resulted in significantly greater organic matter content than any other cover crop. There was no significant affect of cover crops on organic matter in the conventionally tilled plots.

Corn grain yield with the hairy vetch treatment was consistently higher that with bigflower vetch, rye, or corn residue, regardless of tillage and N rates. This was apparently due to the release of biologically fixed N from hairy vetch (1). The familiar response of corn grain yield to tillage and N fertilizer rate, as discussed by Phillips (3), was also apparent in this experiment.

We concluded that during a dry spring, such as 1985 and 1986 in Kentucky, a cover crop may substantially deplete the stored soil water, which may influence early corn growth. However, a few weeks after the cover crops have been killed, soil water in a no-till system tends to be conserved throughout the remaining corn-growing season because of the surface mulch. Chemically killing the cover crop 2 to 3 weeks prior to planting corn may be advantageous during a dry spring. It also was apparent that soil temperature in the row slit of the planter was only slightly lower with no-till than with conventional tillage and may be a better indicator of the soil temperature to which the germinating seed and young seedling are exposed.

REFERENCES

1. Ebelhar, S. A., W. W. Frye, and R. L. Blevins. 1984. *Nitrogen from legume cover crops for no-tillage corn.* Agron. J. 76: 51-55.
2. Frye. W. W., J. H. Herbek, and R. L. Blevins. 1983. *Legume cover crops in production of no-tillage corn.* In W. Lockeretz [ed.] *Environmentally Sound Agriculture.* Praeger Publ., New York. N.Y.
3. Phillips, R. E., R. L. Blevins, G. W. Thomas, W. W. Frye, and S. H. Phillips. 1980. *No-tillage agriculture.* Science 208: 1108-1113.
4. Utomo, M. 1986. *Role of legume cover crops in no-tillage and conventional tillage corn production.* Ph.D. thesis. Univ. Ky., Lexington.

Annual legumes as a fallow substitute in the northern Great Plains of Canada

A. E. Slinkard, V. O. Biederbeck, L. Bailey, P. Olson, W. Rice, and L. Townley-Smith

Tangier flatpea (*Lathyrus tingitanus* L.) and lentil (*Lens culinaris* cv. Indianhead) were grown in paired comparisons for 3 years at five locations on the Brown, Dark Brown, Black, and Gray Wooded soil zones of western Canada. In addition, faba bean (*Vicia faba* cv. Outlook), common flatpea (*L. sativus*), field pea (*Pisum sativum* cv. Trapper), and Austrian winter pea (*P. sativum* cv. Semu S.I. and Poneka) were grown at one or more locations. Top growth was incorporated at early pod set. We collected data on dry matter production and the amount of N in the dry matter and N fixed, as determined by acetylene reduction assay.

Check plots included continuous wheat, continuous wheat plus N, and alternate-year wheat and fallow. Hard red spring wheat (*Triticum aestivum* cv. Neepawa) was seeded in the spring following incorporation of the annual legume green matter crops. We also collected data on seed yield and protein content of the succeeding wheat crop. We replicated all experiments at least four times using a randomized complete block design.

Indianhead lentil and Tangier flatpea averaged 1,698 and 1,666 pounds of dry matter/acre, respectively. The N content of the dry matter was 40 and 36 pounds/acre for Indianhead lentil and Tangier flatpea, respectively,—13 to 14 pounds (33 to 39%) of this N was fixed symbiotically. The other annual legume species—common flatpea, field pea, Austrian winter pea, and faba bean—were tested in about half of the experiments. These species performed similarly in the droughty sites; yields were up to 50% greater at the more humid sties.

Wheat yields have been measured at seven sites to date. Wheat after a green manure crop of Indianhead lentil and Tangier flatpea averaged 1,483 and 1,533 pounds/acre, respectively. This compares with 1,768 pounds/acre for wheat after fallow. Yields of continuous wheat, continuous wheat plus an average of 54 pounds of N/acre, and wheat after an annual legume green manure crop averaged 68%, 77%, and 85% of wheat after fallow. This suggests that incorporation of annual legume crops, containing an average of about 40 pounds of N/acre, increased wheat yields more than the addition of 54 pounds of N/acre on wheat stubble.

However, this is misleading because on the drier sites application of N to wheat stubble sometimes resulted in lower yield than from continuous wheat with no N fertilizer. Thus,

A. E. Slinkard is a senior research scientist at the Crop Development Centre, University of Saskatchewan, Saskatoon, S7N 0W0. V. O. Biederbeck, L. Bailey, P. Olson, W. Rice, and L. Townley-Smith are researchers at the Agriculture Canada Research Stations at Swift Current, Saskatchewan; Brandon, Manitoba; Beaverlodge, Alberta; and Melfort, Saskatchewan, respectively.

incorporation of an annual legume green manure crop always produced higher wheat yields than from continuous wheat plus N. Wheat yield after incorporation of any of these annual legume green manure crops was intermediate between yield of continuous wheat with or without additional N and wheat after fallow. In wet years wheat yield after incorporation of an annual legume green manure crop will approach that from wheat after fallow. While the green manure crop did not significantly affect the protein concentration of the wheat, wheat protein was consistently higher in plots with green manure than from continuous wheat plots.

In summary, (a) dry matter and N yields of these annual legume green manure crops were similar except that pea, faba bean, and common flatpea were more productive at the more humid sites and (b) wheat yield responses were excellent and similar for all annual legumes studied. Thus, the choice of which annual legume species to use should be determined largely by seed cost. The cost per acre for certified seed of Indianhead lentil is $9.00, compared to more than $20 for the peas and more than $25 for common flatpea, faba bean, and Tangier flatpea.

The species of choice, thus, is Indianhead lentil. Indianhead lentil also has the advantage of a low seeding rate, 30 pounds/acre; excellent seedling vigor because it emerges from 3 inches; and a short life cycle, which permits a flexible management system.

In 1986 eleven cooperating farmers in Saskatchewan planted 2-acre plots of Indianhead lentil in their fallow fields. The plots averaged 1,332 pounds of dry matter/acre. Wheat yields will be collected from these plots and the adjacent fallow plots in 1987 to provide a direct comparison under commercial conditions.

Indianhead lentil is being promoted as a fallow substitute in the Dark Brown soil zone of western Canada where over 30% of the land is fallowed each year. It is the only crop studied to date in which the value of the N fixed comes close to equaling seed costs.

Three different management systems appear feasible in the Dark Brown soil zone of western Canada:

▶ Pretill cereal stubble in early spring and seed about 5 days later. If volunteer cereals and weeds are a problem, incorporate all topgrowth about 60 days after seeding.

▶ Pretill cereal stubble in early spring and seed about 5 days later. If volunteer cereals and weeds are not a problem, spray with a low rate of 2, 4-D about 60 days after seeding to prevent seed production. The standing crop will go into the winter, trapping snow and controlling soil erosion.

▶ Pretill cereal stubble in early spring, then till and seed about June 20. This seeding date will result in vegetative growth only and good weed control. The standing crop will go into the winter, again trapping snow and controlling erosion.

Indianhead lentil plants will tolerate 4° to 6° of frost without damage. Colder temperatures may freeze back the topgrowth, but regrowth will be reinitiated from the first two nodes at or below the soil surface. Thus, Indianhead lentil may have potential as a winter legume cover crop in the southern United States.

Nutrient cycling processes in a southeastern agroecosystem with winter legumes

Peter M. Groffman, Paul F. Hendrix, Chun-Ru Han, and D. A. Crossley, Jr.

The capacity of winter legumes to conserve soil and water and to provide N to summer crops is well established. However, relatively little information is known about the basic nutrient cycling processes underpinning winter legume dynamics in southeastern agroecosystems. There is also little information on N losses from winter legume-based systems. Potential problems include high leaching and/or denitrification losses of N under N-fixing legumes, nutrient mineralization that is poorly synchronized with summer crop uptake needs, and high denitrification losses of N during decomposition of legume residues.

Herein is a summary of 3 years of work on basic nutrient cycling and N-loss processes in conventional and no-till systems with either mineral fertilizer or legume-N inputs. Specific studies include analysis of legume decomposition and nutrient release, denitrification, and leaching losses of N. Understanding the basic nutrient cycling processes of winter-legume-based systems is important to the development of efficient strategies of crop production for these systems and to improvement of environmental quality.

We conducted the research at the University of Georgia Horseshoe Bend Experimental Area, near Athens, on the Georgia Piedmont. Soil at this site is a Hiwassee series (typic Rhoddult), a well-drained soil with a sandy clay loam A_p horizon (1). Plowed and no-till treatments have been in place at this site since 1978. Between 1981 and 1984, grain sorghum (*Sorghum bicolor* (L.) Moench] was grown as a summer crop on all plots and winter rye (*Secale cereale* L.) or crimson clover (*Trifolium incarnatum* L.) was grown as split-plot treatments during winter. Plots that had rye received 95 kg N/ha as NH_4NO_3 before sorghum planting.

Legume decomposition and nutrient release

As observed in other studies, residue decomposition was faster in the conventionally tilled plots than in no-till systems for both residue types (Table 1). Clover residue incoporated into conventionally tilled soil decayed nearly twice as fast as the clover remaining on the surface in no-till. The fastest rate (-0.031/day) was observed on the surface of the conventionally tilled plot, but this accounted for only a small fraction of the total residue. Within tillage treatments, clover decomposed faster than rye with conventional tillage. We observed the opposite effect on no-till plots. Surface

Peter M. Groffman was a graduate research assistant, Paul F. Hendrix is an assistant ecologist, Chun-Ru Han was a visiting scientist, and D. A. Crossley, Jr. is a professor at the Institute of Ecology, University of Georgia, Athens 30602. Groffman currently is with the Department of Crop and Soil Science, Michigan State University, East Lansing, 48824; Chun-Ru Han is with the Department of Agronomy, Beijing Agricultural University, Beijing, China.

application of inorganic N fertilizer may have enhanced decomposition of the rye residue on the surface of no-till plots.

Nitrogen release from the residues reflected the pattern of decomposition (Table 2). Largest fluxes were from buried clover residues in conventional tillage and from surface rye residues in no-till. Total N released from the clover residue in conventional tillage was twice the amount released in no-till due to the low biomass of clover in the no-till system (Table 3).

Interestingly, the total N released from the legume was equivalent to or in excess of that applied as inorganic fertilizer in the rye systems (95 kg N/ha). This was especially evident with conventional tillage where buried residues decomposed rapidly. Soil mineral N showed little net change in the legume systems, suggesting that N was readily available to plants.

In addition to accumulating large amounts of N, the winter legume in our study accumulated higher levels of P, K, Ca,

Table 1. Decomposition of rye and clover residues from May through July 1983 in conventional tillage and no-till systems.

Tillage Treatment	Rye and Clover Decomposition			
	Rye		Clover	
	% AFDW Remaining*	-k/day†	% AFDW Remaining	-k/day
Conventional tillage				
Aboveground	35	0.017	12	0.031
Belowground	32	0.017	23	0.021
Total	32	0.017	22	0.022
No-till				
Aboveground	43	0.014	46	0.012
Belowground	48	0.011	59	0.008
Total	44	0.013	51	0.010

* Percent of May ash-free dry weight remaining in July.
† Decay constant from single exponential decay model.

Table 2. Net changes in nitrogen standing stocks from May through July 1983 in conventional tillage and no-till systems.

	Nitrogen Changes (kg/ha)			
	Conventional Tillage		No-Till	
	Rye	Clover	Rye	Clover
Aboveground litter	-2	-9	-111	-87
Belowground litter	-70	-194	-11	-14
Soil mineral pool	-133	-10	-40	-5
Sum	-205	-213	-162	-106
Plant uptake	+141	+144	+145	+96
Balance	-64	-69	-17	-10

Table 3. Biomass and nutrient concentrations in plant tissue in conventional tillage and no-till systems, May 1983 (1, 2).

Tillage Treatment	Biomass (kg/ha)	Nutrient Concentrations (mg/kg)				
		N	P	K	Ca	Mg
Conventional tillage						
Rye	7,135	15,600	1,165	3,460	1,390	465
		*		*	*	*
Clover	7,125	29,500	1,545	11,388	6,328	1,008
No-tillage						
Rye	10,620	14,300	1,105	3,922	1,330	473
	*	*	*	*	*	*
Clover	3,200	29,500	1,492	10,655	5,893	968

* Indicates that the cover crop treatments are signficantly different at p = 0.05.

Table 4. Available nutrients in soil (0-21 cm) of clover-nitrogen and rye/fertilizer-nitrogen treatments in conventional tillage (CT) and no-till (NT) systems, May and June 1983 (1, 2).

Treatment	Available Nutrients (kg/ha)									
	N		P		K		Ca		Mg	
	CT	NT	CT	NT	CT	NT	CT	NT	CT	NT
May										
Rye/fertilizer-N	138	75	143	126	762	561	2,954	2,234	410	294
Clover-N	53	30	149	184	864	868	3,013	2,340	414	307
June										
Rye/fertilizer-N	55	66	74	79	451	546	1,304	1,497	210	213
Clover-N	51	23	90	132	649	623	1,668	1,575	263	200

Table 5. Nitrate concentrations in lysimeters below rye and clover in conventional tillage and no-till systems, spring 1983 and 1984 (1).

Year and Tillage Treatment	Nitrate Concentration (mg/liter)		
	Rye		Clover
1983			
Conventional tillage	0.26	*	1.95
No-till	0.50	*	2.04
1984			
Conventional tillage	0.04	*	1.02
No-till	0.25	*	0.87

*Indicates that cover crop treatments are significantly different at the P = 0.05 level.

and Mg than rye (Table 3). Mowing and/or plowing of these nutrient-rich legume tissues increased levels of P, K, Ca, and Mg in soil in clover-N plots compared to rye/fertilizer-N plots in both May and June (Table 4). Mineral N was higher in plots that were fertilized in May, but this difference disappeared by June. The high release of nutrients from legume tissues relative to rye tissues was associated with higher surface soil pH in the legume plots (2). These data suggest that the legume-N system was able to support high levels of available nutrients relative to the fertilizer-N system. As a result, plant uptake of nutrients tended to be higher in the legume-N system than in the fertilizer-N system (1, 2).

Nitrogen losses

Leaching and denitrification are key processes that control the extent of N losses from agroecosystems. Leaching losses of NO_3-N were higher under clover-N plots than under fertilizer-N plots in the spring of both 1983 and 1984 (Table 5). It is likely that rye is a more effective scavenger for NO_3-N than clover. Although leaching was higher under clover than rye, the magnitude of the N lost to leaching was small, less than 25 kg N/ha/year. Denitrification was higher in fertilizer-N treatments in the spring and fall and was higher in clover-N treatments in summer (1). The extent of N lost to denitrification in our study was low, however, less than 20 kg N/ha/year.

REFERENCES
1. Groffman, P. M., P. F. Hendrix, and D. A. Crossley, Jr. 1987. *Nitrogen cycling in conventional and no-tillage agroecosystems with inorganic fertilizer or legume nitrogen inputs.* Plant and Soil 97: 315-332.
2. Groffman, P. M., P. F. Hendrix, D. A. Crossley, Jr. 1987. *Effects of a winter legume on phosphorus, potassium, calcium and magnesium cycling in a humid subtropical agroecosystems.* Agr. Ecosystems and Environ. (In press).

Early studies on the use of legumes for conservation tillage in Nebraska

Daniel T. Walters

Conservation tillage research in Nebraska developed as a result of the devastating dust bowl of the 1930s. The Soil Conservation Service and the Nebraska Agricultural Experiment Station signed a cooperative agreement on February 23, 1938, for the purpose of conducting studies on soil and moisture conservation. F. L. Duley, senior soil conservationist, and Professor J. C. Russel of the University of Nebraska were project leaders. Through their experiments they sought to reduce runoff and increase infiltration rate by maintaining crop residues on the soil surface. The intent was to lower the erosive power of raindrops and preserve surface structure as well as lower the evaporative loss of water. Development of the V-sweep plow in 1938 proved successful in accomplishing these goals in early studies (3, 5, 18). In 1939 Duley and Kelly (9), using infiltration studies, showed that surface residues, especially alfalfa, markedly improved surface soil structure. Typically, erosion and runoff from dense, subtilled legume plots was minimal compared to oat or wheat stubble (Table 1).

Subsurface tillage or stubble mulch was a great success in its first trial year, 1939, owing to a dry year. The mulch's effects of soil moisture storage and reduced evaporation on corn yields were very apparent.

The principal legumes in use in the 1930s for soil enrichment in rotation with corn were biennial sweetclover and alfalfa. Plowing of the green manure prior to planting corn, however, resulted in a flush of early corn growth due to rapid mineralization of the residue and a concommitant increase in water use by the corn crop. The risk of crop failure due to late season drought then increased. Many farmers reported that land planted in sweetclover "leads to worse erosion after that legume is buried under than had taken place previously." Duley and Russell recognized that subsurface tillage might be successful only to the extent that residue coverage was adequate in "quantity, quality and endurance." As for quantity, it was clear that success depended upon crop sequence and N fertility.

In 1939 Duley and Russel tried seeding small-seeded legumes in wheat-straw stubble. A rotary hoe coupled with a grass seeder was fashioned with the wheels of the rotary hoe reversed so that the teeth faced downward upon leaving the ground. Thus, the teeth were free of trash and disturbed the residue very little. The "mulch treader," as it was coined, is stilled manufactured and used to this day (7). The first plantings of alfalfa were made in the fall of 1940. Alfalfa planted in plowed land winter killed, with only 50% kill on plots seeded into residues.

In 1941 they experimented with subtilling or plowing under sweetclover prior to corn planting. Final corn yield was 11.4

bushels and 18.3 bushels for plowed and subtilled plots, respectively (6). Duley and Russel noted two disadvantages to seeding legumes in residues when spring seeded: delayed emergence because of cool soil temperatures and serious weed infestations. As a result, the practice of subtilling in the fall and spring for weed control prior to seeding resulted in good stands of alfalfa and sweetclover. This practice was adapted for subsequent legume seedings.

In 1942 T. M. McCalla, a soil microbiologist, joined the research team and began investigations on the problems of nitrification, soil microbiological changes, and residue decomposition assocated with stubble-mulch tillage (15, 16, 17). The team found that subtilling legume residues retarded decomposition and nitrification. Thus, the distribution of N could be controlled over two crop years. Tillage also had significant effects on soil fauna. Earthworm casts formed under alfalfa residue were up to three times more stable than under wheat-straw mulch (19). The number of soil fungi, bacteria, and actinomycetes were significantly greater under surface residue. Decomposition byproducts enhanced aggregate stability under stubble mulch (2). When sweetclover was allowed to decompose on the soil surface, 5 to 10 pounds of N/acre were lost as NH_3, with only a trace lost when residue was plowed under (14).

By 1943 the number of subsurface V-sweep tillers in use in Nebrasaka had expanded to 7,000. A consistent problem in the adaptation of stubble-mulch tillage was the detrimental effect of reduced nitrification. Research began in 1945 to examine legumes other than alfalfa and sweetclover and to expand legume acreage farther west. The selection process sought to find a legume that would (a) start in spring, but make most of its growth after the crop was removed; (b) produce seed in late fall and reseed itself; (c) be usable for livestock feed; (d) lend itself to subtillage and produce good ground cover; (e) have enough hard seed to live over several years in the soil; and (f) withstand climatic conditions in the Great Plains.

From 1945 to 1955 experiments were established in both north and south central Nebraska on permanent experiment stations. An additional 21 legume plots were established around the several million acres of sandy land, as these soils were severely deficient in N and highly susceptible to wind erosion when cultivated. The five legumes tested were Korean lespedeza, hairy vetch, partridge pea, annual sweetclover, and biennial sweetclover (10, 11, 12, 13). Lespedeza was a good quality forage, but it did not do well in dry years and did not volunteer well. Partridge pea showed promise because it volunteered well and its short stature in

Table 1. Runoff and erosion from subtilled and plowed field plots, June 1947.*

Crop	Tillage Treatment	Runoff† (inches)	Erosion (tons/acre)
Oats	Plowed	4.80	8.47
	Subtilled	1.23	0.53
Wheat	Plowed	1.26	1.19
	Subtilled	0.77	0.46
Oats	Plowed sweetclover	0.76	0.21
	Subtilled sweetclover	0.45	0.17

* Data from Duley, 1947, annual report.
† Total June rainfall = 11.16 inches.

Daniel T. Walters is an assistant professor agronomy, University of Nebraska, Lincoln, 68583.

Table 2. Corn yields in 1950 after different legumes grown in oats in 1949. *

Legume	Corn Yield (bushels/acre)	
	Subtillage	Plowing
Check - no legumes	58.2	68.9
Lespedeza	78.3	86.1
Partridge pea	79.1	81.7
Annual sweetclover	76.2	83.0
Hairy vetch	93.4	99.0
Biennial sweetclover	97.0	97.5

* Data from Duley, 1950, annual report.

the oat nurse crop allowed for easy oat harvesting. But it was practically impossible to harvest for seed because of its dehiscent nature, and it was a poor forage. Up to 5 years of partridge pea in an oats/corn rotation was realized with only one seeding.

Annual sweetclover was difficult to manage because of its rapid germination and early growth. Biennial sweetclover had to be reseeded each year but contributed most to subsequent corn yields. Hairy vetch was initially spring seeded with oats. Because of its climbing nature, it completely lodged the nurse crop and did not produce seed well. However, its dense cover and N contribution to corn (Table 2) stimulated further management research. Vetch was then fall planted with rye and harvested for seed the following year. The seed yield averaged between 200 and 400 pounds/acre, and at $.30/pound it became very popular with growers. The Madison variety of vetch was certified in 1950 and by 1951 farmers had planted over 25,000 acres.

The use of oats on sandy land gradually became less popular with farmers as a feed crop, and they could produce continuous corn with N fertilizers more profitably. Farmers could manage weed problems easier with new herbicides. Thus, legume rotations with grain crops gave way to N fertilizers. In 1956 legume research plots that had been established 15 years before were discontinued (4).

REFERENCES

1. Colom, J., and T. M. McCalla. 1952. *The decomposition of partridge pea and its influence on nitrification.* Soil Sci. Soc. Am. Proc. 16(2): 208-210.
2. Dawson, R. C., V. T. Dawson, and T. M. McCalla. 1948. *Distribution of micro-organisms in the soil as affected by plowing and subtilling residues.* Bull. 155. Nebr. Agr. Exp. Sta., Lincoln. pp. 1-26.
3. Duley, F. L. 1939. *Surface factors affecting the rate of intake of water by soils.* Soil Sci. Soc. Am. Proc. 4: 60-64.
4. Duley, F. L. 1960. *Yields of different cropping systems and fertilizer tests under stubble mulching and plowing in Eastern Nebraska.* Res. Bull. 190. Nebr. Agr. Exp. Sta., Lincoln. pp. 1-53.
5. Duley, F. L., and J. C. Russel. 1939. *New methods of rainfall and soil moisture conservation.* 31st An. Rpt. Nebr. Crop Growers Assoc., Lincoln. pp. 52-62.
6. Duley, F. L., and J. C. Russel. 1941. *Crop residues for protecting row cropland against runoff and erosion.* Soil Sci. Soc. Am. Proc. 6: 484-487.
7. Duley, F. L., and J. C. Russel. 1942. *Machinery requirements for farming through crop residues.* Agr. Eng. 23(2): 39-42.
8. Duley, F. L., J. C. Russel, T. H. Gooding, and R. L. Fox. 1953. *Soil conservation and management in sandy farm land in northeast Nebraska.* Bull. 420. Nebr. Agr. Exp. Sta., Lincoln. pp. 1-39.
9. Duley, F. L., and L. L. Kelly. 1939. *Effects of soil type, slope and surface conditions on intake of water.* Bull. 112. Nebr. Agr. Exp. Sta., Lincoln. pp. 1-16.
10. Gooding, T. H., F. L. Duley, and J. C. Russel. 1948. *Sweet clover in a stubble mulch system.* Soil Sci. Soc. Am. Proc. 13: 554-557.
11. Gooding, T. H., F. L. Duley, and J. C. Russel. 1948. *Partridge pea in a stubble mulch system.* Soil Sci. Soc. Am. Proc. 13: 558-560.
12. Gooding, T. H., and J. C. Russel. 1952. *Velch for sandy soils in Nebraska.* 43rd An. Rpt. Nebr. Crop Improvement Assoc., Lincoln, pp. 76-84.
13. Gooding, T. H., and J. C. Russel. 1954. *Answers to questions about partridge pea.* Cir. 94. Nebr. Agr. Exp. Sta., Lincoln. pp. 1-8.
14. Gooding, T. H., and T. M. McCalla. 1946. *Loss of carbon dioxide and ammonia from crop residuals during decomposition.* Soil Sci. Soc. Am. Proc. 10: 185-190.
15. McCalla, T. M. 1953. *Microbiology studies of stubble mulching.* Bull. 417. Nebr. Agr. Exp. Sta., Lincoln. pp. 1-14.
16. McCalla, T. M. 1959. *Microorganisms and their activity with crop residues.* Bull. 453. Nebr. Agr. Exp. Sta., Lincoln.
17. McCalla, T. M., and J. C. Russel. 1948. *Nitrate production as affected by sweet clover residues left on the surface of the soil.* Agron. J. 40(5): 411-421.
18. Russel, J. C. 1939. *The effect of surface cover on soil moisture losses by evaporation.* Soil Sci. Soc. Am. Proc. 4: 65-70.
19. Teotia, S. P., F. L. Duley, and T. M. McCalla. 1950. *Effect of stubble mulching on number and activity of earthworms.* Res. Bull. 165. Nebr. Agr. Exp. Sta., Lincoln.

Use of soybeans in an Oklahoma conservation tillage system

Vernon L. Jones

Many arable areas of Oklahoma receive adequate rainfall on an annual basis for good row-crop production. However, the uneven distribution of that rainfall often results in periods of drought during critical stages of crop development. A suitable conservation tillage system potentially could reduce soil erosion and the time required to seed subsequent crops while at the same time conserving soil water (*1, 5, 6*).

Conservation tillage can improve a soil's moisture-holding capacity. Researchers have suggested that some forms of conservation tillage lower soil temperature and increase soil water infiltration capacity (*4, 7*). Declines in soil temperature resulting from conservation tillage have been reported to be great enough to lower the amount of moisture loss from crop canopies (*2, 3*). Such reductions may prolong the period before a crop experiences yield or quality loss from moisture stress.

Conservation tillage practices also save energy. Vaughn and associates (*8*) reported that energy inputs were 10% less for a no-till grain system than for a conventional tillage system.

Study methods

I studied one conventional and two conservation tillage systems. Conventional tillage, as defined herein, refers to moldboard plowing followed by disking. The two conservation tillage treatments were reduced tillage, disking only, and no-till, direct seeding without any prior tillage. Forrest, a maturity group V soybean variety, was sown at 20-inch row spacings in a fine sandy loam soil.

Tillage system and yields

Cloudy, rainy days throughout the month of September and through much of October greatly reduced the amount of sunlight available for photosynthesis at the central Oklahoma study site. This resulted in yield reductions for the 1986 soybean plots over all three tillage treatments. Observations made of pods at harvest revealed incomplete pod-filling. The same was true with soybeans in adjacent fields.

There was no significant difference in yields based on the tillage system used. However, a trend did exist. Yields of soybeans grown under no-till conditions tended to be higher than those grown in reduced or conventional tillage systems (Table 1). Yields of soybeans grown under reduced tillage conditions tended to be equal to those in the conventional tillage treatment. These data suggest that either of the three

Vernon L. Jones is an associate research professor, Langston University, Langston, Oklahoma 73050.

Table 1. Effect of tillage system on soybean seed yields.

Tillage System	Yields (pounds/acre)
No-till	1,110
Reduced tillage	1,010
Conventional tillage	1,009

Table 2. Effect of tillage system on soybean plot heights.

Tillage System	Plot Height (inches)
No-till	36 a*
Reduced tillage	39 b
Conventional tillage	39 b

*Plot height values with different letters indicate that the tillage treatments are significantly different according to the least significant difference (LSD) Test at the 1% level.

tillage systems would have resulted in about the same yields if used by a commercial grower under normal field conditions.

Tillage system and plot heights

I measured heights of soybean plants in each tillage system near the completion of the pod-filling stage (late September). Soybeans grown in the no-till system were significantly shorter than soybeans seeded in reduced tillage or conventional tillage systems (Table 2). Soybeans in reduced tillage and conventional tillage treatments grew to the same height, on the average, across the field.

Summary

No-till and reduced tillage treatments produced yields equal to or slightly higher than the conventional tillage system. Because less time, work, and fuel were used in the no-till and reduced tillage systems, these two conservation tillage systems were more cost-effective than conventional tillage. Conservation tillage should be given consideration as a potential alternative soybean cultural system in Oklahoma.

REFERENCES

1. Aase, J. K., and D. L. Tanaka. 1984. *Effects of tillage practices on soil and wheat spectral reflectances.* Agron. J. 76: 814-818.
2. Ciha, A. J. 1982. *Yield and yield components of four spring wheat cultivars grown under three tillage systems.* Agron. J. 74: 317-320.
3. Finn, Gary A., and William A. Brun. 1980. *Water stress effects of CO_2 assimilation, photosynthate partitioning, stomatal resistance, and nodule activity in soybean.* Crop Sci. 20: 431-434.
4. Gerik, T. J., and J. E. Morrison, Jr. 1984. *No-tillage of grain sorghum on a shrinking clay soil.* Agron. J. 76: 71-76.
5. Ndon, B. A., R. G. Harvey, and J. M. Scholl. 1982. *Weed control in double cropped corn, grain sorghum, or soybeans minimum-till planted following canning peas.* Agron. J. 74: 266-269.
6. Touchton, J. T., and W. L. Hargrove. 1982. *Nitrogen sources and methods of application for no-tillage corn production.* Agron. J. 74: 828-826.

7. Unger, P. W., and B. A. Stewart. 1976. *Land preparation and seedling establishment practices in multiple cropping systems.* In R. I. Papendick, P. A. Sanchez, and G. B. Triplett [eds.] *Multiple Cropping.* Spec. Pub. No. 27. Am. Soc. Agron., Madison, Wisc. pp. 255-273.

8. Vaugh, D. H., and E. S. Smith, and H. A. Hughes. 1977. *Energy requirements of reduced tillage practices for corn and soybean production in Virginia.* In W. Lockeretz [ed.] *Agriculture and Energy.* Academic Press, New York, N.Y. pp. 245-259.

Germplasm resources for legumes in conservation tillage

W. E. Knight

Between 1950 and 1970, research on clovers (*Trifolium* spp.) and special-purpose legumes fell off in the United States. In the last decade, however, interest in using legumes in pastures and in conservation tillage systems has been rekindled (*22, 36, 37, 39, 43, 49*). The energy crisis and subsequent price increases for inorganic N stimulated part of this interest. Emphasis on better quality forage with better seasonal distribution also contributed to renewed interest in forage legumes.

Recent economic conditions have created keen competition for land resources. Commodity crops replaced pastures, resulting in shortages of, and high prices for, legume seed. Erodible land has been planted to row crops, resulting in unacceptable soil losses. There are national concerns about excessive soil loss and the need for improved conservation tillage systems. Legumes for pasture and cover constitute an integral part of these soil-conserving systems.

To date, research on legume cultivars has been directed toward development of forage types, with little emphasis on attributes for conservation tillage. Consequently, this discussion focuses primarily on forage legume germplasm resources, many of which have been adapted to conservation tillage systems, while others are currently being researched for possible adaptation.

Development of adapted legume cultivars has been successful for both annual and perennial species. Ecotypes arose readily after introduction and provided cultivars for immediate use and germplasm for breeding programs (*7*). Several major forage legume species are adapted widely to the climate and soils of the United States. Early efforts in cultivar development identified types for areas with relatively minor climatic and edaphic differences from those of the center of origin of the species.

Geneticists and plant breeders have successfully developed forage legume cultivars with improved characteristics. Development of some of these improved cultivars spanned 25 years. With increased emphasis and more widespread use of legumes, cultivar development efforts by public and private plant breeders must be accelerated to avoid excessive losses from an inevitable buildup of insect and disease pests.

General situation

A diverse array of legume species are available in the United States that fit well into livestock production systems as a source of grazing, hay, silage, and greenchop. Many of these species play a dual role as legumes in conservation tillage systems. The diverse climatic and soil conditions across the seven major continental regions and the tropical areas necessitate this diversity in legume resources. No single legume species is dominant in all the described regions. Alfalfa (*Medicago sativa* L.) is the most widely grown forage legume in the United States, with an estimated 27 to 29 million acres (*1*). Red clover (*Trifolium pratense* L.) is the most widely grown of all the true clovers (*46, 47*), occupying about 12 million acres. It is grown for hay, pasture, and soil improvement and fits well into 3- and 4-year rotations. White clover (*T. repens* L.) is the most important legume in many parts of the temperate zone (*4, 11*). Farmers and ranchers seed a million acres of white clover annually and it volunteers on more than 40 million acres of U.S. grassland. Other important legume species are birdsfoot trefoil (*Lotus corniculatus* L.) (*12*); rose clover (*T. hirtum* All.) (*27*); subterranean clover (*T. subterraneum* L.) (*30*); crimson clover (*T. incarnatum* L.) (*23, 25*); arrowleaf clover (*T. vesiculosum* Savi.) (*25, 31*); vetches (*Vicia* spp.) (*19*); and lespedezas, both annual (*Lespedeza striata* Hook & Arn. and

W. E. Knight is a collaborator, Agricultural Research Service, U.S. Department of Agriculture, Crop Science Research Laboratory, P.O. Box 272, Mississippi State, Mississippi 39762.

L. stipulacea Maxim.) and perennial (*L. cuneata* Don.) (*19*). Still other species are important in various regions of the United States. For example, A. E. Kretschmer in Florida has an extensive program involving over 4,000 accessions of tropical legumes (*37*). These species have great potential for improving tropical pastures in Florida and are basic to grassland agriculture in Hawaii (*38*).

Northeast region

Baylor and Vough (*2*) discussed the role of forage legumes in northeastern agriculture. Ruminant livestock is the primary source of agricultural income. The variable nature of northeastern soils requires careful choice of legume species and cultivars. Low winter temperatures especially affect the choice of perennial species. Alfalfa is the most widely used legume in the Northeast, used commonly in a multiple-harvest system as hay, silage, or greenchop. Red clover is useful in short rotations that often include a small-grain companion crop. Red clover will grow on soils that are either too acid or too wet for alfalfa. Birdsfoot trefoil, grown for hay or silage, is well suited to the less well-drained soils of the northern part of the region. Normally it will outlive red clover by several years. The acreage of ladino clover, used for rotational pasture, declined in recent years but has leveled off at the present time. Farmers use some alsike clover (*T. hybridum* L.) in the region, but its use is declining in favor of birdsfoot trefoil.

In recent years, crownvetch (*Coronilla varia* L.) has shown promise as a pasture and silage legume in the Northeast. Crimson clover is grown as a spring-planted summer annual cover crop in Maine and as a fall-planted winter annual in the coastal plains of New Jersey, Delaware, and Maryland (*26*). There is interest in subterranean clover in these states also. Sweetclovers (*Melilotus officinalis* and *M. alba*) are used for soil improvement and silage. There is some medium red clover and mammoth red clover used in the Northeast region. Geneticists and developers of germplasm at the USDA Pasture Research Laboratory at University Park, Pennsylvania, have active efforts on insect and disease resistance in alfalfa and birdsfoot trefoil. Several clover species are under evaluation for disease and insect resistance at the Geneva, New York, Plant Introduction Station.

North Central region

The eight Central and Great Lakes States include soils formed under both prairie and forest vegetation (*50*). This large region is one of the most fertile areas to be found on the continent. It is well adapted to the production of corn and forage and produces about two-thirds of the U.S. corn and soybeans. The average freeze-free summer period is from 100 days in the north to 200 days in the south.

The major legumes used in crop rotations in the North Central region include soybeans [*Glycine max* (L.) Merr.], peas [*Pisum sativa* (L.) Poir.], alfalfa, red clover, white clover, alsike clover, birdsfoot trefoil, crownvetch, and sweetclover (*Melilotus officinalis* L. Lam. and *M. alba* Medik.) (*45*). Soybeans and peas are annual, self-pollinated species that occur for 1 year in a rotation system. The other legumes are perennial or biennial and usually exist in rotations for more than 1 year.

In the North Central region, farmers plant more land (25 million acres) to soybeans annually than to all other legumes combined. Soybeans usually follow corn or similar grains in the rotation, often being planted on land better suited for perennial legumes. Peas are grown on about 250,000 acres and often precede soybeans in double cropping systems. Both soybeans and peas may be used as forage; however, this use is normally not economical.

Farmers use the remaining legumes generally for forage—silage, hay, or pasture. These biennial or perennial legume species remain in the rotations from 2 to 5 years. Each species has specific attributes that make it adaptable to special environments. Alfalfa and red clover are the most extensive forage legumes. The supply of adapted germplasm is adequate. However, active research programs are developing germplasm with greater N-fixation efficiency, better pest resistance, greater longevity, improved quality, and increased yield. Researchers continually are collecting introduced germplasm to supplement existing germplasm.

White and alsike clover are less desirable forage legume species because they yield less and are less drought tolerant than alfalfa and red clover. These two species, however, are being used in forage mixtures, primarily for pasture seedings. Use of birdsfoot trefoil is increasing, especially on marginal land best suited for pastures. Use of birdsfoot trefoil and crownvetch is limited due to difficulty in establishment and seed production. Farmers are using sweetclover on a limited basis; it has excellent forage quality if harvested in the vegetative stage of growth. One big disadvantage of sweetclover is that most cultivars contain the chemical coumarin, which is converted to dicoumaral if the forage is heated during preservation. Dicoumaral causes bleeding disease in livestock. Sweetclover is available, but not in use, that is essentially void of courmarin.

Other legume species with limited or possible use in the North Central region include lespedezas [*Lespedeza striata*, (Thunb.) H&A. and *L. cuneata* (Dumont) G. Don], vetches (*Vicia villosa* Roth, *V. sativa* L.), sainfoin (*Onobrychis viciifolia* Scop.), kura clover (*T. ambiguum* Bieb.), and crimson clover. In the southern part of the region, farmers are using Korean lespedeza (*L. stipulacea* Maxim.) extensively as pasture and hay on soils with poor to fair productivity. Crop advisory committees are active for all the legumes used in cropping systems in the North Central region. These committees, organized to develop the strategic overview of progress in crop improvement, identify and recommend areas for improvement—needs for germplasm enhancement, maintenance, acquisition, and evaluation.

Humid South

The humid South is characterized by relatively high rainfall (more than 50 inches/year) with erratic distribution between and within years; drought or excessive moisture are common (*5*). The growing season in the upper South ranges from 175 to 200 days; in the lower South, from 210 to 260 days. In the lower South, mild winters are conducive to year-round grazing, while in the upper South little forage is produced in December, January, and February. Cool-season perennial forages reach peak production from March 1 to June 1, with a smaller second peak in September and October. Warm-season perennials produce forage between April 1 and October 1. The relatively high rainfall and warm temperatures in the region have contributed to leached soils that are acid in reaction and low in organic matter and mineral plant nutrients.

In the upper part of the humid South, red clover is the dominant perennial legume species, followed by white clover and alfalfa. In recent years, the use of birdsfoot trefoil has increased. White clover dominates in the lower South. This is particularly true for indigenous stands.

The acreage of alfalfa and alfalfa-grass mixtures declined rapidly in the South with the advent of the alfalfa weevil. Interest in alfalfa production has increased, but substantial increases in acreage have not yet occurred.

In the upper South, ladino clover-grass mixtures generally are more productive than intermediate white clovers. However, farmers use the intermediate white clovers to assure reseeding where drought stress may cause stand failures on light-textured soils. In general, mixtures of dallisgrass-white clover, tall fescue-white clover, and bahiagrass-white clover are well adapted to wetter soils. Bermudagrasses and lespedezas (20), on the other hand, are adapted to well-drained, deep, drier soils.

Winter annual legumes most often used for forage production in the South are crimson, arrowleaf, subterranean, and ball clover, (*T. nigrescens* Viv.) (24). For maximum growth, winter annual clovers must be seeded in late summer or early fall and properly fertilized. They may be grown alone, but usually are mixed with tall fescue, cereals, or annual ryegrass (*Lolium multiflorum* Lam). Hairy vetch (*Vicia villosa* Roth), lupines (*Lupinus L.* spp.), bigflower vetch (*Vicia grandiflora* var. *kitaibeliana W.* Koch), and crownvetch are used as cover crops and provide limited grazing in some management systems.

Summer legumes used in the South include both annual and perennial sericea lespedeza. Alyce clover (*Alysicarpus vaginalis* DC.) is used in the lower South and in Florida.

Southern farmers have attained the greatest success with legumes in cropping systems with vetch, crismon clover, subterranean clover, and winter peas (9, 17). Vetch may have an advantage over crimson clover in the upper part of the region. But the life cycle of crimson clover more nearly meets the requirement for a legume in double-cropping systems over most of the region. In the upper South Chief crimson clover should be used because it is the most winter-hardy of the available crimson clover cultivars (23). Bigbee berseem clover (*T. alexandrinum* L.) appears to have wide adaptation and potential in the lower South.

There are about 76 legume species in public improvement programs in the southeastern United States (37). All 13 states in the region have germplasm evaluation programs, but many of the programs compare species without a program of selection and improvement. General objectives of breeding programs in the South are (a) to develop germplasm and/or cultivars of legumes with improved yield and quality; (b) to improve persistence and reliability by developing germplasm that is winter-hardy, drought-tolerant, and insect- and disease-resistant; and (c) to improve seed production and reseeding. Two research stations are breeding legumes for improved N_2 fixation (22, 44).

Germplasm evaluation is enhanced in the region through distribution of seven regional variety tests—alfalfa, white clover, red clover, annual clovers, trefoil, vetch, and lespedeza. These tests permit regional evaluation of new cultivars and experimentals for adaptation and potential value. One feature of these tests is distribution of specific inoculation cultures with the seed.

Northern Great Plains

This region is characterized by great extremes in climate, soil, and natural vegetation (32). Climate and soil factors largely control the choice of small-seeded grasses and legumes seeded for hay and pasture. The environmental diversity requires special procedures of farming and ranching, including dryfarming methods, grazing management, and irrigation practices. The growing season varies from 123 days in the southeast to 110 days in the northwest. Along the eastern border of the Great Plains, hay and pasture crops are parts of general diversified farming systems. Most farms have fields of alfalfa and smooth bromegrass or natural grasslands used for hay and pasture. Perennial grasses and legumes constitute most of the useful permanent cover in the Great Plains.

In the subhumid to semiarid dryland areas of the region, the use of legumes in crop rotations has been disappointing (42). Crop yields following green-manure fallow generally have been lower than after ordinary fallow. In areas of the semiarid Great Plains with a summer rainfall climate, water is a prime factor in crop production. Water used to grow the green manure crops probably is not replenished sufficiently to grow the following crop.

Alfalfa is the principal legume hay of the region. It usually is grown on the best soils, frequently with irrigation, and is the principal high-protein feed for balancing the hay ration. Its production provides the dominant forage crop for export to other areas in the form of dehydrated alfalfa used in mixed feeds. South Dakota farmers and ranchers harvest over 20 million acres of alfalfa for hay. Sweetclover is seeded in the Great Plains for use either as green manure or as a combination pasture and soil-improving crop. In the Dakotas, it often is used as hay or silage. It is drought resistant and also grows well in parts of the Great Plains receiving as much as 17 inches of rainfall.

Other legumes associated with grasses on low wet soils are red clover and alsike clovers. Birdsfoot trefoil, either grown alone or in grass-legume mixtures, is increasing in importance. Mixtures of birdsfoot trefoil with sainfoin have yielded well in short rotations in Montana. Sainfoin is recommended in Montana for pastures on dryland areas that receive 13 inches or more of annual precipitation.

Intermountain area and Alaska

The Intermountain area is predominantly moutainous terrain, with intensively cropped land in scattered valleys where there is access to water for irrigation (10). Dryfarming is practiced on higher land where the soil is deep and heavy enough to hold moisture. The climate is Mediterranean; precipitation averages about 14 inches/year but ranges from less than 5 inches in the drier valleys to more than 50 inches in the higher mountains.

In the Intermountain States alfalfa is the primary legume crop (3). About 45% of Utah's arable land is in alfalfa. It also is found on rangeland, where it is an important pasture component. Utah, Idaho, Nevada, and Wyoming farmers and ranchers devote about 2.5 million acres of land to alfalfa, the great majority of it under irrigation. However, alfalfa producers in this region have to contend with many production areas located in relatively small mountain valleys, ranging in elevation from 2,000 to 7,000 feet. Moreover, sites at higher elevations frequently receive frost every month of the year. Recommendations as to which alfalfa or other legume variety to grow are difficult without varietal evaluations at each site. Researchers are conducting alfalfa variety trials at some 75 locations in the West. In 1979 Utah researchers had alfalfa variety trials in eight locations throughout the state and added several other locations in 1981. Seeds of the various varieties being tested are readily available through seed companies or their distributors.

In addition to temperature stress, legumes in the Intermountain region must also deal with drought and salinity. The effect of drought stress on plants is particularly pronounced during germination, emergence, and seedling growth. Salinity is a problem in many soils of the semiarid and arid West. Such problems will increase as salinity increases in irrigation waters, water supplies for leaching decline, and planting of alfalfa shifts to marginal soils. Even in relatively trouble-free soils, a short-term buildup of surface soil salinity during germination and early seedling growth can have adverse effects. Thus, successful seedling establishment in these areas hinges on early root initiation and development. One of the criterion, therefore, for evaluating alfalfa varieties should be whether the variety exhibits rapid root extension and growth.

The annual *Medicago* species appear to offer potential (39). These plants effectively compete with weedy species, produce additional forage, and add valuable N to typically N-deficient soils.

Other legumes that offer potential are sainfoin, cicer milkvetch (*Astragalus cicer* L.), faba beans (*Vicia faba* L.), and Utah sweetvetch (*Hedysarum boreale* Nutt.) (19, 40, 49). Because of its nonbloating characteristic and drought- and winter-hardiness, sainfoin is worthy of further study. Faba beans are grown on a limited acreage in Utah as a silage crop. Utah sweetvetch is found on range sites. Milkvetch also is being tried on a limited basis. Each of these species needs further study as a potential crop; Crampton (8) summarized characteristics and potential use of a number of the native range clovers.

Priority areas for future research should include the development of alfalfas with increased water use efficiency, rapid germination, increased N-fixation potential, high degree of compatibility and effectiveness with rhizobial strains, and tolerance to saline/alkaline soils. Researchers also should consider selection of rhizobial strains tolerant of environmental stresses. Alfalfa breeders should consider

a higher proportion of leaves to stems and submerged crowns to increase winter survival and to protect the plants from overgrazing. In addition, plants should be developed with branched rootedness to increase mineral and moisture uptake and the ability to develop and grow faster at low soil temperataures.

In Alaska, farmers are growing forage crops to support the state's dairy, beef, sheep, and swine operations. Growing-season temperatures, although somewhat lower than in midtemperate latitudes, are conducive to cool-season crops. About 5 million acres are suitable for pasture. Perennial and biennial legumes commonly grown in more southern latitudes frequently winterkill. Therefore, grasses are grown principally in monoculture and fertilized with N. Canadian field peas (*Pisum arvense* L.) are grown with oats, cut only once, and preserved as silage.

Alaskans also use a significant amount of native rangeland. Two native legumes among the forbs are the abundant nootka lupine (*Lupinus nootkatensis* Donn.) on uplands and the less common beach pea (*Lathyrus maritimus* L.) along many shorelines.

Southern Great Plains and Southwest

Herbel and Baltensperger (18) described climatic variation in the Great Plains and Southwest. Precipitation not only varies greatly within and among seasons and years but also among locations separated by only a few miles. Because of erratic weather conditions, farming is high risk in all but the eastern portions of the region. However, because of the favorable temperatures, irrigated farming is highly productive where good-quality water is available. A high percentage of the land in the region is used for ranching.

Alfalfa is the most important forage legume in the region; however, it requires a large amount of water for maximum production. Alfalfa and perennial grasses are well adapted for the improvement of desert and semidesert soils for irrigated agriculture. Alfalfa is especially valuable for crop rotation systems in the irrigated parts of the region.

Evaluation of annual legume cultivars from Australia has done much to identify general species adaptation in the southern Great Plains (6, 29, 35, 39, 43). Further improvement to meet the specific requirements of a physiographic locale or management system will come from more intensive scrutiny of additional germplasm. Currently, there are over 500,000 accessions of germplasm in the National Plant Germplasm Program (51, 52). Recent collection trips have added substantially to the germplasm available for evaluation (41).

Pacific Coast

All but a small part of the agricultural land in the Pacific Coast region is essentially arid or semiarid (28). The area west of the Cascade Mountains and the high plateaus in eastern Oregon and Washington and adjacent Idaho are classed as subhumid. Even in the humid and subhumid areas, the climate is dry in summer and supplemental irrigation is required to maintain season-long forage production.

Alfalfa is the principal hay crop in the region. Much of

the hay is harvested from seed fields. The first crop is harvested for hay and the second for seed. Red clover is seeded with ryegrass, and alsike is used with grass on poorly drained land and on acid soils. Oats and vetch or oats and pea mixtures are grown west of the Coast Ranges in Oregon and in western Washington. Birdsfoot trefoil is grown in mixtures with grass for hay in irrigated areas.

Dry peas (*Pisum sativum* L.) and lentils (*Lens culinaris* Medik.) are the predominant grain legume crops in the Palouse region of eastern Washington and northern Idaho (*33*). These crops are included in rotations with cereals for a number of reasons: (a) soil erosion control (the legume replaces summer fallow, a practice known to increase the probability of severe soil erosion); (b) less severe disease infestations in cereals because the legume is not an alternate host for most cereal pathogens; (c) better control of grassy weeds compared with cropping systems with only cereals in the rotation; (d) diversification to exploit the symbiotic association with *Rhizobium* and decrease the consumption of inorganic fertilizer; and (e) broader market opportunities. Commonly used rotations include winter wheat followed by peas or lentils or winter wheat followed by spring barley followed by peas or lentils. Recently, chickpeas (*Cicer arietinum* L.) have shown promise as an additional grain legume crop for use in the Palouse in the place of peas and lentils.

Besides the spring-sown grain legumes, Austrian winter peas [*Pisum sativum* spp. *arvense* (L.) Poir.] have been grown on the Camas prairie area of northern Idaho for either green manure or for seed. Farmers alternate Austrian winter peas with winter wheat, either harvesting seed or using the peas for green manure. Long-term rotations are necessary for this crop because of severe disease problems. The most successful rotations have included 3 years of Alfalfa followed successively by winter wheat, spring barley, and Austrian winter peas. This rotation has been quite successful in supplying N needs to the cereals and for controlling weeds.

In the Columbia Basin (irrigated) of central Washington, grain legumes are grown primarily for seed to be used elsewhere, primarily pea seed and bean seed (*Phaseolus vulgaris* L.). With the variety of legume crops grown in this area, rotations depend upon current crop situations. In the coastal areas of Washington and Oregon, peas are grown for the freezing industry, but not always in rotations with cereals. Some faba beans are used for silage in animal industries. In California dry beans, lima beans (*Phaseolus lunatus* L.), blackeye peas (*Vigna unguiculata* L.), and chickpeas are grown in various rotations. All are grown for grain.

Forage legume seed and hay predominate in the Columbia Basin and the coastal areas of Washington and Oregon. Rotations vary, but yields of potatoes or wheat following alfalfa in the Columbia Basin benefit greatly from the N fixed by the legume.

Legume seed and hay production are extensive in California, again using various rotations. Vetch has been used with limited success in rotation with rice, but weed problems become more severe. In the dry foothill areas of northern California, certain *Medicago* spp. and clovers provide improved, long-term range stands (*21*).

Pacific Northwest and California farmers could use medics

and subterranean clover in rotations with cereals in the dry areas to provide at least a portion of the N requirements of the total rotation and to provide green manure to improve soil organic matter content. Chickpeas (deli types) could be grown in place of summer fallow in the areas that are marginal for annual cropping of cereals. Chickpeas, an upright grain legume, could be harvested if grain production warranted or otherwise used as green manure. Lupines might also be used in these areas. The deep tap root systems of chickpeas and lupines would enable those crops to survive periods of hot, dry weather. Intercropping with sweetclover is another alternative. Such intercropping is not new, but research over 30 years ago showed that planting sweetclover in alternate rows with barley (*Hordeum vulgare* L.) resulted in barley yields that were 85% of those obtained from solid planting. The sweetclover was plowed under for green manure the following summer, and winter wheat was planted in the fall.

Conservation tillage has not become a commonly recommended and accepted practice in cereal farming in California. Thus, the study of the role of annual legumes within this context has not been pursued to any research and development level. For the most part, annual legumes have been studied to a limited extent as green manures and very sparingly as cover crops. W. A. Williams did much of the pioneer work on green manures in rice production in the 1950s and early 1960s (*53*). For the most part, the available annual legume plant materials were somewhat limited and the most successful annual legumes were vetches (*Vicia benghalensis* L., and *V. desycarpa* Jen.).

Use of annual legumes for range improvement in California and most varietal adaptation trials have depended upon imported Australian varieties (*34*). Both subterranean and rose clover varieties have been moderately successful for range seeding. Yet inoculation problems plagued earlier trials of establishment and persistence of these clovers. Only during the 1970s were superior *Rhizobium* strains and inoculation techniques developed to overcome these problems (*21*).

In 1974 researchers initiated testing of the available commercial rose clovers for erosion control purposes in southern California. Hykon rose clover proved to be highly successful for right-of-way erosion control and soil stabilization (*13, 14, 15, 27*).

Plant exploration and exchanges of annual legume germplasm from the Mediterranean region have expanded U.S. annual legume variability and provided resources that are more versatile and potentially adaptable for future inclusion in conservation tillage programs (*41*). In addition to the wide-based germplasm of subclovers, other species that show promise are rose clover, cupped clover (*T. cherleri*), berseem (*T. alexandrinum*), medics (*Medicago* spp.), and sweet vetch (*Hedysarum* spp.).

Legumes in tropical and subtropical areas

Tropical and subtropical areas of the United States include Hawaii, southern Florida, coastal areas on the Gulf of Mexico, Puerto Rico, U.S. Virgin Islands, Guam, and American Samoa. Rotar and Kretschmer (*38*) describe the extreme diversity of climates and forages grown within these areas.

The main centers of origin and diversity for the tropical forage legumes occur in Brazil, Mexico, eastern Africa, and the Sino-Himalayan region. A majority of all commercially available tropical legumes are native to tropical America. Most of the commercial cultivars are in the legume tribes *Indigofereae, Aeschynomeneae, Desmodieae,* and *Phaseoleae.* In a recent review of tropical and subtropical forages, Rotar and Kretschmer (*38*) describe 20 tropical legume germplasm sources under evaluation.

The general approach to forage production in Hawaii has been to use legume-based pastures rather than to rely on fertilizer N for the grasses, as in Puerto Rico. The most productive legume-grass mixtures in Hawaii have been koa haole [*Leucaena leucocephala* (Lam.) deWit.] and guineagrass (*Panicum maximum* L.); greenleaf desmodium (*Desmodium intortum* (Link) DC.) and pangola digitgrass (*Digitaris decumbens* Stent.); kaimi clover [*Desmodium canum* (Gmel.) Schintz & Thellung] and kikuyugrass (*Pennisetum clandestinum* Hochst. ex Chiov.); and white clover and kikuyugrass. Other legumes that show promise are stylo [*Stylosanthes guianesis* (Aubl.) Scv.], glycine [*Neonotonia wightii* (R. Grah. ex Wight & Arn Verdcourt)], centro (*Centrosema pubescens* Benth.), siratro (*Phaseolus atropurpureus* DC.), and big trefoil (*Lotus pedunculatus* Cav.)

In Florida, Kretschmer and Brolmann have an extensive evaluation program involving over 4,000 accessions of tropical legumes representing 12 genera (*36*). Genera in the Florida program include *Arachis, Aeschynomene, Cajanus, Centrosema, Desmanthus, Desmodium, Indigofera, Leucaena, Macroptilium, Stylosanthes, Vigna,* and *Zornia.* Many of the same species are being evaluated in Louisiana. Currently, the use of these species is limited because of seed production problems.

Summary

The range of diverse legume germplasm available for general and special purpose use in the United States indicates that the potential exists to maximize legumes for energy-efficient animal production as well as in conservation tillage systems. Considering the small number of intensive forage-legume breeding programs in the United States, the large number of species available increases the challenge to breeders to provide the public with improved legume germplasm of the best-adapted species. Exploitation of the legume resource available requires an expansion of multidisciplinary research teams. Centers of excellence should be developed in the various regions to enhance the plant and *Rhizobium* germplasm resources available, to develop grazing systems to fully use the potential of legumes in animal production, and to enhance existing species and develop new germplasm sources for use in conservation tillage systems.

REFERENCES CITED

1. Barnes, D. K., and C. C. Shaeffer. 1985. *Alfalfa.* In M. E. Heath, D. S. Metcalf, and R. F. Barnes [eds.] *Forages: The Science of Grassland Agriculture.* Iowa State Univ. Press, Ames. pp. 89-87.

2. Baylor, J. E., and J. R. Vough. 1985. *Hay and pasture seedings for the Northeast.* In M. E. Heath, D. S. Metcalf, and R. F. Barnes [eds.] *Forages: The Science of Grassland Agriculture.* Iowa State Univ. Press, Ames. pp. 338-347.

3. Campbell, W. F. 1983. *Legume germplasm resources for cropping systems in the Intermountain West.* Rpt., Work-Planning Conf. on Legumes in Cons. Tillage Systems. Agr. Res. Serv., U.S. Dept. Agri., Lincoln, Neb. pp. B-78-B-87.

4. Carlson, G. E., P. B. Gibson, and A. A. Baltensperger. 1985. *White clover and other perennial clovers.* In M. E. Heath, D. S. Metcalf, and R. F. Barnes [eds.] *Forages: The Science of Grassland Agriculture.* Iowa State Univ. Press, Ames. pp. 118-127.

5. Chamblee, D. S., and A. E. Spooner. 1985. *Hay and pasture seedings for the humid South.* In M. E. Heath, D. S. Metcalf, and R. F. Barnes [eds.] *Forages: The Science of Grassland Agriculture.* Iowa State Univ. Press, Ames. pp. 359-370.

6. Christiansen, S., and W. L. Graves. 1986. *Screening plant and* Rhizobium *germplasm.* In Proc., Workshop on Annual Legumes. Forage and Livestock Res. Lab., Agr. Res. Serv., El Reno, Okla. pp. 18-19.

7. Cope, W. A., and N. L. Taylor. 1985. *Breeding and genetics.* In N. L. Taylor [ed.] *Clover Science and Technology.* Mono. No. 25. Am. Soc. Agron., Madison, Wisc. pp. 383-404.

8. Crampton, B. 1985. *Native range clovers.* In N. L. Taylor [ed.] *Clover Science and Technology.* Mono. No. 25. Am. Soc. Agron., Madison, Wisc. pp. 579-590.

9. Donnelly, E. D., and J. T. Cope, Jr. 1961. *Crimson clover in Alabama.* Bull. 335. Ala. Agr. Exp. Sta., Auburn. pp. 1-31.

10. Eckert, R. E., Jr., and L. J. Klebesadol. 1985. *Hay, pasture and rangeland of the Intermountain Area and Alaska.* In M. E. Heath, D. S. Metcalf, and R. F. Barnes [eds.] *Forages: The Science of Grassland Agriculture.* Iowa State Univ. Press, Ames. pp. 389-399.

11. Gibson, P. B., and W. A. Cope. 1985. *White clover.* In N. L. Taylor [ed.] *Clover Science and Technology.* Mono. No. 25. Am. Soc. Agron., Madison, Wisc. pp. 471-490.

12. Grant, W. F., and G. C. Marten. 1985. *Birdsfoot trefoil.* In M. E. Heath, D. S. Metcalf, and R. F. Barnes [eds.] *Forages: The Science of Grassland Agriculture.* Iowa State Univ. Press, Ames. pp. 98-108.

13. Graves, W. L., B. L. Kay, and T. Han. 1980. *Rose clover controls erosion in Southern California.* Calif. Agr. 34(4): 4-5.

14. Graves, W. L., B. L. Kay, T. Han, and J. R. Bruce. 1983. *Legumes for erosion control in Southern California.* In Proc., Int. Erosion Control Assoc. Conf. XIV. Int. Erosion Control Assoc., Pinale, Calif. 13 pp.

15. Graves, W. L., B. L. Kay, T. Han, and R. L. Koenigs. 1982. *Stabilizing highway rights-of-way with rose clover in Southern California.* In Proc., Third Symp. on Environ. Concerns in Rights-of-Way Manage. Miss. State Univ., Mississippi State. 8 pp.

16. Heath, M. E., R. F. Barnes, and D. S. Metcalfe. 1985. *Forages: The Science of Grassland Agriculture.* Iowa State Univ. Press, Ames.

17. Henson, P. R., and E. A. Hollowell. 1960. *Winter annual legumes for the South.* Farmers Bull. 2146. U. S. Dept. Agr., Washington, D. C. pp. 1-24.

18. Herbal, C. H., and A. A. Baltensperger. 1985. *Ranges and pastures of the Southern Great Plains and the Southwest.* In M. E. Heath, D. S. Metcalf, and R. F. Barnes [eds.] *Forages: The Science of Grassland Agriculture.* Iowa State Univ. Press, Ames. pp. 380-388.

19. Hoveland, C. S., and C. E. Townsend. 1985. *Other legumes.* In M. E. Heath, D. S. Metcalf, and R. F. Barnes [eds.] *Forages: The Science of Grassland Agriculture.* Iowa State Univ. Press, Ames. pp. 146-153.

20. Hoveland, C. S., and E. D. Donnelly. 1985. *The lespedezas.* In M. E. Heath, D. S. Metcalf, and R. F. Barnes [eds.]

Forages: The Science of Grassland Agriculture. Iowa State Univ. Press, Ames. pp. 128-135.

21. Jones, M. B., J. C. Burton, and C. E. Vaughn. 1978. *Role of inoculation in establishing subclovers on California annual grasslands.* Agron. J. 70: 1081-1085.

22. Knight, W. E. 1984. *Legumes as a nitrogen source.* In Proc., Am. Forage Grasslands Conf. Am. Forage Grasslands Cong., Lexington, Ky. pp. 344-354.

23. Knight, W. E. 1985. *Crimson clover.* In N. L. Taylor [ed.] *Clover Science and Technology.* Mono. No. 25. Am. Soc. Agron., Madison, Wisc. pp. 491-502.

24. Knight, W. E. 1985. *Miscellaneous annual clovers.* In N. L. Taylor [ed.] *Clover Science and Technology.* Mono. No. 25. Am. Soc. Agron., Madison, Wisc. pp. 547-562.

25. Knight, W. E., and C. S. Hoveland. 1985. *Arrowleaf, crimson and other annual clovers.* In M. E. Heath, D. S. Metcalf, and R. F. Barnes [eds.] *Forages: The Science of Grassland Agriculture.* Iowa State Univ. Press, Ames. pp. 136-145.

26. Linscott, D. L. 1983. *Legumes in conservation-tillage Northeastern Region.* Rpt., Work-Planning Conf. on Legumes in Cons. Tillage Systems. Agr. Res. Serv., U.S. Dept. Agr., Lincoln, Nebr. p. 8-6.

27. Love, R. M. 1985. *Rose clover.* In N. L. Taylor [ed.] *Clover Science and Technology.* Mono. No. 25. Am. Soc. Agron., Madison, Wisc. pp. 535-546.

28. Marble, L. V., L. A. Raguse, W. S. McGuire, and D. B. Hannaway. 1985. In M. E. Heath, D. S. Metcalf, and R. F. Barnes [eds.] *Forages: The Science of Grassland Agriculture.* Iowa State Univ. Press, Ames. pp. 400-410.

29. Matches, A. G. 1986. *Annual legume evaluations in the Semi-Arid High Plains.* In Proc., Workshop on Annual Legumes. Forage and Livestock Res. Lab., Agr. Res. Serv., El Reno, Okla. p. 13.

30. McGuire, W. S. 1985. *Subterranean clover.* In N. L. Taylor [ed.] *Clover Science and Technology.* Mono. No. 25. Am. Soc. Agron., Madison, Wisc. pp. 515-534.

31. Miller, J. D., and H. D. Wells. 1985. *Arrowleaf clover.* In N. L. Taylor [ed.] *Clover Science and Technology.* Mono. No. 25. Am. Soc. Agron., Madison, Wisc. pp. 503-514.

32. Moore, R. A., and R. J. Lorena. 1985. *Hay and pasture seedings for the Central and Northern Great Plains.* In M. E. Heath, D. S. Metcalf, and R. F. Barnes [eds.] *Forages: The Science of Grassland Agriculture.* Iowa State Univ. Press, Ames. pp. 371-379.

33. Muehlbauer, F. J. 1983. *Legumes in cropping systems in the Pacific Northwest and California.* Rpt., Work-Planning Conf. on Legumes in Cons. Tillage Systems. Agr. Res. Serv., U.S. Dept. Agr., Lincoln, Nebr. pp. B-14-B-18.

34. Murphy, A. H., M. B. Jones, J. W. Clawson, and J. E. Street. 1973. *Management of clovers on California annual grasslands.* Circ. 564. Div. Agr. Sci., Univ. Calif., Berkeley. 19 pp.

35. Ocumpaugh, W. R. 1986. *Potential for annual legumes in the Semi-Arid Coastal Bend.* In Proc., Workshop on Annual Legumes. Forage and Livestock Res. Lab., Agr. Res. Serv., El Reno, Okla.

36. Pederson, G. A., and W. E. Knight. 1983. *Clover and special purpose legume germplasm resources for the future.* In Proc., 39th South. Pasture Forage Crop Improv. Conf. Agr. Res. Serv., U.S. Dept. Agr., New Orleans, La. pp. 42-55.

37. Pederson, G. A., and W. E. Knight. 1984. *Legume germplasm improvement in the Southeastern U.S.* In. Proc., Am. Forage and Grassland Conf., Am. Forage and Grassland Cong., Lexington, Ky. pp. 256-260

38. Rotar, P. P., and A. E. Kretschmer, Jr. 1985. *Tropical and subtropical forages.* In M. E. Heath, D. S. Metcalf, and R. F. Barnes [eds.] *Forages: The Science of Grassland Agriculture.* Iowa State Univ. Press, Ames. pp. 154-165.

39. Rumbaugh, M. D. 1986. *Evaluation of annual medics for Semi-Arid Steppe and cold desert environments.* In Proc., Workshop on Annual Legumes. Forage and Livestock Res Lab., Agr. Res. Serv., El Reno, Okla. p. 14.

40. Rumbaugh, M. D., and C. E. Townsend. 1985. *Range legume selection and breeding in North America.* In Proc., 30th Ann. Mtg., Soc. Range Manage., Denver Colo. pp. 137-147.

41. Rumbaugh, M. D., and W. L. Graves. 1984. *Foreign travel report: Collecting germplasm in Morocco.* In Proc., 29th Alfalfa Imp. Conf. Agr. Res. Serv., U.S. Dept. Agr., St. Paul, Minn. pp. 92-97.

42. Smika, D. E. 1983. *Legume use in cropping systems in the Northern and Central Great Plains.* Rpt., Work-Planning Conf. on Legumes in Cons. Tillage Systems. Ag. Res. Serv., U.S. Dept. Agr., Lincoln, Nebraska. p. B-26.

43. Smith, G. R. 1986. *Extending forage legumes into the 500-750 mm annual rainfall regions.* In proc., Workshop on Annual Legumes. Forage and Livestock Res. Lab., Agr. Res. Serv., El Reno, Okla. p. 10.

44. Smith, G. R., and W. E. Knight. 1984. *Breeding legumes for improved N_2 fixation.* In G. B. Collins and J. G. Petolino [eds.] *Applications of Genetic Engineering to Crop Improvement.* Martinus Nijhoff/Dr. W. Junk Publishers, Dordrecht, The Netherlands. pp. 1-23.

45. Smith, R. R. 1983. *Use of legumes in cropping systems - germplasm resources in Northcentral Region of U.S.* Rpt., Work-Planning Conf. on Legumes in Cons. Tillage Systems. Agr. Res. Serv., U.S. Dept. Agr., Lincoln, Nebr. pp. B-11-B-13.

46. Smith, R. R., N. L. Taylor, and S. R. Bowley. 1985. *Red clover.* In N. L. Taylor [ed.] *Clover Science and Technology.* Mono. No. 25. Am. Soc. Agron., Madison, Wisc. pp. 457-470.

47. Taylor, N. L. 1985. *Red clover.* In M. E. Heath, D. S. Metcalf, and R. F. Barnes [eds.] *Forages: The Science of Grassland Agriculture.* Iowa State Univ. Press, Ames. pp. 109-117.

48. Townsend, C. E. 1985. *Miscellaneous perennial clovers.* In N. L. Taylor [ed.] *Clover Science and Technology.* Mono. No. 25. Am. Soc. Agron., Madison, Wisc. pp. 563-578.

49. Townsend, C. E., G. O. Hinze, W. D. Ackerman, E. E. Renmenga. 1975. *Evaluation of forage legumes for rangelands of the Central Great Plains.* Gen. Ser. 942. Colo. State Univ. Expt. Sta., Ft. Collins. pp. 1-10.

50. Van Keuren, R. W., and J. R. George. 1985. *Hay and pastures seedings for the Central and Lake States.* In M. E. Heath, D. S. Metcalf, and R. F. Barnes [eds.] *Forages: The Science of Grassland Agriculture.* Iowa State Univ. Press, Ames. pp. 348-358.

51. White, G. A. 1982. *Germplasm acquisition.* In Proc., 27th Grass Breeders Work Planning Conf. Pa. State Univ., University Park. pp. 15-18.

52. White, G. A., and A. J. Oakes. 1979. *Introduction and documentation of forage crop germplasm.* In Proc., 36th South. Pasture and Forage Crop Impr. Conf. Agr. Res. Serv., U.S. Dept. Agr., Washington, D.C. pp. 105-111.

53. Williams, W. A. 1968. *Effects of nitrogen from legumes and crop residues on soil productivity.* Rice J. 77(5): 29-32.

Impact of variability in forage legume root development on seedling establishment

J. F. Gomez, B. L. McMichael,
A. G. Matches, and H. M. Taylor

Differences in rooting characteristics influence the establishment and development of cereal and grain legume plants (*1, 3, 7,10*). At the onset of seedling development, the rate of root elongation is related closely to seed size (*2, 6, 11*) and to cultivar differences (*5, 8, 10, 12, 14*).

The optimum temperature for conducting germination and root development studies of forage legumes has not been defined clearly. The response of root development to temperature differs greatly among species (*1, 2, 9, 10, 13*). Cohen and Tadmor (*2*) measured temperature effects on root elongation in crested wheatgrass [*Agropyron desertorum* (Fish. ex Link) Schult], tall wheatgrass [*A. elgonatum* (Host.) P.B.], bur clover (*Medicago polymorpha* L.), barrel medic (*Medicago truncatula* Gaertn.), and woollypod vetch (*Vicia dasycarpa* Ten.). However, a more complete study on the variability in root development of forage legumes has not been conducted. Our objectives, therefore, were to compare germination and seedling root development of 21 legume entries as a function of temperature and, secondly, to determine relationships between rooting depth in the field with the rooting depth in a laboratory method designed to screen a large number of species for rooting characteristics.

Study methods

We used the laboratory "pouch" procedure (*8*) to study germination and root development of 21 forage legume entries. We added 50 ml of half-strength Hoagland's solution (*4*) to 174-mm-high by 164-mm-wide polyethylene growth pouches (diSPo Seed-Pack[1], Northrup King, Inc.), which contained absorbent paper inserts with a trough near the top for placement of the seed. Perforations in the bottom of the trough allowed roots to grow downward. We suspended the growth pouches upright on wooden racks that were placed in germination chambers held at constant temperatures of 10°, 15°, 20°, 25°, and 30°C.

We placed two seeds of different entries into each pouch, and planted a total of six seeds for each entry at each temperature. Length of the main root axes and lateral roots were measured daily for 10 days. We also made daily counts of the number of lateral roots on each main axis. In addition, daily germination percentages at each temperature were determined on 100 seeds/entry.

The experimental field was located at the Texas Tech University campus farm on a Acuff loam soil. We planted

J. F. Gomez is a postdoctoral research associate, B. L. McMichael is a plant physiologist, A. G. Matches is Thornton professor, and H. M. Taylor is Rockwell professor, Department of Plant and Soil Science, Texas Tech University, and Agricultural Research Service, U.S. Department of Agriculture, Lubbock, 79409.

entries on March 6, 1986, in plots of five rows placed 17 cm apart. The experimental design was a randomized complete block with four replications. At 32, 54, and 74 days after planting, we recorded rooting depth using the soil-core method. Soil cores, 6.5 cm in diameter, were removed over individual plants. We measured the root length or rooting depth by breaking the soil core and locating the root tip.

Results

Generally, germination of the 21 forage legumes in pouches was highest at 15° and 20°C; the greatest reduction in germination occurred at 30°C. Main root axes length varied significantly with time, temperature, and entry. The length of the main root axes increased linearly with time and quadratically with temperature. Main root axes length was generally greatest at 20°C. The shallowest rooted entry was Yuchi arrowleaf clover (21 mm), while the deepest rooted entry was Austrian winter pea (151 mm), both of which occurred at 10 days after planting and 20°C.

Of the 21 forage legume entries, we found lateral roots on 15 entries. There was lateral root growth at 15°, 20°, and 25°C and as early as 5 days after planting. The number of lateral roots and total lateral root length were significantly different among the entries. Generally, the number of lateral roots tended to be greater with lower temperature. The total lateral root length, on the other hand, was greater with higher temperature. Also, the number and total length of lateral roots for seven entries were positively correlated with main axes length.

Overall, higher temperatures within the range investigated accelerated root elongation, while lower temperatures retarded root elongation. The root elongation rate (mm/day) increased linearly with seed weight at 10°C and increased quadratically at 15°, 20°, 25°, and 30°C.

We found significant, positive linear relationships between the main axes length determined by the laboratory method and the rooting depth in the field. The highest coefficients of determination between the field study and the pouch runs at 10°, 15°, 20°, 25°, and 30°C were 0.30, 0.41, 0.66, 0.57, and 0.60, respectively. Overall, rooting depth in the field at 32 days after planting was significantly predicted by main axes length in the pouches by 5 days after planting. Pouch studies thus appear to have promise for preliminary screening for differences in rooting characteristics of forage legumes.

---

[1]Mention of a company name or trademark is for the reader's benefit and does not constitute endorsement of a particular product over others that may be commercially available.

REFERENCES

1. Arndt, C. H. 1945. *Temperature-growth relations of the roots and hypocotyls of cotton seedlings.* Plant Physiol. 20: 200-220.
2. Cohen, Y., and N. H. Tadmor. 1969. *Effects of temperature on the elongation of seedling roots of some grasses and legumes.* Crop Sci. 9: 189-192.
3. Gregory, P. J. 1983. *Response to temperature in a stand of pearl millet (*Pennisetum typhoides S&H). II. Root development.* J. Exp. Bot. 34: 744-756.
4. Hoagland, D. R., and D. I. Arnon. 1938. *The water culture*

method for growing plants without soil. Circ. 347. Calif. Agr. Exp. Sta., Berkeley.

5. Kaspar, T. C., C. D. Stanley, and H. M. Taylor. 1978. *Soybean root growth during the reproductive stages of development.* Agron. J. 70: 1105-1107.
6. Kittock, D. L., and J. K. Patterson. 1959. *Measurement of relative root penetration of grass seedlings.* Agron. J. 51: 512.
7. Klepper, B., R. K. Belford, and R. W. Rickman. 1984. *Root and shoot development in winter wheat.* Agron. J. 76: 117-122.
8. McMichael, B. L., J. J. Burke, J. D. Berlin, J. L. Hatfield, and J. E. Quisenberry. 1985. *Root vascular bundle arrangements among cotton strains and cultivars.* Environ. and Exp. Botany. 25: 23-30.
9. Sprague, V. G. 1943. *The effects of temperature and day length on seedling emergence and early growth of several pasture species.* Soil Sci. Soc. Am. Proc. 8: 287-294.
10. Stone, J. A., and H. M. Taylor. 1983. *Temperature and the development of the taproot and lateral roots of the four intermediate soybean cultivars.* Agron. J. 75: 613-618.
11. Tadmor, N. H., and Y. Cohen. 1968. *Pre-emergence seedling root development of Mediterranean grasses and legumes.* Crop Sci. 8: 416-419.
12. Taylor, H. M., E. Burnett, and G. D. Booth. 1978. *Taproot elongation rates of soybeans.* Z. Acker-Pflanzenb. 146: 33-39.
13. Taylor, H. M., M. G. Huck, and B. Klepper. 1972. *Root development in relation to soil physical condition.* In D. Hillel [ed.] *Optimizing the Soil Physical Environment Toward Greater Crop Yields.* Academic Press, New York, N.Y.
14. Vincent, C., and P. J. Gregory. 1986. *Differences in the growth and development of chickpea seedling roots (Cider arietinum).* Expl. Agr. 22: 233-242.

Fall-seeded legume nitrogen contributions to no-till corn production

A. M. Decker, J. F. Holderbaum, R. F. Mulford, J. J. Meisinger and L. R. Vough

We conducted six field experiments at Piedmont and Coastal Plain locations from 1982 through 1985 to evaluate the N-supplying potential of 14 fall-seeded legumes for no-till corn production systems. We found that hairy vetch, Austrian winter peas, and crimson clover had the greatest potential, especially on the lighter textured Coastal Plain soils where the winter growing season is considerably more mild than in the central Piedmont. Subterranean clover may have a place in the lower Coastal Plain region as a naturally reseeding winter annual legume.

Winter cover crops in these studies were seeded after fall corn harvests and allowed to grow until corn planting time the following spring—late April to early May. About 10 days prior to corn planting, paraquat plus residual herbicides were applied. Application of the knock-down herbicide prior to corn planting is necessary, especially on heavy-textured soils during wet springs, to allow the soil to dry adequately for good seed coverage and good seed-soil contact. In studies through 1985, we applied 80 pounds of N/acre to half of each legume plot when corn was at the 5- to 7-leaf stage. Starting in 1986, we increased the number of N rates to four to identify more accurately the actual N contributions of these fall-seeded legumes. Rates varied from 40 to 240 pounds of N/acre depending upon species and location.

Only data on vetch, peas, and crimson clover are discussed here (Table 1). Responses varied depending upon the soil N status; the specific legume; and the climatic conditions, especially rainfall, during the periods of fall legume establishment and spring corn planting and emergence. The longer the legume was allowed to grow in the spring, the larger the amount of N fixed, but the greater the risk of inadequate soil moisture for good corn emergence and early growth. This was dramatically illustrated by 1986 data. Research is being initiated in 1987 to address the optimum herbicide application and corn planting dates.

Results clearly show that fall-seeded legumes can replace much of the applied N fertilizer so essential for maximum economic corn yields. These findings are in agreement with work reported by Mitchel and Teel (*4*), Ebelhar and associates (*1*), and Hargrove (*2*). The percent N in the top

A. M. Decker is a professor, Agronomy Department, University of Maryland, College Park, 20742; J. F. Holderbaum is a graduate assistant, Agronomy Department, University of Florida, Gainesville, 32611, and formerly a graduate research assistant, University of Maryland; R. F. Mulford is manager of the Poplar Hill Research Farm, Quantico, Maryland 21856; J. J. Meisinger is a soil scientist with the Agricultural Research Service, U.S. Department of Agriculture, Beltsville, Maryland 20705; and L. R. Vough is an associate professor, Agronomy Department, University of Maryland, College Park, 20742. Contribution from Maryland Agricultural Experiment Station. Work was supported by USDA/SAE/ARS Grant No. 58-32U4-2-424, 9-14-82 through 9-30-86.

growth of these legumes varied considerably among species and their stage of growth when the knock-down herbicide was applied. But the single, most important factor affecting legume N contributions to no-till corn production was legume dry matter production. Legume covers can have a significant effect on subsequent corn N uptake. Small grains, either as pure stands or included with legume cover crops, reduced corn N uptake as well as corn grain yields relative to no cover controls and pure legume covers, respectively. These results contrast those of Moschler and associates (5), who reported small grain cover crops to be superior to legumes. When large amounts of soil N are available, legume N fixation declined dramatically, as illustrated by the 1985 Piedmont location data. Moustafs and associates (6) and Heichel (3) also found this to be true. Over 30% of total N in the top growth of many of the legumes in our study appeared to be available to the subsequent corn crop. This is substantially more than the 15% to 25% suggested by Heichel (6).

Our tests led to four conclusions:

▶ Fall-seeded legumes can supply significant amounts of N to no-till corn; these responses are greatest at Coastal Plain locations with more favorable winter growing conditions and generally lighter textured soils with lower N-supplying capabilities.

▶ Small grains seeded with legumes reduced N uptake by the corn crop and subsequent grain yields.

▶ No-till drilling of fall-seeded legumes after corn harvest was consistently more successful than broadcast seeding prior to corn harvest.

▶ The most reliable legume species at both Piedmont and Coastal Plain locations was hairy vetch.

REFERENCES
1. Ebelhar, S. A., W. W. Fry, and R. L. Blevins. 1984. *Nitrogen from legume cover crops for no-tillage corn.* Agron. J. 76: 51-55.
2. Hargrove, W. L. 1986. *Winter legumes as a nitrogen source for no-till grain sorghum.* Agron. J. 78: 70-74. .
3. Heichel, G. H. 1985. *Nitrogen recovery by crops that follow legumes.* In *Forage Legumes for Energy-Efficient Animal Production.* Proc., Trilateral workshop. Nat. Tech. Info. Serv., Springfield, Va.
4. Mitchell, W. H., and M. R. Teel. 1977. *Winter annual cover crops for no-tillage corn production.* Agron. J. 69: 569-73.
5. Moschlar, W. W., G. M. Shear, D. L. Hallock, R. D. Sears, and G. D. Jones. 1967. *Winter cover crops for sod-planted corn: Their selection and management.* Agron. J. 59: 547-51.
6. Moustafa, E., R. Ball, and T. R. O. Field. 1969. *The use of acetylene reduction to study the effects of nitrogen fertilizer and defoliation on nitrogen fixation by field-grown white clover.* N. Z. J. Agr. Res. 12: 691-96.

Table 1. Effects of selected winter cover crops on nitrogen production, corn nitrogen uptake, and grain yields.

Location and year	Cover Crop (pounds)	Nitrogen Taken Up by Corn Crop		Grain Yield	
		0-N	80-N	0-N	80-N
		—pounds/acre—		—bushels/acre—	
Coastal Plains 1984					
Hairy vetch	313	156	189	168.0	167.0
Peas	133	125	163	142.8	158.2
Crimson	202	122	147	123.2	144.1
Barley	62	64	125	64.8	136.9
No cover	—	73	115	66.4	127.2
LSD (.05)	52	19		14.2	
Coastal Plain 1985					
Hairy vetch	257	157	193	86.1	83.3
Hairy vetch-wheat	227	116	173	77.0	80.4
Peas	191	156	193	96.0	91.7
Peas-wheat	170	125	171	86.0	82.8
Crimson	206	111	189	82.8	89.9
Wheat	41	41	113	18.5	79.0
No cover	—	47	125	28.3	89.6
LSD (.05)	38	23		12.7	
Piedmont 1984					
Hairy vetch	144	—	—	182.4	196.1
Peas	30	—	—	172.7	186.4
Barley	62	—	—	118.3	165.7
No cover	—	—	—	165.6	160.8
LSD (.05)	22			14.0	
Piedmont 1985					
Hairy vetch	169	157	193	164.4	177.8
Hairy vetch-wheat	186	105	131	119.6	145.2
Peas	171	169	186	172.2	174.3
Peas-wheat	176	107	157	123.5	145.0
Wheat	91	70	118	70.7	130.4
No cover	—	144	182	147.6	167.8
LSD (.05)	26	32		21.0	

Soil drainage, acidity, and molybdenum effects on legume cover crops in no-till systems

B. N. Duck and D. D. Tyler

Agronomists and farmers have assumed for many years that most legumes are better adapted to well-drained soils with neutral to slightly acid reaction. However, legume species differ in adaptation to soil conditions. Practical considerations make it imperative that legumes used as cover crops be reasonably suited to conditions under which the primary crops in the planting sequence are grown. Most information about adaptation of legumes used as cover crops was developed in conventional tillage systems. We conducted a field experiment to evaluate effects of internal soil drainage, molybdenum supplementation, and soil acidity on performance of five cool-season legumes used as cover crops for corn in a no-till system.

Treatments were arranged in a split-plot design (Table 1). The poorly drained Routon silt loam (fine silty, mixed, thermic, Typic Ochraqualfs, 0-2% slopes) and well-drained Lexington silt loam (fine silty, mixed, thermic, Typic Paleudalfs, 2-5% slopes) were located about 200 feet apart in a field that had been planted to row crops with conventional tillage for many years. Molybdenum was applied to seed as an aqueous solution of sodium molybdate in an inoculation slurry with a gel adhesive and strains of *Rhizobium* suitable for the respective legumes.

We established soil acidity levels by determining the pH of each sub-subplot in the 0- to 6-inch depth, then applying agricultural lime to achieve the desired pH level. Liming was done in August 1983. The sites were planted to no-till corn in 1984. Legumes were planted in corn residue with a no-till drill on October 1, 1984. Samples were taken for forage productivity and N content (micro-Kjeldahl) measurements in May 1985.

Soil drainage and acidity and legume species influenced dry matter production (Table 2). The main effects of each of these factors were significant when averaged across all levels of the other factors. As anticipated, mean dry matter production was greater at pH 6.7 and 6.2 than at pH 5.7. However, average forage yield on the poorly drained site (1.74 tons/acre) was greater than on the well-drained site (1.40 tons/acre). This is contrary to usual perceptions of legume adaptation.

Partitioning of interaction effects revealed adaptational differences among species (Table 3). Crimson clover produced more forage than other legumes on the poorly drained soil. Its productivity, however, was not different from that of hairy vetch on the well-drained site. Similarly, crimson clover produced more forage than other legumes at pH 6.2 and 6.7, but its productivity was not different from that of hairy vetch

and sub clover at pH 5.7. Closer examination of the drainage/acidity/species interaction showed marked response of crimson clover and caley pea productivity to higher pH levels on the well-drained site. Winter pea, on the other hand, showed a similar response to increasing pH on the poorly drained soil.

Percent N content of forage differed among species when averaged across all other factors (Table 2) and was higher for legumes grown on the poorly drained soil than for those on the well-drained soil. Analysis of the drainage/species interaction showed that hairy vetch, winter pea, and caley pea differed significantly in N content on the poorly drained soil but not on the well-drained site. Molybdenum supplementa-

Table 1. Treatment factors and levels and their arrangement in a split-plot design.

Design Categories	Treatments	
	Factors	Levels
Main plots	Soil drainage	Poorly drained, well-drained
Sub-plots	Molybdenum	0.0 and 1.0 ounces/acre
Sub-sub-plots	Soil acidity (pH)	5.7, 6.2, and 6.7
Sub-sub-sub-plots	Legume species	Caley pea (*Lathyrus hirsutus*)
		Crimson clover (*Trifolium incarnatum*)
		Hairy vetch (*Vicia villosa*)
		Sub clover (*T. subterraneum*)
		Winter pea (*Pisum sativum arvense*)

Table 2. Influence of various factors and interactions on response variables forage yield and nitrogen content.

	Probabilities of Larger F-Values	
Factors*	Forage Yield	N Content
DR	.021	.006
MO	.296	.586
PH	.035	.262
CO	.0001	.0001
DR/MO	.160	.763
DR/PH	.529	.610
DR/CO	.002	.0001
MO/PH	.913	.402
MO/CO	.309	.436
PH/CO	.006	.883
DR/MO/PH	.290	.572
DR/MO/CO	.155	.972
DR/PH/CO	.016	.340
MO/PH/CO	.033	.883

*DR = soil internal drainage, MO = molybdenum rate, PH = soil acidity (pH), and CO = legume species

Table 3. Separation of interactions of legume species with soil internal drainage and acidity. Values are forage dry matter yields, tons/acre.

	Forage Dry Matter Yields (tons/acre)				
	Soil Internal Drainage		Soil pH Levels		
Legume	Poorly Drained	Well-Drained	5.7	6.2	6.7
Caley pea	1.49 c*	1.26 bc	1.22 c	1.29 d	1.63 b
Crimson clover	2.05 a	1.66 a	1.63 a	1.97 a	1.96 a
Hairy vetch	1.71 b	1.56 a	1.62 a	1.74 b	1.54 b
Sub clover	1.80 b	1.15 c	1.48 ab	1.48 cd	1.45 b
Winter pea	1.64 bc	1.36 b	1.38 bc	1.57 bc	1.54 b

*Values within a column not accompanied by the same letter are different at the p = .05 level.

B. N. Duck is a professor of plant science, University of Tennessee, Martin, 38238, and D. D. Tyler is an associate professor of soil science, University of Tennessee, Jackson, 38301.

tion and soil acidity level did not influence N content of legume forage.

These data show that soil internal drainage and acidity affect forage production of legumes grown as cover crops in no-till systems and that species respond differently to these factors. They also show that soil drainage affects N content of legume forage and that response to this factor differs among species.

Soil temperature and the growth, nitrogen uptake, dinitrogen fixation, and water use by legumes

J. A. Zachariassen and J. F. Power

Our study sought to investigate the effects of root/soil temperature and plant age on (a) plant dry matter production, (b) N uptake, (c) dinitrogen fixation, and (d) water use and water use efficiency of eight species of legumes and three non-nodulated species. The study, conducted in soil temperature tanks in the greenhouse, consisted of three separate split-plot experiments at soil temperatures of $10°$, $20°$, and $30°C$. Main plots were plant species split by five sampling dates, with three replicates. Water use was taken as the weight difference between one weighing and the next. We used ^{15}N isotope dilution (ID) to determine the percent plant N derived from the atmosphere (%Ndfa), where:

$$atm.\%^{15}N \ ex. = atom \ \%^{15}N \ excess$$

and

$$\%Ndfa = \left[1 - \frac{atm.\%^{15}N \ ex.(fs)}{atm.\%^{15}N \ ex.(nfs)} \right] \times 100$$

This calculation of %Ndfa is independent of yield, requiring only that the fixing system (fs) and nonfixing system (nfs) take the same proportions, but not the same amounts, of N from the soil and/or fertilizer. From the %Ndfa we calculated the absolute amount of N_2 fixed at each sampling date for each legume. This figure represents a cumulative or integrated value for N_2 fixation up to each sampling date. Winter wheat and Kentucky bluegrass were used as a non-fixing control species for annuals and perennials, respectively.

Hairy vetch, non-nodulating soybeans, nodulating soybeans, and faba bean produced greatest amounts of dry matter at $10°C$; faba bean, non-nodulating soybeans, nodulating soybeans, and lespedeza produced the most dry matter at $20°C$; and nodulating soybeans, non-nodulating soybeans, and lespedeza produced the most dry matter at $30°C$ (Table 1). For the $10°C$ and $20°C$ soil temperature treatments, the legumes accumulated more dry matter/pot than the nonlegumes, winter wheat, and bluegrass throughout the growing period. At $10°$, the annuals crimson clover, faba bean, field pea, lespedeza, and nodulating soybean generally produced more dry matter than the biennials hairy vetch, sweetclover, and white clover.

Generally, N uptake followed similar trends as the dry mat-

J. A. Zachariassen is a graduate research assistant, Agronomy Department, Colorado State University, Fort Collins, 80523, and J. F. Power is a research leader with the Agricultural Research Service, U.S. Department of Agriculture, and adjunct professor, Agronomy Department, University of Nebraska, Lincoln, 68583.

ter production. Nitrogen uptake at 10 °C was greatest for hairy vetch; at 20 °C and 30 °C it was greatest for nodulated soybeans. Usually, N uptake by annuals was greater than that by biennials. At 10 °C and 20 °C, N uptake by the legumes was greater than that by the nonlegumes. Among the non-nodulated species, N uptake by non-nodulated soybeans was greater than by winter wheat and bluegrass at all samplings and temperature treatments. At most sampling dates N uptake at 10 °C was also high for hairy vetch, faba bean, white clover, nodulated soybeans, and crimson clover; at 20 °C it was high for faba bean, non-nodulated soybeans, and hairy vetch; and at 30 °C it was high for non-nodulated soybeans and lespedeza.

Hairy vetch, faba bean, white clover, and crimson clover fixed the greatest amounts of N_2 at 10 °C; fixation by lespedeza was virtually nil at this temperature throughout the growth period (Table 1). For the 20 °C temperature treatments fixation was greatest by nodulated soybeans, faba bean, white clover, and sweetclover; values for faba bean were consistenly high throughout the growth period. At 30 °C N_2 fixation was greatest by nodulated soybeans and almost doubled between day 63 and day 105 in lespedeza and sweetclover. Fixation by crimson clover and field peas generally decreased over time at this temperature treatment. There was not any clear distinction in amount of N_2 fixed between the annuals and biennials. Among the annuals, generally fixation was greatest by nodulated soybeans and faba bean (at 10° and 20° for faba bean), and by hairy vetch among the biennials. Overall N_2 fixation followed similar trends observed for dry matter production and N uptake.

Nodule count data were more variable than dry matter production and N uptake. Greatest nodule counts were observed in hairy vetch and sweetclover at 10 °C. But for sweetclover this was not reflected by greater dry matter production and N uptake. Lespedeza often had greatest nodule numbers at 20° and 30 °C. Among the annuals, faba bean had the greatest number of nodules at 10° and 20 °C, but at 30 °C nodule count decreased toward the end of the growth period. Likewise, at 20° and 30 °C nodule counts decreased in field peas and crimson clover over time. Nodule count was well correlated with N uptake for nodulated soybeans, lespedeza, white clover, and crimson clover at 30 °C and for sweetclover at 10 °C. Some nodules were found on roots of non-nodulated soybeans, especially at 30 °C, although in considerably fewer numbers than for nodulated soybeans.

Water use generally increased with increased soil temperature. At 10 °C, water use was greatest for faba bean. At 20° and 30 °C, nodulated soybeans used the most water. Water use efficiency tended to be highest at the earlier samplings, then declined (Table 1). A notable exception was for lespedeza. Water use efficiency was usually greatest at 10 °C, presumably because a greater portion of the water used was lost through transpiration rather than through evaporation. Several annual species, faba bean and field peas in particular, approached maturity and began senescence, so water use effenciency was not calculated for the later sampling dates. Water use efficiency has little relationship to dry matter, and differences between species did not follow easily iden-

Table 1. Dry matter production, nitrogen fixation, and water use efficiency of legume species as affected by soil temperature and time.

Species and Temperature	Dry Matter Production (g/pot) by time (days)				Nitrogen Fixation (mg N fixed/pot) by time (days)				Water Use Efficiency (mg DM/g H_2O) by time (days)			
	42	63	84	105	42	63	84	105	42	63	84	105
Hairy vetch												
10	2.32	3.05	5.33	6.39	113.39	122.99	167.21	253.04	2.77	1.75	2.10	2.08
20	1.41	2.56	2.65	4.26	52.41	100.40	77.07	138.85	1.36	1.26	0.88	1.10
30	1.16	1.71	1.79	2.10	16.96	9.90	14.68	34.62	0.92	0.90	0.70	0.69
Sweetclover												
10	1.33	1.67	2.42	2.41	24.53	53.21	76.53	75.10	2.34	1.66	1.54	1.15
20	1.85	3.15	3.95	4.24	59.26	97.95	145.64	139.01	1.54	1.57	1.47	1.18
30	1.85	2.63	2.97	4.43	47.99	52.39	57.36	118.61	1.30	1.00	0.93	1.01
Faba bean												
10	4.60	4.57	4.15	4.95	175.24	148.81	153.33	138.46	3.03	2.10	*	*
20	5.01	5.47	5.82	6.53	154.53	141.66	138.41	152.15	1.96	1.45	*	*
30	2.43	2.62	3.07	2.80	25.75	14.09	12.88	4.76	1.16	1.09	*	*
Lespedeza												
10	0.59	0.70	1.55	2.10	0.00	5.28	3.99	0.00	3.58	1.65	2.45	2.46
20	1.19	2.43	4.41	7.84	10.43	33.76	106.27	164.64	1.96	1.87	2.00	2.25
30	1.57	3.04	4.94	10.10	18.35	25.69	68.88	200.27	1.42	1.68	1.74	2.31
Field pea												
10	2.03	2.92	2.40	2.90	41.47	54.97	32.91	58.69	3.14	3.38	*	*
20	2.04	2.09	2.31	2.07	43.50	24.01	14.06	9.72	2.07	1.75	*	*
30	1.68	1.22	1.59	1.31	14.13	11.81	6.85	0.00	1.58	0.94	*	*
White clover												
10	1.12	1.89	3.22	4.65	22.85	62.03	99.98	184.74	3.08	2.04	2.15	2.23
20	1.63	2.81	3.47	5.20	44.63	88.93	123.00	174.46	2.36	2.18	1.74	1.70
30	1.36	2.43	3.00	3.39	3.20	0.70	45.71	43.66	1.17	1.43	1.31	1.06
Nodulated soybeans												
10	3.81	3.43	4.07	5.00	36.11	37.47	42.55	39.23	4.65	3.27	2.79	2.55
20	4.01	7.40	10.50	14.63	131.70	244.25	357.64	471.24	2.44	2.15	2.14	2.29
30	4.01	7.91	11.19	16.68	107.55	185.89	295.46	330.41	1.50	1.73	1.77	1.96
Crimson clover												
10	2.05	2.37	3.66	4.79	49.26	49.76	82.12	122.15	2.93	2.37	2.44	2.71
20	2.50	3.78	4.71	3.37	61.35	89.89	97.98	63.70	1.86	1.94	2.01	1.31
30	1.74	1.75	2.14	2.00	10.66	2.35	15.90	6.70	1.28	1.12	1.04	0.78

*Plants dead.

tifiable patterns. However, nodulated soybeans were notable in that, for almost any time and temperature comparison, they had among the highest water use efficiency values of any species.

These results show that different species are best adapted for different situations. When seeking a legume cover crop for a short growing period (less than 60 days) and at low temperature (early spring), hairy vetch or faba bean would be the best choices. With warmer temperatures (early summer), nodulated soybeans would fix the most N. If the growing period is longer for cool-season growth, the above species plus white clover or crimson clover could be considered. At warmer temperatures with a long growth period, nodulated soybeans and lespedeza would be the best candidates.

Interseeding in corn

Chris Nanni and C. S. Baldwin

Interseeding of legumes into standing row crops is not a new concept. Rather, it is one that has given way to the constraints of time and money. However, with renewed emphasis on soil degradation because of erosion and compaction, losses of organic matter, and rising fertilizer costs, the practice of interseeding may again become commonplace.

Pilot-type interseeding experiments began at the Ridgetown College of Agricultural Technology in 1981. It was not until 1984 that a fully documented research project was initiated. The purpose of the project is to study the effects of various interseeded crop species on a number of soil parameters, such as bulk density, soil structure, and organic matter additions. As well, we hope to determine the effect that interseeded crop species will have on the growth and yield of cash crops, in our case, corn, and assess the ability of these same interseeds to reduce erosion.

There are six trials associated with the research. Three of the trials compare various interseed species, with respect to their growth and development coupled with the previously listed parameters, sown into row corn grown on three different soil types—Berrien sand, Brookston clay, and Brookston clay loam. Two other trials compare various interseed species sown in three different corn row widths, 30-, 45-, and 60-inch rows, on two soil types—Haldimand loam and Brookston clay loam. The last trial compares the rate and time of interseeding various interseed species sown in standing corn on a Berrien sandy loam soil.

Trials are located on the Ridgetown campus and a number of on-farm locations. All trials are sprayed with eradicane, a preemergent herbicide, prior to planting. Corn is planted in 30- or 38-inch rows, depending on the co-operators' equipment. Once the corn is 6 to 8 inches high, cultivation and postemergent herbicide are used to eliminate weeds. Herbicide selection is such that no residual elements of the herbicide retard or restrict interseed growth.

Interseeding begins when the corn is 12 to 16 inches high. The ground is cultivated, for the last time in the growing season, to a depth of 4 inches. A second lighter cultivation, 1 to 1.5 inches, follows. With this second cultivation interseed is distributed behind the cultivator tines. Rolling packers attached to the cultivator unit press the soil down, affording good seed-soil contact and preventing moisture from evaporating from the soil. The trials are then left until harvest in mid-autumn.

Results to date show no significant differences in corn yields among treatments, except in the two trials in which corn was planted in three various row widths. However, in those two trials significant differences in corn yield did not exist among treatments having the same row widths. Results on the dry matter contributions of the various interseeds also

Chris Nanni is a lecturer and researcher and C. S. Baldwin is head of the Soils Section at Ridgetown College of Agricultural Technology, Ontario Ministry of Agriculture and Food, Ridgetown, Ontario, Canada N0P 2C0.

showed no significant differences among treatments. Only in the various row widths and in the rate and time trials did differences in dry matter contribution occur. This was especially evident in the 60-inch-row corn plots where significant differences also occurred among treatments, that is, among interseed varieties, and where rate and times of interseeding varied in the trial on the sandy loam.

Possibly the most important aspect of the interseeding project was the contribution of dry matter from the interseed despite the droughty conditions that prevailed during the germinating and emergence phase. This occurred on all of the trials.

Initial observations from interseed data point to a high degree of success, in terms of interseed biomass production, in the wider spaced rows, specifically in the 60-inch rows. This increased growth of interseed in the wide rows was, and will continue to be, expected as sunlight penetration appears to be a limiting factor of interseed growth in the narrow rows. Soil improvement capabilities from the increased interseed biomass in the 60-inch row plots have not yet been documented. But that will undoubtedly reduce soil erosion. Again, inferences relating to row width cannot, and indeed should not, be made on the observations of a single year's data. One way of controlling soil erosion, employing the use of interseeds, is accumulating biomass in the soils. This is achieved through proper tillage practices. Accordingly, it would take a number of years for this accumulation to become appreciable and, in fact, measurable.

Bulk density and water infiltration rate analysis both show no signficiant differences among treatments to date. The relatively low biomass contribution, both top and root growth, can account for this significance.

The interseeding research is expected to last until the spring of 1988. The final results will help us develop recommendations on the use of legume interseeds in corn production.

Evaluation of cool-season legumes for conservation tillage

C. M. Owsley and E. D. Surrency

Since the fall of 1983, the Americus Plant Materials Center of the Soil Conservation Service has conducted a screening program of cool-season annual legumes for use in conservation tillage systems. *Lathyrus, Trifolium, Vicia, and Medicago* are the genera included in this screening test.

The legumes are assembled from foreign as well as naturalized populations. All foreign accessions came through the Plant Introduction System. The naturalized legumes were collected and processed by SCS personnel in the Southeast. These naturalized populations originated from legume cover crop plantings grown during the middle of the century.

We evaluated each accession, a documented and numbered legume, for adaptability, growth, vigor, winter hardiness, stand, reseeding ability, flowering date, seed production, disease resistance, and insect resistance. Each accession was compared to all known standards. The evaluation area is located on Orangeburg sandy loam in Americus, Georgia.

Since the screening program began, over 900 cool-season annual legume accessions have been assembled, planted, and evaluated. The following superior accessions have been selected for seed increase and further evaluation:

Legume	Identification Number	Common Name	Origin
Vicia villosa	PI-222177	Hairy vetch	France
Vicia villosa	PI-383803	Hairy vetch	Turkey
Medicago hispida	9039823	Bur clover	Lee Co., Ala.
Medicago hispida	9039842	Bur clover	Lee Co., Ala.
Medicago orbicularis	PI-199258	Button clover	Greece
Medicago orbicularis	PI-289311	Button clover	Hungary

The Americus Plant Materials Center has also increased 87 caley pea (*Lathyrus hirsutus*) accessions that were forwarded to Auburn University for future evaluation and testing in the Black Belt area.

The superior legumes will be studied in cooperation with the University of Georgia. The legumes will be compared against existing varieties in a replicated grain sorghum conservation tillage research test. The standards of comparison will include commercially available crimson clover (*Trifolium incarnatum*), sub clover (*T. subterraneum*), common vetch (*Vicia sativa*), bigflower vetch (*V. grandiflora*), caley pea (*Lathyrus hirsutus*) and hairy vetch (*V. villosa*). The legumes will be evulated for ground cover and erosion control potential. The legumes will also be analyzed for N and dry matter content. The conservation-tillage planted grain sorghum will be evaluated for yield and protein content.

This and future evaluations could provide documentation to support a new varietal release for use as a legume cover crop in conservation tillage systems.

C. M. Owsley is manager of the Americus Plant Materials Center, Soil Conservation Service, U.S. Department of Agriculture, Americus, Georgia 31709. E. D. Surrency is a plant materials specialist with SCS/USDA, Athens, Georgia 30601.

A NITROGEN SOURCE

Legumes as a source of nitrogen in conservation tillage systems

G. H. Heichel

Conservation tillage has been defined as "any tillage sequence that reduces loss of soil or water relative to conventional tillage; often a form of noninversion tillage that retains protective amounts of residue mulch on the surface" (*34*). Residue is often interpreted to be living or dead components of legume crops.

There are many types of conservation tillage systems. These range from traditional crop rotations to the more contemporary use of intercropping (*9*) and living mulches (*8*). The importance of conservation tillage systems varies geographically throughout the United States. It is evident, however, that legume crops figure prominently in many types of conservation tillage systems (Table 1). Inclusion of forage or grain legumes in cropping systems often has a beneficial effect on the growth or yield of interseeded or succeeding crops by providing N and through such non-N effects as reduced soil erosion; control of weeds, insects, or pathogens; or improved soil structure and soil moisture.

Herein, I look at how much symbiotically derived N might be contributed by forage and grain legumes in conservation tillage systems and at research needed to enhance the use of legume N in conservation tillage systems.

Economic value of symbiotic nitrogen

Crop legumes symbiotically remove, or fix, nutritionally inert, gaseous dinitrogen (N_2) from the earth's atmosphere and convert it into amino acids that are nutritionally useful to crops and subsequently to farm animals or humans. The capability to fix atmospheric N_2 decreases the requirements

G. H. Heichel is a plant physiologist with the Agricultural Research Service, U.S. Department of Agriculture, St. Paul, Minnesota 55108. This paper is a joint contribution of the USDA-ARS and the Minnesota Agricultural Experiment Station, Scientific Journal Series No. 15241.

of crop legumes for soil N mineralized from organic matter or for application of commercial N fertilizer. The genetic endowment of legumes to fix N_2 is a major factor in their cost-effective production.

Farmers and others seldom appreciate the potential economic value of partly substituting biologically renewable, fixed N_2 for commercial N fertilizer. What, indeed, is the value of the total N_2 fixation of all U.S. crop legumes in terms of the volatile cost of N in anhydrous ammonia? Currently, anhydrous ammonia costs about 10 cents/pound in the upper Midwest. In this case, the total N_2 fixation of U.S. crop legumes (16 billion pounds annually) has a potential value of about $1.6 billion.

Once crop legumes are harvested for hay or grain, much of the fixed N_2 is removed from the land and, depending upon marketing or management practices, may not be returned in crop residue or animal wastes (*16*). Crop residue or stockpiled crop growth left on the land contain fixed N_2 that can become available to interseeded or succeeding nonlegume crops grown in conservation tillage systems, thus substituting a renewable nutrient source for a manufactured one.

For example, farmers annually incorporate about 8.9 million of the 27 million acres of alfalfa grown in the U.S. into the soil using some form of conventional or conservation tillage. Alfalfa is then followed by a nonlegume crop. If the alfalfa returns 85 pounds/acre of fixed N_2 to the soil, excluding the soil-derived N in the legume, the value of the fixed N_2 is about $8.50/acre, or about $89 million annually for that portion of the U.S. alfalfa crop incorporated by tillage. The other major legume crops listed in table 1 would substantially increase the value of legume-fixed N_2. Thus, the potential economic value of the fixed N_2 contributed by legumes grown in conventional or conservation tillage systems is quite large. The value of the N alone represents a major cost savings to producers, discounting any other economic or noneconomic benefits of conservation tillage that might accrue.

Information Base

With the exception of some recent publications (eg. *7, 9, 13, 23*) and a recently initiated regional research project,[1] information needed to use and manage legumes effectivly as a source of net N input in contemporary conservation tillage systems on a national scale is sparse. Simply stated, there is inadequate knowledge of the environmental and

Table 1. Examples of regional use of legumes in cropping or conservation tillage systems.

Region	Legume Species	Cropping or Tillage System
Southeast	Crimson clover, hairy vetch	Winter cover crop— no-till corn
	Bigflower vetch, crownvetch, alfalfa, lupine, arrowleaf clover, red clover	Winter cover crops preceding grain sorghum and cotton
Northeast	Alfalfa, birdsfoot trefoil, red clover	Legumes grown for hay or silage in crop rotations that include conventional or no-till corn as feed grain or silage; also used as living mulches
North Central	Soybean, pea	Grown in 1-year rotation with nonlegume, possibly using conservation tillage methods; peas may precede soybeans in a double-cropping system
	Alfalfa, red clover, white clover, alsike clover	Grown for 2 years or more in 3- to 5-year rotations with small grains or corn, possibly by use of conservation tillage methods
	Birdsfoot trefoil, crownvetch, sweetclover	Used for forage, silage, or pasture
Great Plains	Native legumes (Void in genetically adapted material and economically compatible enterprises; water is the limiting factor)	Rangeland for grazing
Pacific Northwest	Dry pea, lentil, chickpea	Rotation or double-cropped with grains
	Austrian winter pea	Green manure or alternated with winter wheat
	Alfalfa	Grown in rotation with winter wheat, spring barley, and winter peas
	Faba bean	Grown in rotation for silage
California	Dry bean, lima bean, blackeye pea, chickpea	Grown for grains in various rotations
	Alfalfa	Grown for seed on irrigated land and for erosion control and forage on steeply sloping soils
	Subterranean clover	Rangeland for grazing

biological factors regulating legume N_2 fixation and the transfer of fixed N within conservation tillage systems. We need to determine the N_2-fixation capability of legumes adapted for use in conservation tillage systems. We need to develop host/rhizobial combinations for the most cost-effective performance in enterprises that use conservation tillage. We need to determine mechanisms and quantities of N transfer from legumes to interseeded or succeeding nonlegumes grown in conservation tillage systems. Finally, we need to determine the fate of fixed N_2 within the soil-plant-animal system and how the fate of fixed N_2 is mediated by other sources of N, by soil organic matter, or by C from crop residues.

While some research issues that follow are generic to the use of legumes as a net source of N in conservation tillage systems, each research issue may vary with cropping system and geographic region.

Nitrogen nutrition of legumes

Legumes derive N from three principal sources: the soil, through commercial fertilizer or manure application; the soil, by mineralization of indigenous organic matter; and the atmosphere, by symbiotic N_2 fixation. Hay and pasture legumes grown on a soil with the same initial N concentration in the profile derive different amounts of N_2 from symbiosis (Table 2). Birdsfoot trefoil may derive less N_2 from symbiosis than the other hay or pasture legumes because of its lower productivity and lack of persistent nodules, which must regenerate after each harvest. The amount of N_2 fixed also varies with growth stage (Table 2) and year of stand (Table 3), which may reflect the dynamic balance between the availability of soil N to the plant and the activity of the root nodules.

In addition to species, growth stage, and inherent soil fertility, legume N_2 fixation can be influenced by crop management practices, life form (annual or perennial habit), and location or environment. In general, the environmental factors that promote high yields are conducive to high rates and high seasonal totals of N_2 fixation of legumes nodulated by effective bacteria. Such factors include optimum mineral nutrition at a pH slightly below neutrality, a long growing season, low concentration of plant-available soil N, optimum water availability, and freedom from insects or pathogens.

Selection of annual legume cultivars for rapid emergence after planting, rapid leaf area expansion, high productivity, and maturation before the end of the growing season would favor highest N_2 fixation. Also, management practices and cultivar selection that enhance stand persistence would favor N_2 fixation of perennial legumes.

Although these principles can be listed, evidence to support each is inferential because field experiments on N_2 fixation per se are seldom performed in which cultivar, management, and environmental variables are not confounded. However, with better methods of measuring N_2 fixation in the field values representative of hay and pasture

[1]Regional Project No. S-201, "Role of Legume Cover Crops in Conservation Tillage Production Systems," coordinated by F. C. Boswell, Agronomy Department, Georgia Agricultural Experiment Station, Experiment.

Table 2. Variation of dinitrogen fixation capacity with legume species, legume productivity, and initial soil nitrogen concentration (19, 36).

Species	N from Symbiosis by Harvest (%)				Dry Matter Yield (pounds/acre)
	1	2	3	Mean	
Hay and pasture legumes					
Alfalfa*	49	81	58	63	6,809
Red clover*	51	79	65	65	6,230
Birdsfoot trefoil*	27	67	25	40	4,880
Harvest at Grain Maturity					
Grain legumes					
Soybean†			76		2,494
Soybean‡			52		7,837

*Established in soil with 3.7% organic matter and an initial NO_3^--N concentration of 12 ppm (0- to 6-inch depth).
†Established in soil with 1.8% organic matter and an initial NO_3^--N concentration of 12 ppm (0- to 8-inch depth).
‡Established in soil with 4.8% organic matter and an initial NO_3^--N concentration of 31 ppm (0- to 8-inch depth).

Table 3. Seasonal dinitrogen fixation by forage and grain legumes (19, 36).

Crop	Seasonal N_2 Fixation by Year of Stand (pounds/acre N)			
	1	2	3	4
Alfalfa	142	102	143	199
Red clover	118	61	77	68
Birdsfoot trefoil	100	44	87	97
Soybean 1*	13-75			
Soybean 2†	68-135			

*Grown on high-N soil (see table 2).
†Grown on low-N soil (see table 2).

legumes (Table 4) and seed and pulse legumes (Table 5) grown in North American cropping systems can be summarized. These measurements were made using either of two procedures known to be the most accurate for determining the contribution of symbiotic N_2 fixation to the total N economy of a legume crop community: the difference method and the isotope dilution method (21). Unfortunately, few of the measurements were in conservation tillage cropping systems. Measurements on long-term stands of hay and pasture species fit the concept most closely.

The values of N_2 fixation in the total crop (Tables 4 and 5) include only N_2 removed from the earth's atmosphere and

sequestered within the plant; N derived from soil organic matter or from fertilizer sources has been excluded. Dinitrogen fixed by legumes can be viewed as a free resource because it occurs as a consequence of capture of the energy of continuously renewable sunlight in the photoassimilates of green plants. Substituting fixed legume N for fertilizer N results in production of economic plant products with less dependence upon commercial N fertilizers. However, a net enrichment or renewal of the soil resource by fixed N in legumes can only occur when the legume is grown and managed with attention to returning fixed N to the soil rather than permanently exporting it from the land (18).

Legume nitrogen and companion plants

Researchers often have inferred that N from legumes is transferred to nonlegumes. Such conclusions come from measurements made in controlled and field environments (9, 16). Increased use of N tracer methodology has expedited measurements of legume-nonlegume N transfer and has allowed some assessments of its significance in the field. Pot-grown ryegrass received 27% of its N from associated white clover over a 17-week period, while field-grown ryegrass on the same soil obtained less than 6% of its N from the associated white clover (14). In another experiment, ryegrass obtained 12% of its N from associated white clover, with transfer occurring only at the fourth harvest of a 105-day season (15). In similar experiments ryegrass grown in the field obtained up to 80% of its N from associated ladino clover, but no transfer was observed in the glasshouse (3).

Recent field results from mixtures of reed canarygrass with alfalfa or with birdsfoot trefoil showed that 64% to 79% of the N in the grass was originally symbiotically fixed by the legume (Table 6). This N represented 13% of the N_2 fixed by birdsfoot trefoil and 17% of that fixed by alfalfa. The results indicated that N transfer occurred over a distance of 8 inches, with maximum N transfer in areas of high legume/grass ratio (4).

In all such N transfer experiments, the method(s) and mechanism(s) of transfer are unclear. Although direct excretion or degradation of nodule and root tissue are the usual explanations (27), no one has attempted to develop a budget to account for N transfer from various sources.

Table 4. Seasonal total of dinitrogen fixation by hay and pasture legumes and legume-grass swards, measured by the difference or isotope dilution methods.

Species	N_2 Fixation (pounds/acre N/growing season)	Location	Measurement Method	Reference
Alfalfa	189	Lexington, Ky	Difference	(24)
Alfalfa	102-199	Rosemount, Minn.	Isotope dilution	(20)
Alfalfa-orchardgrass sward	13-121	Lucas Co., Iowa	Isotope dilution	(39)
Alsike clover	19	New Jersey	Difference	(35)
Birdsfoot trefoil	44-100	Rosemount, Minn.	Isotope dilution	(17)
Crimson clover	57	New Jersey	Difference	(35)
Hairy vetch	99	New Jersey	Difference	(35)
Ladino clover	147-168	Maryland	Difference	(37)
Red clover .	61-101	Rosemount, Minn.	Isotope dilution	(17)
Subterranean clover	52-163	Hopland, Calif.	Isotope dilution	(29)
Subterranean clover-soft chess sward	19-92	Hopland, Calif.	Isotope dilution	(29)
Sweetclover	4	New Jersey	Difference	(35)
White clover	114	Lexington, Ky	Difference	(24)

Despite the emphasis on root and nodule degradation, other mechanisms may include stolon decay in some forage legumes, abscission of leaves and petioles in soybean-small grain intercrops, loss of leaves to pests or pathogens in other legume species, or death of individual legume plants within the intercrop or legume-nonlegume sward (16). Except for the emerging quantitative evidence on N transfer between forage legumes and grasses (4), little is known of the quantity or significance of N transfer in conservation tillage systems.

Management effects

It is now clear that several factors influence the proportion of legume N that is symbiotically fixed and that the soil N contribution to legume N content represents a temporary storage until the N is cycled back into the soil N pool. The discussion below applies equally to the recovery by nonlegumes of soil and symbiotically fixed N_2 contained in legume residues. However, only recovery of symbiotically fixed N_2 represents a substitution for commercial fertilizer N or a net N addition to the plant-soil system unless, of course, the legume recovers soil N that would be permanently lost to leaching (16). Legume conservation of soil N typically occurs in environments where precipitation events leach mineralized N below the rooting zone of crops or even to the water table. In this case the legume functions as a trap crop for N conservation (16, 28).

Table 5. Seasonal total of dinitrogen fixation by seed or pulse legumes, measured by the difference or isotope dilution methods.

Species	N_2 Fixation (pounds/acre N/ growing season)	Location	Measurement Method	Reference
Chickpea	21-75	Alberta	Isotope dilution	(32)
Common bean	11-108	Alberta	Isotope dilution	(31)
	2-98	Alberta	Difference	(30)
Faba bean	159-224	Alberta	Isotope dilution	(32)
Field pea	155-175	Alberta	Isotope dilution	(32)
Lentil	149-169	Alberta	Isotope dilution	(32)
Soybean	12-67	Iowa	Difference	(38)
	20-71	Minnesota	Difference	(12)
	49-143	Arkansas	Difference	(2)
	234-277	Washington	Difference	(1)
	49-115	Minnesota	Difference	(11)
	93	Nebraska	Difference	(6)

Table 6. Maximum nitrogen transfer from legumes to grass. The legume/grass ratio was 2:3 for the reed canary-grass-alfalfa community, and 3:2 for the reed canary-grass-trefoil community (4).

Community	N in Grass from Legume (%)*		
	Harvest 1	Harvest 2	Harvest 3
Grass-alfalfa	64	68	68
Grass-trefoil	68	66	79

*Percent (proportion x 100, mass basis) of accumulated grass N received from legume via N transfer.

In conservation tillage systems the amount of commercial fertilizer N that legume N can replace depends upon (a) the quantity of legume residue returned to the soil by tillage or other transfer mechanisms, (b) the content of symbiotically fixed N_2 in the residues, and (c) the availability of N in the legume residue to the companion or succeeding nonlegume.

Nitrogen budgets for soybeans developed from N tracer experiments (16) illustrate the partitioning of fixed N_2 in a crop with different amounts of symbiotic activity (Table 7). In the midwestern United States soybeans may fix 40% and obtain 60% of their N from the soil either from mineralization of organic matter or from carryover fertilizer applied to a preceeding crop. When these crops are harvested for grain, soil N export in the harvested grain may exceed symbiotic N_2 return in the residue, which leads to a calculated soil N deficit of 74 pounds/acre N. If the same crop is grown under conditions where it fixes 90% of its N, for example, in the Piedmont region, the return of fixed N_2 in the residue may exceed soil N export in grain by 22 pounds/acre N.

Typically, only part of the symbiotically fixed N_2 in a hay or pasture legume is returned to the soil for use by a succeeding crop. This is because a portion of the fixed N_2 is removed from the land when the legume is harvested, with the balance remaining in unharvested roots and crowns. The N available for incorporation into the soil depends on the time of the season when incorporation occurs and the proportion of the plant that is N-rich herbage compared with relatively N-poor crown and roots. The N budget in table 8 illustrates the net N return to the soil that is possible when alfalfa is incorporated by either fall or late-summer moldboard plowing (18). Comparable data for noninversion tillage practices are not available. In related experiments, tillage method had no consistent effect on the N or dry matter yield of sudangrass following alfalfa (10).

If two herbage harvests are taken, followed by herbage regrowth before moldboard plowing on October 20, the early season N deficit is nearly replaced by the late season N_2 fixation, with an inconsequential loss of 4 pounds/acre N. A seasonal N deficit of 38 pounds/acre N would occur if a third herbage harvest was taken after the first frost, but before the October 20 tillage operation, so that only roots and crowns were incorporated. In contrast, removal of one herbage harvest followed by plowdown of a lush regrowth on August 30 allowed a net input of 48 pounds/acre N. Clearly, the benefit of growing a forage or grain legume before a subsequent nonlegume is influenced by how the legume is managed for return of N to the soil. Such knowledge is urgently needed for all types of conservation tillage systems.

For example, in living mulch tillage systems forage legumes can be inhibited by sublethal applications, or destroyed by lethal applications, of herbicides before seeding of a subsequent crop with reduced tillage methods (5, 8). In such situations, losses of volatile N compounds from plant material undoubtedly occur, but the extent and significance of these losses to the N balance of the cropping system is poorly understood. Similarly, the yield stimulation of a nonlegume interseeded into a chemically suppressed living mulch has not been ascribed quantitatively to the transfer of soil N, symbiotically fixed N_2, or stimulation by some

Table 7. Nitrogen budget of soybeans illustrating the allocation of soil and symbiotic N among plant components and the net return of N to the soil with 40% and 90% of plant nitrogren from symbiosis. (18).

Crop Component	Dry Matter Content	Total Reduced N Content	Content of Symbiotic N		Soil N Export in Grain		Symbiotic N Return in Residue		Loss (-) or Gain (+) or N	
			40% N Symbiosis	90% N Symbiosis	40% N	90% N	40% N	90% N	40% N	90% N
			pounds/acre		N					
Grain	2,100	151	61	136	90	15	—	—	—	—
Residue*	3,424	40	16	37	—	—	16	37	—	—
Total Plant	5,524	191	77	173	—	—	—	—	-74	+22

*Pod walls, leaves, stems, roots and nodules; incomplete grain harvest would increase this value.

non-N factor. The results ascribing yield stimulation to N transfer (8) are equivocal at best.

Nonlegume yield response

The N fertilizer equivalent of a legume preceding a nonlegume is traditionally measured as the rate of N fertilizer application to a nonlegume that gives a grain yield equivalent to that obtained by preceding the nonlegume by a legume. This method requires three assumptions: that the entire increased nonlegume yield is due to N, that the entire N effect is due to additions through symbiotic N_2 fixation, and that fertilizer N and residue N are equally available (16, 18).

In short-rotation studies of legumes and nonlegumes (Table 9), incorporation of alfalfa managed in a one-cut system gave corn grain yields equal to those expected from 100 to 200 pounds/acre N. The incorporation of alfalfa managed in a three-cut management gave a corn grain yield response equal to the application of 50 pounds/acre N. Yields were greater with incorporation of more residual plant N in the one-cut system compared to the three-cut system. Corn yields following wheat were also greater, suggesting that there are benefits of changing species in crop rotations that are not solely attributable to the N in legumes. New analytical methods now allow discrimination between N and non-N effects of legumes preceding nonlegumes in crop rotations (33).

Advances in N tracer methology have refined the understanding of the recovery of legume N by subsequent nonlegume crops (25). Recent experiments in Australia (Table 10) indicated that the first crop of wheat grown after incorporation of medic residues recovered an average of 23% of the legume N, while the second crop recovered an additional 4% (26). Comparable results from experiments done on conservation tillage systems in the United States are lacking. Nevertheless, the message is clear that only a portion of the traditional fertilizer N equivalence attributed to legumes in cropping systems is actually due to symbiotically fixed legume N. The challenge to understand N_2 fixation, N cycling, and N use efficiency in conservation tillage systems is ripe for investigation as these systems become more widely adopted throughout the United States.

Perspective and Challenge

More knowledge is needed to understand and quantify the resource conservation and economic consequences of using legumes as N sources in conservation tillage systems. Much

is known about how much N_2 legumes fix (Tables 2 to 5), especially in conventional cropping systems employing inversion tillage. These data illustrate the range of N_2 fixation anticipated from legumes used in conservation tillage systems but do not reduce the urgency of obtaining quantitative information on legumes used in or adapted to conservation tillage systems.

Regardless of the amount of N in the forage crop, cover crop, intercrop, or living mulch, only that N_2 fixed by the plant and returned to the soil represents a net N addition to ameliorate the export of soil N in the crop product. Thus, the genetic endowment of legumes to fix large amounts of free N_2 is foreshadowed by the necessity to manage the

Table 8. Nitrogen budget for seeding-year alfalfa illustrating the net return of nitrogen to the soil with two moldboard plowing practices (18).

N From	Plant Part	N (pounds/acre) at Harvest on		
		July 12	August 30	October 20
N_2 fixation	Total	51	91	30
	Herbage	46	66	20
	Root and crown	5	25	10
Soil	Total	54	22	23
	Herbage	48	16	14
	Root and crown	6	6	9
N budget with plowing on August 30				
N return/harvest		-43*	+91	
Cumulative N		-43	+48	
N budget with plowing on October 20				
N return/harvest		-43	+9	+30
Cumulative N		-43	-34	- 4

*N return/harvest = fixed N in roots and crowns returned by soil (5 pounds/acre) - soil N removed in herbage (48 pounds/acre) = -43 pounds/acre.

Table 9. Corn grain yields as influenced by previous crop and fertilizer nitrogen rate at the Southern Experiment Station, Waseca, Minnesota (22).

N Rate (pounds/acre)	Corn Grain Yield* (bushels/acre) Following Previous Crop of				
	Corn	Soybeans	Wheat	3-Cut Alfalfa	1-Cut Alfalfa
0	50	58	57	65	108
50	65	90	99	112	138
100	100	122	128	123	142
150	103	138	127	126	137
200	100	140	144	133	140

*Short rotation—1 year of corn following each crop.

Table 10. Recovery of nitrogen in residues of medic (tops and roots, topsoil-field) (26).

N Addition (pounds/acre)	Assay Crop		Legume N Recover (% of addition)	
	Species	Growth period	Crop 1	Crop 2
43	Wheat (whole plant)	8 months	27.8	—
	Wheat (tops)	8 months		4.8
43	Wheat (whole plant)	8 months	20.2	—
21	Wheat (tops)	8 months,	22.4	4.3
43	(tops)	8 months	24.9	4.1
86	(tops)	8 months	21.1	3.3

legume properly, either by harvest, tillage, or hitherto undiscovered practices, to allow a net N return to the soil-plant system. As the various types of conservation tillage systems are regionalized (Table 1), so must be the principles of tillage and harvest management.

Conservation tillage practices produce soil rhizophere conditions that are often considerably different from those existing in conventionally tilled cropping systems. We need better knowledge of legume-rhizobial germplasm combinations adapted to conservation tillage systems. While the rhizosphere ecology of conventionally tilled cropping systems is still in its infancy, the rhizosphere ecology of conservation tillage systems is even less well developed.

Finally, the mechanisms of legume-nonlegume N transfer in conservation (and conventional) tillage systems are poorly understood. Once the net N input to the soil-plant system is quantified, the fate of the added N must be understood. Crop recovery of legume N is held to be large, but non-N rotation effects stimulate yields by unknown mechanisms. Does some of the net N added by legumes escape to groundwater or become immobilized in the soil organic matter pool? Do living mulch systems really contribute N to an interseeded crop, or are the results confounded by N volatilization or by other growth-limiting factors?

Clearly the knowledge base on the use of legumes as sources of net N input into conservation tillage systems is meager, and the research challenges are numerous. Because of the growing adoption of conservation tillage systems, the proposed research may enhance the self-sufficiency of conservation tillage systems for N, facilitate the conservation of finite resources of soil N in regions prone to severe erosion or organic matter oxidation, provide strategies to enhance the managerial skills of producers and the economic viability of enterprises, and preserve the land base while maintaining or increasing its productivity without the need for costly purchased inputs.

REFERENCES

1. Bezdicek, D. F., D. W. Evans, B. Abede, and R. E. Witters. 1978. *Evaluation of peat and granular inoculum for soybean yield and N fixation under irrigation.* Agron. J. 70: 865-868.
2. Bhanghoo, M. S., and D. J. Albritton. 1976. *Nodulating and non-nodulating Lee soybean isolines response to applied nitrogen.* Agron. J. 68: 642-645.
3. Broadbent, F. E., T. Nakashima, and C. Y. Chang. 1982. *Estimation of nitrogen fixation by isotope dilution in field and greenhouse experiments.* Agron. J. 74: 625-628.
4. Brophy, L. S., G. H. Heichel, and M. P. Russelle. 1987. *Nitrogen transfer from forage legumes to grasses in a systematic planting design.* Crop Sci. 27 (In Press).
5. Cardina, J., and N. L. Hartwig. 1981. *Influence of nitrogen and corn population on no-tillage corn yield with and without crownvetch.* Proc. N.E. Weed Sci. Soc. 35: 27-31.
6. Diebert, E. J., M. Bijeriego, and R. A. Olson. 1979. *Utilization of ^{15}N fertilizer by nodulating and non-nodulating soybean isolines.* Agron. J. 71: 717-723.
7. Ebelhar, S. A., W. W. Frye, and R. L. Blevins. 1984. *Nitrogen from legume cover crops for no-tillage corn.* Agron. J. 76: 51-55.
8. Elkins, D., D. Frederking, R. Maraski, and B. McVay. 1983. *Living mulch for no-till corn and soybeans.* J. Soil and Water Cons. 37: 431-433.
9. Elmore, R. W., and J. A. Jackobs, 1986. *Yield and nitrogen yield of sorghum intercropped with nodulating and non-nodulating soybeans.* Agron. J. 78: 780-782.
10. Groya, F. L., and C. C. Sheaffer. 1985. *Nitrogen from forage legumes: Harvest and tillage effects.* Agron. J. 77: 105-109.
11. Ham, G. E. 1978. *Use of ^{15}N in evaluating N_2 fixation of field-grown soybeans.* In *Isotopes in Biological Dinitrogen Fixation.* Int. Atomic Energy Agency, Vienna, Austria. pp. 151-162.
12. Ham, G. E., I. E. Liener, S. D. Evans, R. D. Frazier, and W. W. Nelson. 1975. *Yield and composition of soybean seed as affected by N and S fertilization.* Agron. J. 67: 293-297.
13. Hargrove, W. L. 1986. *Winter legumes as a nitrogen source for no-till grain sorghum.* Agron. J. 78: 70-74.
14. Haystead, A., and C. Marriott. 1978. *Fixation and transfer of nitrogen in a white clover-grass sward under hill conditions.* Ann. Appl. Biol. 88: 453-457.
15. Haystead, A., and C. Marriott. 1979. *Transfer of legume nitrogen to associated grass.* Soil Biol. Biochem. 11: 99-104.
16. Heichel, G. H. 1987. *Legume nitrogen: Symbiotic fixation and recovery by subsequent crops.* In *Energy in World Agriculture Handbook, Vol. II, Energy in Plant Nutrition and Pest Control.* Elsevier Sci. Publ., Amsterdam, The Netherlands. (In Press).
17. Heichel, G. H., C. P. Vance, D. K. Barnes, and K. I. Henjum. 1985. *Dinitrogen fixation, and N and dry matter distribution during 4-year stands of birdsfoot trefoil and red clover.* Crop Sci. 25: 101-105.
18. Heichel, G. H., and D. K. Barnes. 1984. *Opportunities for meeting crop nitrogen needs from symbiotic nitrogen fixation.* In D. Bezdicek and J. Power [eds.] *Organic Farming: Current Technology and its Role in a Sustainable Agriculture.* Spec. Pub. No. 46, Am. Soc. Agron., Madison, Wisc. pp. 49-59.
19. Heichel, G. H., D. K. Barnes, C. P. Vance. 1981. *Nitrogen fixation of alfalfa in the seeding year.* Crop Sci. 21: 330-335.
20. Heichel, G. H., D. K. Barnes, C. P. Vance, and K. I. Henjum. 1984. *Dinitrogen fixation, and N and dry matter partitioning during a 4-year alfalfa stand.* Crop Sci. 24: 811-815.
21. Henson, R. A., and G. H. Heichel. 1984. *Dinitrogen fixation of soybean and alfalfa: Comparison of the isotope dilution and difference methods.* Field Crops Res. 9: 333-346.
22. Hesterman, O. B., C. C. Sheaffer, D. K. Barnes, W. E. Lueschen, and J. H. Ford. 1986. *Alfalfa dry matter and N production, and fertilizer N response in legume - corn rotations.* Agron. J. 78: 19-23.
23. Huntington, T. G., J. H. Grove, and W. W. Frye. 1985. *Release and recovery of nitrogen from winter annual cover crops in no-till corn production.* Commun. Soil Sci. Plant Anal. 16: 193-211.
24. Karraker, P. E., C. E. Bartner and E. N. Fergus. 1950. *Nitrogen balance in lysimeters as affected by growing Kentucky bluegrass and certain legumes separately and together.* Bull. 557. Ky. Agr. Exp. Sta., Lexington. 16 pp.

25. Ladd, J. N., J. H. A. Butler, and M. Amato. 1986. *Nitrogen fixation by legumes and their role as sources of nitrogen for soil and crop.* Biol. Agr. Hort. 3: 269-286.
26. Ladd, J. N., J. M. Oades, and M. Amato. 1981. *Distribution and recovery of nitrogen from legume residues decomposiong in soils sown to wheat in the field.* Soil Biol. Biochem. 13: 251-256.
27. Mulder, E. G., T. A. Lie, and A. Houwers. 1977. *The importance of legumes under temperate conditions.* In R. W. F. Hardy and A. H. Gibson [eds.]. *A Treatise on Dinitrogen Fixation. Section IV. Agronomy and Ecology.* Wiley-Interscience, New York, N.Y. pp. 221-242.
28. Olson, R. A., W. R. Rawn, Y. S. Chun, and J. Skopp. 1986. *Nitrogen management and interseeding effects on irrigated corn and sorghum and on soil strength.* Agron. J. 78: 856-862.
29. Phillips, D. A., and J. P. Bennett. 1978. *Measuring symbiotic nitrogen fixation in rangeland and plots of* Trifolium subterraneum *L. and* Bromus mollis *L.* Agron. J. 70: 671-674.
30. Rennie, R. J. 1984. *Comparison of N balance and* ^{15}N *isotope dilution to quantify* N_2 *fixation in field-grown legumes.* Agron. J. 76: 785-790.
31. Rennie, R. J. and G. A. Kemp. 1983. N_2*-fixation in field beans quantified by* ^{15}N *isotope dilution. I. Effect of strains of* Rhizobium phaseoli. Agron. J. 75: 640-644.
32. Rennie, R. J., and S. Dubetz. 1986. *Nitrogen-15 determined nitrogen fixation in field-grown chickpea, lentil, faba bean, and field pea.* Agron. J. 78: 654-660.
33. Russelle, M. P., O. B. Hesterman, C. C. Sheaffer, and G. H. Heichel. 1987. *Estimating N and "rotation" effects in legume-corn rotations.* In *The Role of Legumes in Conservation Tillage Systems.* Soil Cons. Soc. Am., Ankeny, Iowa. pp. 41-42.
34. Soil Conservation Society of America. 1982. *Resource Conservation Glossary.* Ankeny, Iowa. 33 pp.
35. Sprague, H. B. 1936. *The value of winter green manure crops.* Bull. 609. New Jersey Agr. Expt. Sta., New Brunswick. 19 pp.
36. Vasilas, B. L. 1981. *A tracer technique for the quantitative measurement of symbiotic dinitrogen fixation by soybeans in field studies.* Ph.D. Diss. Abstr. No. DA821159. Univ. of Minn., St. Paul. 80 pp.
37. Wagner, R. E. 1958. *Legume nitrogen versus fertilizer nitrogen in protein production of forage.* Agron. J. 46: 233-237.
38. Weber, C. R. 1966. *Nodulating and non-nodulating soybean isolines. II. Response to applied nitrogren and modified soil conditions.* Agron. J. 58: 46-49.
39. West, C. P., and W. F. Wedin. 1985. *Dinitrogen fixation in alfalfa-orchardgrass pasture.* Agron. J. 77: 89-94.

Timing effects of cover crop dessication on decomposition rates and subsequent nitrogen uptake by corn

M. G. Wagger

As well as providing potential soil and water conservation benefits, decomposing residues of winter annual legumes may be able to contribute a significant portion of the required N to nonleguminous summer crops, such as corn and sorghum. With respect to corn production in North Carolina, cover crop management relative to corn planting operations can be particularly important. Patterns of dry matter accumulation and water use by cover crops change rapidly during what is considered a relatively narrow window for optimum corn planting dates. An understanding of residue N release rates with regard to residue composition, corn demand for N, and environmental conditions is necessary to maximize the efficient use of residue N.

In this study I sought to evaluate timing effects of cover crop burndown on dry matter accumulation and chemical composition of several winter annual cover crops and to determine the effect of cover crop management on residue N release patterns in a no-till corn system.

Study methods

The experiment sites included an Enon fine sandy loam (fine, mixed, thermic Ultic Hapludalfs) and a Pacolet sandy loam (clayey, kaolinitic, thermic, Typic Hapludults) in 1984 and 1985, respectively. This study was part of a larger, split-split plot experiment consisting of cover crop burndown/corn planting combinations (early kill/early plant, early kill/late plant, and late kill/late plant) as main plot treatments; cover crop type (fallow, rye, crimson clover, and hairy vetch) as subplot treatments; and fertilizer N rates (0, 90, and 180 pounds/acre) as sub-subplot treatments. Plot dimensions were 10 feet by 35 feet, with four replications. Corn was planted in 30-inch rows. There was about a 2-week interval between the early and late combinations, corresponding to corn planting dates of April 18 and May 1 in 1984 and April 19 and May 7 in 1985. Because the study reported here focused primarily on cover crop decomposition rates with regard to N turnover and subsequent corn N uptake, plots receiving fertilizer N were not used.

To monitor cover crop decomposition, I placed nylon mesh bags containing air-dried cover crop material on the surface of the no-till plots about 1 week after corn planting. I retrieved the bags at 1, 2, 4, 8, and 16 weeks after placement. Samples were dried, weighed, and analyzed for total N concentration. To minimize the effects of any soil contamination, I calculated residue weight remaining on an ash-free weight basis by determining the ash content on a subsample from each bag. I collected whole-plant corn samples

M. G. Wagger is an assistant professor, Department of Crop Science, North Carolina State University, Raleigh, 27695.

and soil samples at various times during the growing season to estimate apparent residue N availability within the cover crop management/corn planting combinations.

Results and discussions

Table 1 shows the changes in cover crop composition resulting from a 2-week delay in dessication, averaged over both years of the study. Late burndown resulted in nearly a 45% increase in top growth dry matter for rye and crimson clover and about a 60% increase for hairy vetch. The accompanying increase in total N content ranged from 13% for rye to 43% for hairy vetch. In turn, these changes altered the C:N ratio of the residues, most notably for rye. Perhaps of greater consequence were the composition differences that occurred in some of the fiber constituents because their relative proportions can govern the rate of decomposition. Sharp increases in the cellulose fraction were associated with the late burndown for all residues, while the hemicellulose fraction remained nearly constant. Within rye, the magnitude of change in the lignin fraction was relatively small compared to the legumes.

Figure 1 shows residue N release patterns for 1984, expressed as a percentage of the initial residue N, for both the early and late burndown. Across all residue/burndown combinations, an exponential equation reflected the decline in residue N over time. However, differences in N disappearance patterns were readily apparent between species and residue burndown dates. On average the decline in residue N was on the order of hairy vetch > crimson clover > rye. These results were not unexpected, considering the C:N ratio of each residue. The N content or C:N ratio of plant residues has been used frequently as a tool to predict the rate of residue decomposition, yet they are not the sole determinants.

Nitrogen release patterns between the early and late burndown dates differed more for the legumes than for rye, particularly during the early stages of decomposition. For example, after 4 weeks in the field, crimson clover contained 31% of the initial N when killed early compared to 64% of the initial N with a late burndown. These differences were still pronounced at 8 weeks and corresponded with the approximate corn tasseling-silking period. I observed a similar pattern for hairy vetch. By 16 weeks, however, the N remaining between the early and late burndown dates for each cover crop residue approached nearly the same level.

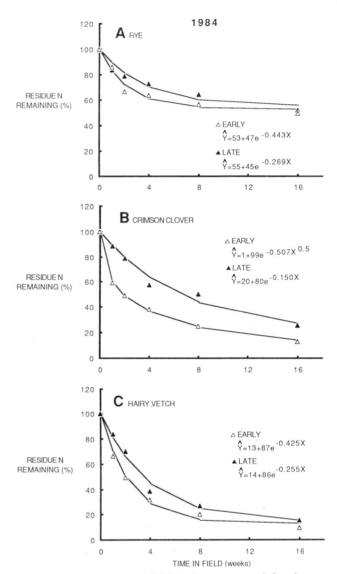

Figure 1. Percentage of initial nitrogen remaining in rye (top), crimson clover (middle), and hairy vetch (bottom) residue at various times during the 1984 growing season for the early and late cover crop burndown dates.

The considerably faster rate of N turnover associated with the early burndown for the legumes most likely was related to residue composition changes that occurred with the late burndown. Other investigators have indicated that, in addition to the C:N ratio of residue, the lignin:carbohydrate ratio is also important, with lignin being considerably more resistant to decomposition than carbohydrates. Environmental conditions, mainly such weather-related factors as temperature or precipitation, can influence the rate of residue decomposition. These factors did not appear to explain the observed differences in my study because there was no evidence of any temperature extremes during the first 8 to 10 weeks and rainfall amounts and distributions were nearly identical for the two periods.

In contrast, climatic conditions exerted a strong influence on N disappearance patterns during the 1985 growing season

Table 1. Cover crop residue characteristics as affected by time of dessication, 2-year average.

Cover Crop and Burndown	Dry Matter (tons/acre)	N Content (pounds/acre)	C:N Ratio	Cellulose (%)	Hemi-cellulose (%)	Lignin (%)
Rye						
Early	2.6	71	32:1	30.9	27.6	7.0
Late	3.8	80	38:1	34.9	28.2	7.7
Crimson clover						
Early	1.7	101	14:1	19.4	12.5	6.7
Late	2.4	124	16:1	25.1	13.9	9.8
Hairy vetch						
Early	1.5	120	10:1	17.4	8.4	8.4
Late	2.4	171	11:1	23.1	8.0	10.7

(Figure 2). Residue N remaining at 16 weeks for the early burndown was 59%, 43%, and 35% for rye, crimson clover, and hairy vetch, respectively. By comparison, corresponding values in 1984 were 53%, 20%, and 13%. Relatively dry conditions persisted during most of the 1985 growing season, which probably was a prominent factor limiting the release of N. The rapid disappearance of N observed in 1984 during the first 4 weeks did not occur in 1985; rather there was a more steady rate of decline.

It is also important to obtain some index of N availability from decomposing crop residues with respect to N demand by the succeeding summer crop. Estimates of residue N availability indicated that optimal synchronization varied by year (environmental conditions). In the rye cover crop system all burndown/corn planting combinations depicted a net N immobilization. Synchronization of N release between the legume residues and corn demand for N influenced early season growth but had no apparent effect on grain yield. Under the conditions that existed during the course of this study, the soil moisture status appeared to be a more important determinant of residue N availability with respect to corn demand for N.

These results indicate that although the potentially available legume N pools differed with respect to cover crop management, residue N was released in a timely manner so as not to limit corn yields. It would also appear that legume cover crop systems offer greater flexibility with regard to timing of supplemental N fertilization.

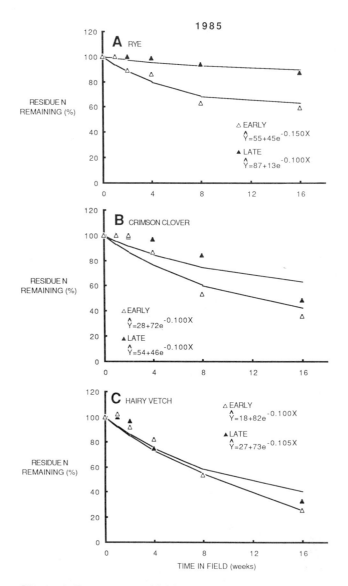

Figure 2. Percentage of initial nitrogen remaining in rye (top), crimson clover (middle), and hairy vetch (bottom) residue at various times during the 1985 growing season for the early and late cover crop burndown dates.

Fertilizer use efficiency in a sorghum-soybean rotation

R. D. Lohry, M. D. Clegg, G. E. Varvel, and J. S. Schepers

Use of manure and legumes dates back to 2000 B.C. Emphasis on production efficiency has renewed interest in rotations providing some necessary plant growth factors.

Baldock and Musgrave (1), studying the effects of manure and mineral fertilizer in various crop rotations, concluded that 2 years of legumes contributed the same amount of N as the application of 136 kg/ha of mineral N. The usual method of assessing the N-supplying ability of a legume has been to compare yields of unfertilized nonlegume crops to those fertilized with various levels of N fertilizer. Whether the observed nonlegume yield response from rotation is due solely to fixed N has been questioned (4). Cook and associates (3) suggest that a likely explanation for yield decreases in continuous cropping practices is a progressive buildup of plant root pathogens that damage roots and subsequently impair nutrient uptake. Bock (2) hypothesized that higher ammonium/nitrate ratios may provide better growth.

The divergent hypotheses surrounding the observed yield response of nonlegume crops following legumes prompted the initiation of a production system to evaluate the effects of crop rotation and fertilizer N rate on soybeans [*Glycine max.* (L.)] and sorghum [*Sorghum bicolor* (L.)]. The plots, established in 1975, have been used to study potentially mineralizable N, microbial biomass, and N contribution of soybeans under the rotation. The study presented here uses tagged N fertilizer (^{15}N depleted) to assess the N-use efficiency by sorghum in rotation with soybeans compared to monocropped sorghum. The study also compared dry matter yields across cropping systems.

Study methods

The study site was at the University of Nebraska research farm new Mead. The soil, a Sharpsburg silty clay loam (fine montmorillonitic, mesic typic Argiudoll), is highly base saturated and slightly acid or neutral in the natural condition. Mineralization of abundant feldspar materials gives rise to ample K. The exchange complex is saturated with Ca and Mg.

Sorghum hybrid Dekalb 'DKE59' and soybean 'Amsoy 71' were used. The plots were 12 m long and 6 rows wide, with each row spaced 0.76 m apart. Grain sorghum was seeded in excess and thinned to a final population density of about 125,00 plants/ha. The seedbed was prepared by disking twice and then harrowing. Plots received chemical weed control before seedling emergence, were cultivated when necessary, and received no irrigation.

R. D. Lohry is a graduate assistant and M. D. Clegg is an associate professor, Department of Agronomy, University of Nebraska, Lincoln, 68583. G. E. Varvel and J. S. Schepers are associate professors, University of Nebraska, and Agricultural Research Service, U.S. Deparment of Agriculture, Lincoln, Nebraska.

The experimental design was a randomized complete block with four replications. Treatments consisted of continuous sorghum and sorghum following soybeans with four levels of N fertilizer applied to the sorghum crop. Double-labelled, depleted ^{15}N ammonium nitrate (NH_4NO_3) fertilizer was surface applied as solution on the row at the four-leaf stage in 1985 and 1986. The four N fertilizer levels were 0, 56, 112, 168 kg/ha.

We analyzed grain and stover samples for total N by modified micro-Kjeldahl digestion (5), followed by N determination with a Technicon Autoanalyzer. The isotope ratios of ^{15}N to ^{14}N in plant and soil samples were determined by mass spectrometry at the Agricultural Research Service's horticultural and special crops lab in Peoria, Illinois.

Percent N derived from fertilizer (PNDF) was calculated using the following formula:

$$PNDF = [(A-B)/(C-B)] \times 100$$

where A = sample atom %^{15}N, B = control atom %^{15}N, and C = fertilizer atom %^{15}N.

The amount of fertilizer N recovered (FNR) was calculated as follows:

$$FNR = PNDF \times E \times N$$

where E = the amount of material (kg/ha) and N = percent N in the material.

Percent fertilizer N recovered (PFNR) and apparent N use efficiency (ANUE) were calculated as follows:

$$PFNR = FNR/R$$
$$ANUE = (T-K)/R$$

where T = the N removed in the fertilized treatment, K = N removed in the check, and R = rate of fertilizer N applied (kg/ha).

Results

Isotopic N data for the 1986 season is, at this writing, being analyzed; thus, only yield data are presented for both years. The grain yields from the continuous cropping system were consistently lower than in the rotation system. Grain yields in plots receiving no N fertilizer were significantly lower than all other treatments in both years (Table 1). Grain yield at the 56-kg/ha rate in the rotation was significantly higher than in the continuous system. The maximum yield was at the 168-kg/ha rate in the rotation system. However, there were no increases beyond the 56-kg/ha rate in the rotation system. We observed no treatment differences in any cropping system at the 112-kg/ha rate or to the next higher

Table 1. Mean sorghum grain dry weight yields, 1985-1986, Mead, Nebraska.

N Rate (kg/ha)	Sorghum Yield (Mg/ha)		P>F
	Continuous	Rotation	
0	3.21	5.41	.0001
56	5.08	5.97	.0021
112	5.95	6.09	.6204
168	6.06	6.31	.3861
LSD (.05)	.84	.80	

University of Winnipeg, 515 Portage Ave, Winnipeg, Manitoba, Canada R3B 2E9

Table 2. Total nitrogen recovered by sorghum, 1985, Mead, Nebraska.

N Rate (kg/ha)	N Recovered (kg/ha)		
	Continuous	Rotation	P>F
Grain N Removed			
0	37.7	73.3	.04
56	63.0	87.4	.07
112	87.0	107.9	.11
168	81.5	96.0	.27
LSD(0.5)	33.34	22.6	
Total N Removed			
0	60.0	118.0	.01
56	112.0	148.2	.04
112	157.5	191.2	.05
168	170.0	179.4	.57
LSD(.05)	33.4	45.0	

Table 3. Mean sorghum apparent nitrogen-use efficiency, 1985-1986, Mead, Nebraska.

N Rate (kg/ha)	Sorghum ANUE (%)		
	Continuous	Rotation	P>F
56	92.9	100.0	NS
112	86.5	93.8	NS
168	33.02	45.0	NS

Table 4. Fertilizer nitrogen recovered by sorghum, 1985, Mead, Nebraska.

N Rate (kg/ha)	Fertilizer N Recovered (kg/ha)		
	Continuous	Rotation	P>F
Grain FNR			
56	15.25 (27.2)*	17.47 (31.2)	.5320
112	21.32 (19.0)	25.70 (22.9)	.2241
168	25.53 (25.5)	30.31 (30.3)	.1855
(LSD .05)	6.91	9.54	
Total FNR			
56	29.71 (53.1)	26.69 (47.7)	.6378
112	45.81 (40.9)	51.19 (45.7)	.4036
168	48.99 (29.2)	54.47 (32.4)	.3954
(LSD .05)	16.35	13.57	

*Value in parentheses indicate PFNR.

REFERENCES

1. Baldock, Jon O., and R. B. Musgrave. 1980. *Manure and fertilizer effects in continuous corn and rotational crop sequences in central New York.* Agron. J. 72: 511-518.
2. Bock, B. R., and K. R. Kelley. 1986. *Nitrate and ammonium nutirition of grain sorghum: I. Yield and nutrient composition.* Agron. Abstr. p. 194.
3. Cook, R. James, M. G. Boosalis, and B. Doupnik. 1984. *Influence of crop residues on plant diseases.* In W. R. Oswald [ed.] *Crop Residue Management Systems.* Am. Soc. Agron., Madison, Wisc.
4. Heichel, G. H., and D. K. Barnes. 1984. *Opportunities for meeting crop nitrogen needs from symbiotic nitrogen fixation.* In D. F. Bezdicek [ed.] *Organic Farming: Current Technology and its Role in a Sustainable Agriculture.* Am. Soc. Agron., Madison, Wisc.
5. Schuman, G. E., M. A. Stanley, and D. Knudson. 1973. *Automated total nitrogen analysis of soil and plant samples.* Soil Sci. Soc. Am. Proc. 37: 480-481.

increment of N fertilizer.

The maximum amount of N removed was at the 112-kg/ha rate, with a decline in the N content at the highest rate (Table 2). PFNR was similar at corresponding N rates across cropping systems, indicating that the previous legume crop did not enhance fertilizer uptake at any N rate. In contrast, ANUE (Table 3) indicated a much higher rate of N recovery. The amount of fertilizer N recovered was not different across cropping systems (Table 4). Given the data, how do we ascertain the N-use efficiency with the isotope or the nonisotope data?

The correct interpretation of the difference between the results in the equations lies in the magnitude of the contribution of indigenous soil N. The higher value of ANUE reflects the higher amount of indigenous soil N removed. The higher amount of N removed in the rotation system was due to increased mobilization or availability of soil N as a result of the previous soybean crop.

Legume nitrogen transformation and recovery by corn as influenced by tillage

J. J. Varco, W. W. Frye, M. S. Smith, and J. H. Grove

Increasing use of conservation tillage by farmers has brought about a renewed interest in growing winter legume cover crops. Farmers traditionally plowed legume cover crops under as green manure, providing an important source of organic matter and introducing a significant quantity of N into the soil through biological N fixation.

Little is known about the decomposition and N dynamics of legume cover crops left on the soil surface. Therefore, in 1984 we initiated several field experiments to determine the effects of tillage on dynamics of fertilizer and legume N. We used the stable ^{15}N isotope as a tracer, allowing direct determination of recovery of fertilizer and legume N in different soil N fractions and recovery of legume N by corn (*Zea may* L.).

Study methods

We obtained 2-inch x 8-inch soil cores from a 7-year no-till corn cover crop experiment (*1*). Soil cores removed from plots with a history of hairy vetch (*Vicia villosa* Roth.) winter cover were treated with ^{15}N-enriched vetch, either surface-applied for no-till or incorporated for conventional tillage. Soil cores removed from plots with a history of corn stalk residue cover were treated with surface-applied, ^{15}N-enriched fertilizer (ammonium nitrate in 1984, ammonium sulfate in 1985) for both simulated no-till and conventional tillage. On a larger scale, 7.5 x 9.8 feet, we replaced in situ, unlabeled vetch with ^{15}N-depleted vetch. The labeled vetch was either left on the surface or plowed under. We treated soil cores from a 4-year stand of alfalfa (*Medicago sativa* L.) with surface-applied, ^{15}N-enriched alfalfa with and without 107 pounds/acre N fertilizer. All soil cores were incubated under field conditions. The experiments lasted for 2 years.

Results

Incorporated vetch residue decomposed and released N quicker than surface-applied vetch. Within 15 days, 60% of the vetch residue weight was lost with conventional tillage, while only 20% was lost with no-till. For the same period of time, surface-applied alfalfa lost about 27% of its original weight. At 120 days, 20% of the vetch residue remained with conventional tillage and about 40% remained with no-till.

The more rapid decomposition of vetch residue with con-

ventional tillage resulted in a greater accumulation of plant-available N (ammonium and nitrate) in the soil than with no-till. Plant available N levels with incorporated vetch approached those of the fertilized corn residue cores with conventional tillage and no-till. In 1985, 47% of the vetch N was recovered as plant-available N at 15 days with conventional tillage. In contrast, no more than 10% of the added vetch N was recovered as plant available N with no-till.

More N was recovered as soil organic N when applied as vetch than as fertilizer with both tillage systems. For example, at day 75, 30% of the added vetch N was found as soil organic N with both tillage treatments, while only about 16% of the added fertilizer N was soil organic N. The addition of fertilizer N in combination with alfalfa residue decreased the quantity of alfalfa N immobilized in the soil organic fraction. Thus, fertilizer N added, along with legume N, could improve the crop recovery of legume N by diluting the pool of N from which soil microbes are immobilizing.

Total recovery of added N for the periods studied was two to three times greater from vetch than from fertilizer. Although the soil cores were only 8 inches in length and, therefore, do not represent the total rooting depth, it is obvious that potential N losses are greater with fertilizer N than legume N.

The percentages of vetch residue ^{15}N recovered by the corn stover and grain were 17% and 31% for no-till and conventional tillage, respectively, in 1984. In 1985 percentages were 22% and 32% for no-till and conventional tillage, respectively. Vetch incorporation resulted in about 10% greater recovery than when left on the soil surface. The total N accumulated by corn with vetch winter cover in excess of the corn residue check treatment ranged from 45 to 60 pounds/acre depending upon tillage and year. This resulted in apparent vetch N recovery of 52% with no-till and 57% with conventional tillage. It is apparent that recovery of N directly from the vetch is less than total recovery, which includes residual vetch N. Although more vetch N was recovered with conventional tillage than no-till, no significant difference in yield occurred either year, apparently due to greater water use efficiency with no-till than with conventional tillage.

Summary

Tillage enhances legume residue decomposition and N mineralization. In our study a greater proportion of the vetch N was recovered as plant-available soil N and subsequently by the crop plant. Vetch N immobilization was two to three times greater than fertilizer N immobilization. It appears that potential N losses from vetch are less than for fertilizer N. Further work in this area is needed. Although more N is recovered when vetch is plowed under, the disadvantages of plowing, both monetary and environmental, may favor using no-till practices and adding some additional N fertilizer rather than plowing. Also, the greater water use efficiency of the no-till system is an important consideration.

J. J. Varco is an assistant professor, Mississippi State University, Mississippi State, 39762. W. W. Frye is a professor, M. S. Smith is an associate professor, and J. H. Grove is an assistant professor, Department of Agronomy, University of Kentucky, Lexington, 40546-0091.

REFERENCE
1. Ebelhar, S. A., W. W. Frye, and R. L. Blevins. 1984. *Nitrogen from legume cover crops for no-tillage corn.* Agron. J. 76: 51-55.

Estimating nitrogen and rotation effects in legume-corn rotations

M. P. Russelle, O. B. Hesterman, C. C. Sheaffer, and G. H. Heichel

For centuries farmers have known that yields of non-legume crops usually increase when those crops are grown simultaneously or in rotation with legumes. Researchers usually attribute yield increases of a nonlegume crop, such as corn (*Zea mays* L.), solely to the N contribution by the legume (*1, 7*). But positive "rotation" effects not directly associated with N have been observed (*2, 3*). These rotation effects include improved soil physical properties and elimination of phytotoxic substances (*3*), addition of growth-promoting substances (*6*), and reduced disease incidence (*4*). Estimating these rotation effects has been problematic. Generally, they have been restricted to cases in which rotation systems continue to produce higher yields than monocrop systems after N limitations to yield have been removed by fertilization (*2*). The assumption in this approach is that rotation effects are constant and not affected by changes in yield potential due to N availability. This assumption has not been tested to our knowledge.

Methods are needed to discern between N and rotation effects at any rate of applied N. Then scientists can make reliable improvements in crop management systems. Presumably, increased N contribution can be realized by use of appropriate legume harvest management (*5*) or with specially developed cultivars, such as Nitro alfalfa (*Medicago sativa* L.) (D. K. Barnes, personal communication). Without an understanding of the magnitude of rotation effects, little progress in rotation improvement can be expected. Herein, we report one method to estimate N and rotation effects in crop rotations.

This method is based on the relationship between grain yield and total N accumulation in the shoots of the nonlegume crop. As an example, we cite data from an experiment conducted at four locations in Minnesota in 1982 and 1983, with corn as the nonlegume grown in an annual rotation with either alfalfa or corn. We developed yield-N relationships for continuous corn on an irrigated sandy loam soil and on three nonirrigated, fine-textured soils (Figure 1). These relationships should be developed for the same year

M. P. Russelle is a soil scientist, U.S. Dairy Forage Research Center, Agricultural Research Service, U.S. Department of Agriculture, and an assistant professor, Department of Soil Science, University of Minnesota, St. Paul, 55108; O. B. Hesterman, is an assistant professor, Department of Crop and Soil Sciences, Michigan State University, East Lansing; C. C. Sheaffer is a professor, Department of Agronomy and Plant Genetics, University of Minnesota; and G. H. Heichel is a plant physiologist, Agricultural Research Service, U.S. Deparament of Agriculture, and professor, Department of Agronomy and Plant Genetics, University of Minnesota. A contribution of USDA-ARS and the Agricultural Experiment Stations of the University of Minnesota and Michigan State University.

in which estimates of N and rotation effects are calculated to minimize confounding effects of environmental factors.

The total effect of a rotation is the difference in grain yield between corn-after-legume and corn-after-corn at any specified rate of applied N. For a given legume treatment, 'a

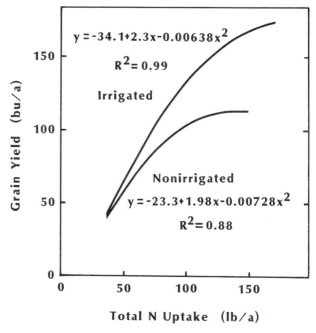

Figure 1. Response of continuous corn grain yield to total (stover + grain) N uptake at physiological maturity under irrigated and nonirrigated conditions in Minnesota, 1983.

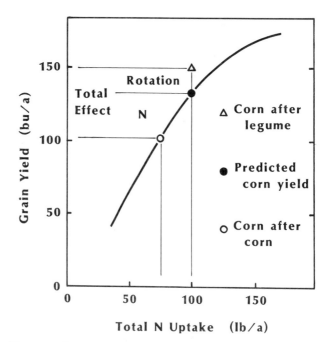

Figure 2. Relationship among total effects, N effects, and rotation effects in a hypothetical case at the irrigated site, Minnesota, 1983.

predicted value of corn grain yield is calculated from the total N accumulated in that treatment and the appropriate yield-N response function. The difference between the predicted yield and the yield of corn-after-corn is attributed to the improved N supply in the legume-corn rotation, and serves as an estimate of the N effect. The difference between the predicted yield and the actual grain yield in the legume-corn rotation is interpreted as the rotation effect.

For example, assume that yields of continuous corn and corn-after-alfalfa at a particular rate of applied fertilizer N at the irrigated site were 103 and 150 bushels/acre, respectively, and that N accumulations in these treatments were 75 and 100 pounds/acre, respectively (Figure 2). The predicted yield of corn grown in rotation with alfalfa would be 132 bushels/acre, based on N accumulation alone. The total effect of rotation would be 47 bushels/acre (150-103), the N effect would be 29 bushels/acre (132-103), and the rotation effect would be 18 bushels/acre (150-132). The rotation effect represents 38% of the total effect in this example.

Fertilizer N rate had an important effect on the relative importance of the rotation effect at the irrigated and nonirrigated locations (Figure 3). At the irrigated site, the N effect of alfalfa accounted for the entire corn yield response at low rates of applied fertilizer N because soil N supply was very small. As N became less limiting to yield, rotation effects became more important. Rotation effects accounted for about one-fourth of the yield response of unfertilized corn on the fine-textured soils and were the dominant factor at higher N rates.

There are two assumptions implicit in this proposed approach. First, we assume the effects of N and non-N factors are additive and do not interact. Second, the relationship between N accumulation in corn shoots and roots is either constant across rotation systems or does not significantly influence the estimates. Included in the N effects are all impacts of the cropping sequence on soil and crop residue N mineralization; indirect effects on N absorption by the non-legume roots, for example, improved soil aeration; altered fertilizer N availability; and others. Our method cannot provide a measure of these individual processes, but we assert that the net effect of these direct and indirect effects on N availability is described most completely by the N accumulated in the nonlegume. Also, the proposed method cannot differentiate among the various processes affecting the rotation component.

The proposed approach is the first that allows discrimination between N and non-N effects in crop rotations, without the use of N isotopes. It should be tested under a wide variety of conditions to determine its general utility, but it should be applicable to any tillage system and to most crop rotation combinations.

REFERENCES

1. Baldock, J. O., and R. B. Musgrave. 1980. *Manure and mineral fertilizer effects in continuous and rotational crop sequences in central New York.* Agron. J. 72: 511-518.
2. Baldock, J. O., R. L. Higgs, W. H. Paulson, J. A. Jackobs, and W. D. Shrader. 1981. *Legume and mineral N effects on crop yields in several crop sequences in the upper Mississippi Valley.* Agron. J. 73: 885-890.
3. Barber, S. A. 1972. *Relation of weather to the influence of hay crops on subsequent corn yields on a Chalmers silt loam.* Agron. J. 64: 8-10.
4. Curl, E. A. 1963. *Control of plant diseases by crop rotation.* Bot. Rev. 29: 413-479.
5. Hesterman, O. B., C. C. Sheaffer, D. K. Barnes, W. E. Lueschen, and J. H. Ford. 1986. *Alfalfa dry matter and N production, and fertilizer N response in legume-corn rotations.* Agron. J. 78: 19-23.
6. Ries, S. K., V. Wert, C. C. Sweeley, and R. A. Leavitt. 1977. *Triacontanol: A new naturally occurring plant growth regulator.* Science 195: 1339-1341.
7. Voss, R. D., and W. D. Shrader. 1979. *Crop rotations: Effect on yields and response to nitrogen.* Pm-905. Coop. Ext. Serv., Iowa State Univ., Ames.

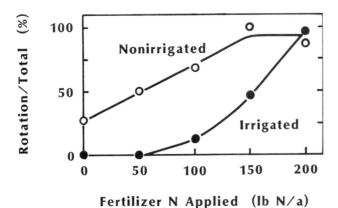

Figure 3. Relationship between fertilizer N application rate and the relative importance of rotation (non-N) effects of a 1-year-old alfalfa crop on corn grain yield under irrigated and nonirrigated conditions, Minnesota, 1983.

Fertilizer nitrogen recovery by no-till corn as influenced by a legume cover crop

S. J. Corak, W. W. Frye, M. S. Smith, J. H. Grove, and C. T. MacKown

Hairy vetch (*Vicia villosa* Roth), a winter annual legume, has proven to be an agronomically viable cover crop for no-till corn (*Zea mays* L.) production in Kentucky. When broadcast overseeded into standing corn in mid-September, hairy vetch overwinters as small plants and then accumulates substantial dry matter and N by the time of corn planting in mid-May. That portion of vetch N derived from biological N fixation is assumed to represent a significant input of N into the cropping system. The combination of hairy vetch and 90 pounds/acre of fertilizer N has consistently resulted in both greater corn yields (1) and higher net profits (2) than any other combination of cover crop and fertilizer N rate studied. Yield increases with hairy vetch are undoubtedly attributable to interactions of increased N fertility; a water-conserving mulch; and possibly improvements in other physical, chemical, and biological factors.

We initiated a study to determine the effect of a hairy vetch cover crop on the uptake of fertilizer N by corn plants and the final recovery of that N in grain. This information should prove useful for refining fertilizer N recommendations for the hairy vetch/no-till corn cropping system.

Study methods

The study site is located in Lexington, Kentucky, on a Maury soil (fine-silty, mixed, mesic, Typic Paleudalfs). Treatments include fertilizer N applied at rates of 0, 95, 190, and 285 pounds/acre. We are evaluating each N rate with and without a hairy vetch cover crop. Hairy vetch seed was inoculated with *Rhizobium* and broadcast overseeded into standing corn on assigned plots in mid-September of 1985. Beginning in 1986, and to be repeated in 1987, isotopically labeled, ^{15}N-depleted fertilizer N was applied to a subplot within each main plot. Unlabeled fertilizer N was applied to the remainder of the plot. Corn (FRB73 x PA81) was no-till planted on June 3, 1986. Herbicides were applied at planting to kill the vetch and control weeds. We removed six whole corn plants (aboveground) from each plot 6 weeks after planting, at silking, and at harvest; three were from the isotopically labeled subplot and three were from the main plot. We estimated dry matter production from the average plant dry weight multiplied by the number of plants/acre. Each three-plant subsample was analyzed separately for N. Atom-% ^{15}N of the isotopically labeled subsample was determined with a mass spectrometer. We used this result to

S. J. Corak is a graduate assistant, W. W. Frye is a professor, M. S. Smith is an associate professor, J. H. Grove is an assistant professor, and C. T. MacKown is an associate professor adjunct in the Department of Agronomy, University of Kentucky, Lexington, 40546-0091.

calculate the percentage of total plant N derived from the current year's applied fertilizer.

Results and discussion

A combination of factors resulted in a severe loss of the hairy vetch stand during the winter of 1985-1986, the first such event in 10 years of evaluation of cover crops at this location. The cover crop was replanted with a no-till drill in March. Initial growth was slow. At 4.97 inches, rainfall for April and May was 3.27 inches below the 30-year average. Corn planting was delayed until June 3 to allow the cover crop to make significant growth; this would not be a recommended management practice under normal circumstances. During the dry spring, hairy vetch depleted soil water prior to corn planting. June rainfall was 2.87 inches, 3.38 inches below normal. We observed severe wilting of corn plants on vetch plots beginning June 21. We irrigated with 2 inches of water to the entire area on June 24.

Because of the dry spring and resultant depletion of soil water, hairy vetch inhibited dry matter accumulation, total N, and fertilizer N uptake by corn as measured at 6 weeks after planting and at silking (Table 1). Fertilizer N did not increase dry matter at any sampling date, although it increased total N uptake throughout the season. Although the differences between cover treatments narrowed considerably during the season, stover dry matter was lower on vetch plots at harvest. Vetch had no effect on total N or fertilizer N in the stover at harvest. Neither N, nor vetch, nor their interaction had any effect on grain yield. Fertilizer N increased both total and fertilizer N content of the grain. At no time did vetch affect the percentage of total plant N derived from fertilizer.

These data suggest that under the adverse conditions prevailing during the summer of 1986, hairy vetch had little positive influence upon the N nutrition of the corn crop. Low rainfall during much of the season was undoubtedly a major cause of the overall poor response to both fertilizer N and hairy vetch N. The average grain yield of 83.5 bushels/acre was about half of the highest yield obtained at this site dur-

Table 1. Percentage of labeled fertilizer nitrogen recovered by corn as affected by fertilizer nitrogen rate and a hairy vetch cover.

N Rate and Cover	Fertilizer N Recovered (%)				
	Whole Plant		Harvest		
	6 Weeks	Silking	Stover	Grain	Stover + Grain
95 pounds/acre					
Without vetch	31.1	39.6	27.0	25.8	52.8
With vetch	12.0	37.8	22.4	22.0	44.4
190 pounds/acre					
Without vetch	20.8	36.7	22.1	17.9	40.0
With vetch	10.5	28.7	28.5	18.5	47.0
285 pounds/acre					
Without vetch	18.0	35.8	18.8	14.7	33.5
With vetch	11.5	25.3	15.3	15.2	30.5
ANOVA*					
N	B	B	B	B	B
HV	B	NS	NS	NS	NS
NxHV	B	NS	NS	NS	NS

*Significance of effect by analysis of variance (ANOVA): NS = not signficant (P > 0.05), A = significant (P ≤ 0.05), B = highly significant (P ≤ 0.01).

ing 1985. It is noteworthy, however, that the rate of crop growth and N uptake were greater on vetch plots from 6 weeks after planting until harvest. Clearly, growing vetch had further depleted the already low supply of soil water, but later in the season the mulch conserved a significant amount of the water from intermittent rainfall. The percentage of fertilizer N recovered was not affected by vetch or the interaction of N and vetch, except at the 6-week sampling date (Table 1). The percentage recovery of fertilizer N in stover and grain was generally lower with increasing N rates.

REFERENCES

1. Ebelhar, S. A., W. W. Frye, and R. L. Blevins. 1984. *Nitrogen from legume cover crops for no-tillage corn.* Agron. J. 76: 51-55.
2. Frye, W. W., W. G. Smith, and R. J. Williams. 1985. *Economics of winter cover crops as a source of nitrogen for no-till corn.* J. Soil and Water Cons. 40: 246-249.

Nitrogen fertilizer requirements for corn as affected by legume cropping systems and rotations

L. J. Oyer and J. T. Touchton

Early maturing winter legumes can be used as the sole N source for summer crops that have a low N requirement or that have a relatively late optimum planting date. These legumes, however, do not provide sufficient N for corn, a crop that must be planted early and that has a high N requirement.

Soybeans provide about one-third of the total N needed by subsequent corn crops. However, comparing current legume seed and seeding costs to commercial N prices, the legume must provide about 80 pounds/acre N to cover production costs. If reseeding legumes can be used, production costs may be reduced.

The major objective of our study was to determine if legume systems can be used to reduce N fertilizer requirements for corn.

Study methods

The experimental design was a split plot within a randomized complete block, replicated four times. Treatments for the whole plots included (a) continuous corn with no winter crops, (b) soybean-corn rotation with no winter crops, (c) continuous corn with fall-planted crimson clover, and (d) soybean-corn rotation with reseeded crimson clover. Subplots consisted of N treatments of 0, 60, 120, and 180 pounds/acre N.

The study, initiated in 1984, is at two sites: an Appalachian Plateau Wynnville sandy loam at the Sand Mountain Substation and a Coastal Plain Dothan fine sandy loam at the Wiregrass Substation. The two soils have an organic matter content of about 1.1%, cation exchange capacity of 4 meq/100 g, and pH of 6.1. Both soils are high in P (60 and 99 pounds/acre), K (114 and 193 pounds/acre), Ca (823 and 1,018 pounds/acre), and Mg (42 and 223 pounds/acre), respectively, at the Sand Mountain and Wiregrass Substations.

At the Sand Mountain Substation, RA 1502 corn was planted during mid April 1985 in 30-inch rows; irrigation was not available. At the Wiregrass Substation, Dekalb T1230 corn was planted in late March 1985 using Twin 7 on 36-inch centers.

At both study locations, we seeded the plots with an in-row subsoiler. We also applied starter fertilizer at the time of seeding.

At corn planting we determined the weight and N content of clover and mulch and soil N to a depth of 6 inches. At

L. J. Oyer is a graduate research assistant and J. T. Touchton is a professor, Department of Agronomy and Soils, Auburn University, Auburn, Alabama 36849.

silking, we measured ear leaf N. Finally, at maturity we measured grain yield and N.

Results

Seeded clover after corn at Sand Mountain weighed 3,198 pounds/acre. After soybeans, the reseeded clover produced 4,237 pounds/acre. The N contents were similar (2.9%) but more total N at corn planting was produced after soybeans (121 pounds/acre) than after corn (93 pounds/acre) due to greater growth period.

At Wiregrass, seeded clover after corn produced 1,103 pounds/acre; after soybeans the reseeded clover produced 2,425 pounds/acre. Percent N after corn (4.2%) was higher than after soybeans (3.8%) because the clover was less mature.

Mulch samples—all material laying on the soil surface prior to corn planting—taken at the Wiregrass Substation showed a greater contribution of dry matter and total N with the continuous corn (3,619 pounds/acre and 38 pounds/acre N) and clover-corn treatments (3,308 pounds/acre and 49 pounds/acre N) than with the soybean-corn rotation (3,187 pounds/acre and 29 pounds/acre N) and the soybean-clover-corn rotation (1,879 pounds/acre and 28 pounds/acre N). However, percent N was insignificant, equal to or less than 1.5%, and of little contribution to the subsequent corn crop.

At Sand Mountain total soil N at corn planting ranged from 230 ppm for continuous corn to 270 ppm for soybean-clover-corn. At Wiregrass the total soil N increased signficantly from 370 to 470 ppm with a winter clover crop on continuous corn and from 380 to 490 ppm with a clover crop on a soybean-corn rotation.

At Sand Mountain ear leaf N at silking significantly differed with cropping systems when no N was applied: 1.69%, 1.89%, 2.44%, and 3.02% for continuous corn, soybean-corn, clover-corn, and soybean-clover-corn, respectively. At higher N rates we found no significant differences due to cropping systems (average = 3.5%). At Wiregrass signficant differences in ear leaf N due to cropping systems occurred with 0 up to 120 pounds/acre N; ear leaf N for the soybean-clover-corn rotation was significantly higher.

At Sand Mountain corn grain yields (Table 1) peaked at 120 pounds/acre N. Cropping systems changed yield levels significantly but had little effect on N fertilizer requirements.

At Wiregrass, in contrast, corn grain yields (Table 1) were affected little by rotation when N was at optimum levels, 180, 180, 180, and 120 pounds/acre N, respectively, for the cropping systems.

Corn grain from Sand Mountain analyzed for percent N showed a primary effect due to N applied, not due to cropping system.

Conclusions

▶ Reseeded clover behind soybeans resulted in 53% more N in clover tissue at corn planting.

▶ At Wiregrass total soil N was 105 ppm higher when clover was grown in winter.

▶ At Sand Mountain the highest yielding cropping system was the soybean-clover-corn rotation with 155 bushels/acre, compared to 110 bushels with continuous corn. Cropping systems had no effect on N fertilizer requirements; 120 pounds/acre N were required regardless of cropping system. However, the soybean-clover-corn system set a yield potential.

▶ At Wiregrass the legumes replaced some but not all of the N fertilizer requirement for corn. Fertilizer N requirements for the continuous corn, soybean-corn, clover-corn and soybean-clover-corn systems were 180, 180, 180 and 120 pounds/acre respectively.

Table 1. Corn grain yield as affected by treatment.

Location and Cropping System	Corn Grain Yield* by N Treatment (pounds/acre N)			
	0	60	120	180
Sand Mountain Substation	bushels/acre			
Continuous corn	12	67	110	110
Soybean-corn	39	102	123	135
Clover-corn	53	104	132	131
Soybean-clover-corn	81	135	156	155
Wiregrass Substation				
Continuous corn	61	138	155	186
Soybean-corn	89	125	165	171
Clover-corn	85	139	152	164
Soybean-clover-corn	139	170	182	163

*Int. FLSD = 26.

Soybeans as a source of nitrogen for conservation tillage systems in the southeastern Coastal Plain

P. G. Hunt and T. A. Matheny

Nitrogen accumulation and fixation are very important to the growth and yield of soybeans (*Glycine max* L.). Total N accumulations of more than 200 kg/ha are common (*1, 5*). Net N returned to the soil—N accumulated from N_2 fixation minus N removed in seed—is important to crops that follow soybeans in rotation. Corn yields generally increase following soybeans as opposed to continuous corn. Researchers have attributed these yield increases to both net soil N increases and rotational effects. In the Midwest or other areas with high N-content soils, the annual balance of N returned to the soil from soybeans is likely to be negative (*13*). However, Patterson and LaRue (*10*) found N_2 fixation to account for 50% to 84% of accumulated N as measured by both ^{15}N dilution and difference methods when large amounts of cabonaceous materials were used to immobilize soil N. In the southeastern Coastal Plain N_2 fixation has accounted for more than 75% of accumulated N in soybeans grown on a loamy sand under irrigated conditions (*9*). Thurlow and Hiltbold (*11*) also found that 70% of the soybean N frequently came from N_2 fixation. Under these conditions of high N_2 fixation and total N accumulation, the annual return of N to the soil is positive.

Soil N levels, acidity, and temperature are known to affect nodulation and N_2 fixation (*1, 6, 7, 12*). These soil conditions differentially affect growth and symbiotic performance of various *Rhizobium japonicum* strains (*7, 8*). Different populations of *R. japonicum* strains likely cause different nodule occupancy, and this, in turn, may affect the N accumulation and/or soybean yield (*2, 5, 14*).

We grew three determinate soybean cultivars on a Norfolk loamy sand (Typic Paleudult) with either conservation or conventional tillage. Plots were split for inoculation with strain 3Ilb 110 of *R. japonicum*. We estimated N_2 fixation by the difference between total shoot N for nodulating and non-nodulating Lee isolines. We estimated net N returned to the soil by the difference between N_2-fixed N and seed N. Drought was significant the first year; only 19.5 cm of rain fell during the podfill period. Rainfall and soil water were adequate the second year. Tillage, cultivar, and inoculation did not consistently affect the percentage of nodules formed by particular rhizobial strains individually, but specific treatment combinations significantly affected nodular occupancy by certain strains.

Inoculated Coker 338 had the highest shoot total N accumulation in both years; however, the difference between years overshadowed tillage, cultivar, or inoculation effects. For the 2-year period, we estimated percentages of N supplied by N_2 fixation were 58% to 67% and 49% to 65% under conservation and conventional tillage, respectively. Yields generally did not differ significantly for tillage or inoculation; however, Coker 338 soybeans had the lowest seed yield even though it had the highest total N accumulation in the shoots. Estimates of net N returned to the soil varied more between years than between tillage treatments. Values ranged from 14 to 40 kg/ha in 1980 and from 57 to 123 kg/ha in 1981. An expanded discussion of this research has been undertaken by Hunt and associates (*4*).

We confirmed the large dry matter accumulations of determinate soybeans by measurements in a separate experiment. Dry matter production by Davis, Braxton, and Coker soybeans were measured in large plots, greater than 20 m²/sample. Mean dry matter accumulation was 6.72 Mg/ha with a coefficient of variation of 0.088 (*3*). Maximum accumulation of ground litter was 3.37 Mg/ha, and the maximum N accumulation was 293 kg/ha. Seed N mean was 130 kg/ha. If 70% to 50% of the N was supplied from N_2 fixation, this study would also show 75 to 17 kg/ha net N returned to the soil. These values are similar to those obtained in the tillage experiments. Thus, the crop rotation and nonpoint pollution control aspects of determinate soybeans are substantial for the southeastern Coastal Plains.

REFERENCES

1. Bhangoo, M. S., and D. J. Albritton. 1976. *Nodulating and non-nodulating Lee soybean isolines response to applied nitrogen.* Agron. J. 70: 322-326.
2. Hunt, P. G., A. G. Wollum II, and T. A. Matheny. 1981. *Effects of soil water on* Rhizobium japonicum *infection, nitrogen accumulation, and yield in Bragg soybeans.* Agron. J. 73: 501-505.
3. Hunt, P. G., K. P. Burnham, and T. A. Matheny. 1987. *Precision and bias of various soybean dry matter sampling techniques.* Agron. J. (In Press).
4. Hunt, P. G., T. A. Matheny, and A. G. Wollum II. 1985. Rhizobium japonicum *nodular occupancy, nitrogen accumulation, and yield for determinate soybean under conservation and conventional tillage.* Agron. J. 77: 579-584.
5. Hunt, P. G., T. A. Matheny, D. C. Reicosky, A. G. Wollum II, R. E. Sojka, and R. B. Campbell. 1983. *Effect of irrigation and* Rhizobium japonicum *strain 110 upon yield and nitrogen accumulation and distribution of determinate soybeans.* Comm. Soil Sci. Plant Anal. 17: 223-238.
6. Keyser, H. H., and D. N. Munns. 1979. *Effects of calcium, manganese, and aluminum on growth of rhizobia in acid media.* Soil Sci. Soc. Am. J. 43: 500-503.
7. Lindemann, W. C., and G. E. Ham. 1979. *Soybean plant growth, nodulation, and nitrogen fixation as affected by root temperature.* Soil Sci. Soc. Am. J. 43: 1134-1137.
8. Mahler, R. L., and A. G. Wollum II. 1981. *The influence of soil water potential and soil texture on the survival of* Rhizobium japonicum *and* Rhizobium leguminosarum *isolates in the soil.* Soil Sci. Soc. Am. J. 45: 761-766.
9. Matheny, T. A., and P. G. Hunt. 1983. *Effects of irrigation on accumulation of soil and symbiotically fixed N by soybeans grown on a Norfolk loamy soil.* Agron. J. 75: 719-722.
10. Patterson, T. G., and T. A. LaRue. 1983. *Nitrogen fixation by soybeans: Seasonal and cultivar effects and comparisons of estimates.* Crop Sci. 23: 488-492.
11. Thurlow, D. L. and A. E. Hiltbold. 1985. *Dinitrogen fixation by soybeans in Alabama.* Agron. J. 77: 432-436.

P. G. Hunt and T. A. Matheny are soil scientists at the Coastal Plains Soil and Water Conservation Research Center, Agricultural Research Service, U.S. Department of Agriculture, Florence, South Carolina 29502-3039.

12. Weber, C. R. 1966. *Nodulating and non-nodulating soybean isolines: I. Agronomic and chemical attributes.* Agron. J. 58: 43-46.
13. Welch, L. F., L. V. Boone, C. G. Chambliss, A. T. Christiansen, D. L. Mulvaney, M. G. Oldham, and J. W. Pendelton. 1973. *Soybean yields with direct and residual nitrogen fertilization.* Agron. J. 65: 547-550.
14. Williams, L. E., and D. A. Phillips. 1983. *Increased soybean productivity with a* Rhizobium japonicum *mutant.* Crop Sci. 23: 246-250.

Wheat-red clover interseeding as a nitrogen source for no-till corn

C. F. Ngalla and D. J. Eckert

For many years Ohio farmers have seeded red clover (*Trifolium pratense* L.) into winter wheat (*Triticum aestivum* L.) in the spring for use as a green manure crop following wheat harvest. The clover generally is plowed under in the fall following wheat harvest. If corn (*Zea mays* L.) is planted the next spring, the green manure crop provides about 60 to 80 pounds/acre N to the corn crop (*1*).

The growing acceptance of no-till corn production in Ohio has raised the question of whether plowing is necessary to realize the N benefit from the clover. Triplett and associates (*2*) demonstrated that corn receives large amounts of N from a killed alfalfa sod under no-till conditions in Ohio. But no data exist regarding red clover, which is common in the cash-grain producing regions of the state. Thus, we sought to evaluate the N-supplying power of red clover to no-till corn.

Wheat was planted in the fall of 1982, 1983, and 1984 at the Northwest Branch of the Ohio Agricultural Research and Development Center, located on a poorly drained Hoytville silty clay loam about 30 miles south of Toledo. In 1983, 1984, and 1985, 10 pounds/acre of redclover was drilled either into standing wheat in April or following wheat harvest in July. Corn was planted on the plots in late April in each of the succeeding years, and different plots received 0, 100, or 200 pounds/acre N as anhydrous ammonia. Though we took numerous soil and plant measurements during each corn year, herein we present only plant stand and grain yield data.

Wheat yields averaged 65 bushels/acre over the 3-year study. Clover drilling generally decreased wheat yields by 3 bushels/acre. Drilling is not necessary to establish clover, but was done to ensure quality stands. Clover coverage averaged 90% or more in October on all plots. Summer-seeded plants were smaller at this time than spring-seeded ones. Clover generally ceased top growth by late October.

Table 1 shows the pertinent corn performance data. In 1984 corn plant stands were lower on spring-seeded plots than on no-clover plots. The cover crop had been killed after corn planting and we assumed that the stand problem was due to excessive residue. Thereafter, we killed clover in the October preceding corn planting to reduce residue cover the next spring. The Hoytville soil does not require high levels of residue for optimum no-till corn performance. Fall-killing was not expected to affect the N-supplying power of the clover because legumes fix little N between late October and optimum corn planting time in northern Ohio. In 1985 we noted no consistent stand reductions due to clover; however, in 1985 we noted very consistent stand losses on clover plots. At this time, we feel that the stand losses were due to infestations of corn root webworm (*Crambus Caliginosellus*

C. F. Ngalla was a former graduate student and D. J. Eckert is an associate professor, Department of Agronomy, Ohio State University, Columbus, 43210.

Table 1. Corn plant stands and grain yields as affected by clover and nitrogen treatment, 1984 to 1986.

Year and Clover Treatment	Corn Stand (1,000 plants/acre) by N Rate (pounds/acre)				Yield (bushels/acre) by N Rate (pounds/acre)		
	0	100	200	Av.	0	100	200
1984							
None	26.8	26.0	27.2	26.7	121	193	196
Spring	24.2	22.3	22.5	23.0	156	180	200
Summer	28.5	25.3	24.9	26.2	113	176	191
LSD.10		2.4		1.4		20	
1985							
None	24.3	24.1	22.2	23.5	124	177	153
Spring	22.9	25.2	24.1	24.1	156	173	173
Summer	24.1	21.7	22.4	22.7	125	155	160
LSDS.10		1.6		NS		17	
1986							
None	20.6	19.1	21.3	20.3	85	176	196
Spring	17.5	17.3	17.9	17.6	118	156	176
Summer	16.2	17.3	19.1	17.5	102	158	174
LSD.10		2.2		1.3		24	

Clemens), a pest species identified as causing stand losses in corn following clover in production fields in northwestern Ohio.

Interpretation of yield data is confounded by differences in stands. Some general conclusions can be drawn. The summer seeding was not beneficial to the succeeding corn crop in any year. The short time between planting in July and cessation of growth in late October did not allow for sufficient N fixation to affect corn yields. The presence of spring-seeded clover residue, however, increased corn yields on plots receiving no N each year. The magnitude of the yield response was probably attenuated by stand reductions, but it did appear that the clover supplied the equivalent of 50 to 60 pounds/acre N as commercial fertilizer. Tissue analysis of corn plants indicated that the response to clover treatment was indeed an N response. Finally, there was no evidence that spring-seeded clover residue increased corn yields when supplemental N was applied, perhaps because 100 pounds/acre N was near optimum for corn production in the 3 study years. (This rate is normally regarded as below optimum; 200 pounds/acre N is normally recommended.)

Presently, the use of red clover as an N source for no-till corn is not recommended in Ohio due to the relatively high cost of N compared to commercial N fertilizer. The yield difference between no clover and spring-seeded clover treatments on plots receiving no supplemental N, 33 bushels/acre over 3 years, would certainly pay for the clover seeding and possible use of a soil insecticide to control webworms ($20 to $25/acre total). However, producers could achieve much larger yield increases (72 bushels/acre over 3 years) and returns by using commercial N sources at 100 pounds/acre N ($10 to $20/acre depending on source). Nitrogen credits are given when no-till corn follows a wheat-red clover interseeding, but we cannot yet consider it as a major alternative for supplying N to corn in Ohio.

REFERENCES

1. Schmidt, W. H., D. Myers, and R. W. Van Keuren. 1974. *Value of legumes for plowdown nitrogen.* Agron. Tip Misc-4. Ohio Coop. Ext. Serv., Columbus.
2. Triplett, G. B., Jr., F. Haghiri, and D. M. Van Doren, Jr. 1979. *Plowing effect on corn yield response to N following alfalfa.* Agron. J. 71: 801-803.

Nitrogen contribution of winter legumes to no-till corn and grain sorghum

C. L. Neely, K. A. McVay, and W. L. Hargrove

Winter legume cover crops may provide significant quantities of biologically fixed N while conserving soil and water resources. In this study we sought to determine the N contribution of various winter cover crops to grain production by no-till corn (*Zea mays* L.) and grain sorghum [*Sorghum bicolor* (L.) Moench].

We conducted field experiments in 1985 and 1986 in the Limestone Valley region of Georgia on a Rome gravelly clay loam (Typic Hapludult) and in the Coastal Plain on a Greenville sandy clay loam (Rhodic Paleudult). At the Limestone Valley location corn followed crimson clover (*Trifolium incarnatum* L.), hairy vetch (*Vicia villosa* Roth), winter pea (*Psium arvense*), wheat (*Triticum aestivum* L.), or no cover crop. Nitrogen rates applied to the corn crop were 0, 25, 50, 100, or 200 pounds/acre. In the Coastal Plain grain sorghum followed crimson clover, hairy vetch, berseem clover (*Trifolim alexandrinum* L.), wheat, and no cover crop. Nitrogen rates applied to the grain sorghum were 0, 20, 40, 80, or 160 pounds/acre. The experimental design and methodology were very similar to that described by Hargrove (1).

Results and discussion

Dry matter production and N content of cover crops. Table 1 shows total dry matter, N concentration, and N content. At both locations crimson clover produced more dry matter than hairy vetch. However, total N content was comparable because hairy vetch has a higher N concentration. Dry matter production and N concentration for wheat was much less than for crimson clover or hairy vetch. Dry matter production of berseem clover and winter pea was comparable to that of wheat. Nitrogen concentration in berseem clover and winter pea, however, was similar to hairy vetch, resulting in intermediate N content. Berseem clover and winter pea are not as well adapted to the region and dry matter production was limited compared to the other covers. Dry matter production in 1986 was considerably less than in 1985 due to lower rainfall in the winter and spring of 1986.

Yields. Table 2 shows yields for corn grain and grain sorghum as influenced by cover crops, with corresponding regression equations listed in table 3. Corn yields following wheat or fallow increased greatly with added fertilizer N. Corn yields following hairy vetch tended to be greater than those following crimson clover and winter pea. The fertilizer N replaced by legumes ranged from 75 to 100 pounds/acre N when compared to fallow and wheat treatments.

C. L. Neely and K. A. McVay are graduate research assistants, and W. L. Hargrove is an associate professor, Agronomy Department, University of Georgia, Agricultural Experiment Station, Experiment, 30212.

Table 1. Dry matter, nitrogen concentration, and nitrogen content of cover crops.

	1985			1986			Mean		
Cover Crop	Dry matter (pounds/acre)	N Conc. (%)	N Content (pounds/acre)	Dry Matter (pounds/acre)	N Conc. (%)	N Content (pounds/acre)	Dry Matter (pounds/acre)	N Conc. (%)	N Content (pounds/acre)
Limestone Valley									
Wheat	2,055b	1.79c	37c	1,500c	2.22c	33b	1,178	2.01	35
Winter pea	2,195b	4.13a	91b	650d	4.98a	32b	1,423	4.56	61
Hairy vetch	3,137a	3.91a	124a	1,915b	5.32a	101a	2,526	4.62	113
Crimson clover	3,353a	3.28b	110ab	2,313a	4.05b	94a	2,833	3.67	102
Coefficient of									
variation (%)	14.5	8.8	20.7	9.5	7.0	9.6			
LSD .05	625	0.46	30	242	0.46	10			
Coastal Plain									
Wheat	1,669b	1.89c	32c	734c	2.06b	15c	1,202	1.98	23
Hairy vetch	4,306a	3.23a	139a	2,951b	4.13a	122a	3,629	3.68	130
Crimson clover	4,017a	2.20bc	88c	4,619a	2.25b	103a	4,318	2.23	96
Berseem clover	1,248c	3.04ab	38c	1,433c	2.87b	41b	1,341	2.96	40
Coefficient of									
variation (%)	7.1	20.7	30.9	18.2	18.7	23.3			
LSD .05	318	0.86	37	710	0.85	26			

Table 2. Influence of cover crop on grain yield.

	Yield by Year and N rate (pounds/acre)					
	1985		1986		Mean	
Cover	0	200	0	200	0	200
Corn, Limestone Valley			bushels/acre			
Fallow	55	191	71	130	63	161
Wheat	15	193	50	152	32	121
Crimson clover	159	197	128	147	143	172
Hairy vetch	181	188	131	147	156	168
Winter-pea	147	179	117	150	132	165
Sorghum, Coastal Plain						
Fallow	90	97	38	86	64	92
Wheat	64	97	38	64	51	80
Crimson clover	91	95	74	67	82	81
Hairy vetch	103	112	83	53	93	83
Berseem	88	99	52	65	70	82

Table 3. Regression equations for yield as a function of fertilizer nitrogen rate for each cover crop.

Site, Year, and Cover Crop	Regression Equation	$P_R > F$	R^2
Corn, Limestone Valley			
1985			
Winter pea	$Y = 149.9 + 0.692x - 0.002x^2$	0.0003	0.6217
Hairy vetch	$Y = 178.6 + 0.296x - 0.001x^2$	0.036	0.3236
Crimson clover	$Y = 157.1 + 0.223x - 0.0001x^2$	0.0165	0.3838
Wheat	$Y = 18.8 + 2.146x - 0.006x^2$	0.0001	0.9141
Fallow	$Y = 53.3 + 1.668x - 0.004x^2$	0.0001	0.9131
1986			
Winter pea	$Y = 122.1 + 0.386x - 0.001x^2$	0.0769	0.2604
Hairy vetch	$Y = 132.8 + 0.457x - 0.0019x^2$	0.0287	0.3408
Crimson clover	$Y = 121.7 + 0.486x - 0.001x^2$	0.0519	0.3091
Wheat	$Y = 52.8 + 1.144x - 0.003x^2$	0.0001	0.8766
Fallow	$Y = 76.2 + 0.833x - 0.002x^2$	0.0009	0.5606
Sorghum, Coastal Plain			
1985			
Berseem	$Y = 89.3 + 0.299x - 0.001x^2$	0.0115	0.4086
Crimson clover	$Y = 89.6 + 0.120x - 0.0005x^2$	0.4819	0.0823
Fallow	$Y = 89.1 + 0.241x - 0.001x^2$	0.0615	0.2797
Hairy vetch	$Y = 101.9 + 0.099x - 0.001x^2$	0.0575	0.2853
Wheat	$Y = 64.6 + 0.713x - 0.003x^2$	0.0001	0.8217
1986			
Berseem	$Y = 51.6 + 0.763x - 0.004x^2$	0.0006	0.5828
Crimson clover	$Y = 75.4 + 0.077x - 0.0008x^2$	0.1132	0.2260
Fallow	$Y = 40.8 + 0.859x - 0.003x^2$	0.0001	0.8463
Hairy vetch	$Y = 82.4 + 0.056x - 0.001x^2$	0.0058	0.4545
Wheat	$Y = 38.9 + 0.988x - 0.005x^2$	0.0001	0.9189

Grain sorghum yields following wheat in 1985 and 1986 responded to applied fertilizer N. Where grain sorghum followed fallow in 1985, there was no response to additional fertilizer N. Grain sorghum yields after crimson clover showed no response to additional fertilizer N. Where vetch was the cover, grain sorghum yields showed no response to additional N in 1985.

In 1986 yields responded negatively to additional N. Grain sorghum yields following berseem clover responded moderately to additional N. In general, when grain sorghum followed any cover or fallow treatment, additional fertilizer N rates above 80 pounds/acre N either did not influence yields or did so in a negative way. The amount of fertilizer N replaced by the legumes ranged from 60 to 70 pounds/acre N when compared to wheat and fallow treatments, respectively. In 1986, at N rates greater than 80 pounds/acre N, there was a greater incidence of anthracnose and Fusarium-induced foliar/stalk rot symptoms as well as weed pressure. These factors probably contributed to the reduced yields.

In summary, grain yields following hairy vetch or crimson clover with no applied fertilizer N were equal to that following wheat or fallow with 60 to 100 pounds/acre N. Under the environmental conditions of this study, a well-adapted legume can replace all the fertilizer N requirement of grain sorghum and about 100 pounds/acre N for corn production.

REFERENCE

1. Hargrove, W. L. 1986. *Winter legumes as a nitrogen source for no-till grain sorghum*. Agron. J. 78: 70-74.

Estimating response curves of legume nitrogen contribution to no-till corn

D. D. Tyler, B. N. Duck, J. G. Graveel, and J. F. Bowen

Cropping systems involving winter cover crops followed by a no-till grain crop are a good way to reduce soil erosion to acceptable levels and still maintain farm profit. If the winter cover is a legume, part of the N recycled from the legume after it is chemically killed is fixed from the atmosphere. This provides an additional source of N fertility.

One approach to estimating the N contributed by a legume cover has been to compare yield response curves at various N rates with and without a legume.

We initiated an experiment in 1985 to compare corn yields with 0, 50, 100, 150, and 200 pounds/acre N planted with no-till practices into previous crop residue or chemically killed wheat, vetch, Austrian winter pea, and crimson clover. Nitrogen rates constituted the main plots and cover the subplots. The soil type on the experimental area was a Memphis silt loam (fine-silty, mixed, thermic, Typic Halpudalf).

In the first year of the experiment, there was no residue from a previous crop. Figure 1 shows no-till corn yields from the no-cover treatment compared to those of corn planted in killed vetch or wheat. We derived the curves from calculated regression equations having R^2 values of 0.81, 0.78, and 0.95 for the vetch, no cover, and wheat, respectively. Corn yields in no cover and wheat were influenced differently by N fertilizer rates. Yields of corn planted into killed wheat were lower than yields of corn in the no-cover treatment, up to about 175 pounds/acre N. The lower yields for corn planted into wheat and receiving less than optimum N fertilization could be explained by immobilization of mineralizable and fertilizer N during microbial decomposition of the wheat residue. Analysis of residues of the cover crops showed C:N ratios of 32:1, 18:1, 10:1, and 13:1 for wheat, crimson clover, vetch, and winter pea, respectively. Researchers generally recognize that C:N ratios as high as that of the wheat residue may result in N immobilization during residue decay.

Yields of corn planted into killed vetch were higher at all N rates than those of the other two covers. Our results were similar with Austrian winter pea and crimson clover. Corn yields in the no cover treatment peaked at about 150 pounds/acre N. Yields also were similar with 75 pounds/acre N when corn was planted into killed vetch. When corn was planted into killed wheat, yields did not peak until 200 pounds/acre N. Based on response curve comparisons, we estimated the N equivalency of the legume to be about 75

D. D. Tyler is an associate professor of plant and soil science, West Tennessee Experiment Station, Jackson, Tennessee 38301; B. N. Duck is a professor of plant and soil science, University of Tennessee, Martin, 38237; and J. G. Graveel is an assistant professor and J. F. Bowen is a graduate student, Plant and Soil Science, University of Tennessee, Knoxville, 37996.

pounds/acre N using no cover and about 125 pounds/acre N with wheat cover.

Rainfall distribution was favorable during the 1985 growing season, but the 1986 season was much drier and corn yields were about half of those obtained in 1985 (Figure 2). R^2 values for the curves are 0.51, 0.83, and 0.93 for the vetch, corn stalks, and wheat, respectively. Yield comparisons using maximums for corn stalk and wheat residue treatments indicated N equivalencies of about 100 pounds/acre N from vetch compared to corn stalks and about 120 pounds/acre N compared to wheat.

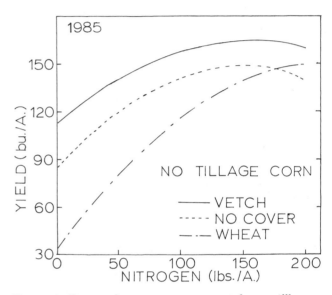

Figure 1. Regression response curves for no-till corn yields versus nitrogen rates for three covers, Milan, Tennessee, 1985.

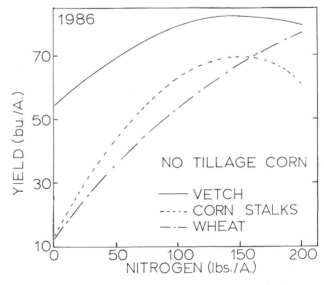

Figure 2. Regression response curves for no-till corn yields versus nitrogen rates for three covers, Milan, Tennessee, 1986.

Higher corn yields following a legume cover compared to the two nonlegume covers make direct yield comparisons for N equivalency very difficult. Thus, where a legume is used in a conservation tillage system, a question still remains on the recommended N fertilization: Should farmers reduce the amount of fertilizer N or, conversely, apply the optimum N and potentially produce a higher yield with the legume? Comparing the cost of the cover crops, N, and labor will help answer this question.

Nitrogen from legume cover crops for no-till corn and grain sorghum

J. H. Herbek, W. W. Frye, and R. L. Blevins

Legumes in crop rotations have provided N for nonlegume crops since the early history of agriculture. Recently, there appears to be a renewed interest in the use of cover crops, along with the adoption of no-till or other conservation tillage production systems. Cover crops fit well with conservation tillage systems because they produce plant material that can be chemically killed to provide a surface mulch for the subsequent crop. Legume cover crops offer advantages because they not only provide an erosion control benefit and a surface mulch similar to small grain cover crops but they also biologically fix N that can be used to meet a significant portion of the N requirement of the subsequent grain crop. Because of the large quantity of N needed for production of corn and grain sorghum, it is reasonable to consider use of legumes as a management strategy to reduce N fertilizer expenses as well as provide soil and water conservation benefits.

Research shows that several legumes perform well as winter cover crops for no-till corn or grain sorghum and provide significant levels of N to the grain crop (*1, 2, 3, 5, 6, 7*). Estimated values of the N contribution range from 50 to 180 pounds/acre.

We designed a study to determine the amount of biologically fixed N that can be obtained from different legume cover crops in continuous production of corn and grain sorghum under no-till practices.

Study methods

We began our experiment in the fall of 1979 at the West Kentucky Research and Education Center in Princeton. The soil is a Zanesville silt loam (fine silty, mixed, mesic, Typic Fragiudalfs). We planted cover crops of hairy vetch (*Vicia villosa*) and bigflower vetch (*Vicia grandiflora*) into corn each fall. Beginning in 1985 we added grain sorghum (*Sorghum bicolor* L. Moench) as a variable in place of a legume seeding variable. Seeding rates for the two vetches were 35 pounds/acre. We compared cover crop treatments to a cover of corn stalk residue from the previous year. Initial N fertilizer rates were 0, 45, and 90 pounds/acre N as ammonium nitrate. But in 1984 we increased the N rates to 0, 75, and 150 pounds/acre. The corn and grain sorghum was planted with a no-till planter in late May each year. After planting, we sprayed the plots with a tank mixture of 2 pints/acre of Paraquat, 2.5 pounds/acre of Bladex, 2.5 pounds/acre of Lasso, and a surfactant in 45 gallons of water/acre.

J. H. Herbek is an associate extension professor, Department of Agronomy, Research and Education Center, Princeton, Kentucky 42445, and W. W. Frye and R. L. Blevins are professors, Department of Agronomy, University of Kentucky, Lexington, 40546-0091.

Table 1. No-till corn yields, 1980-1986, and grain sorghum yields, 1985-1986, as affected by winter cover crops and nitrogen rates, Princeton, Kentucky.

Cover Treatments	Grain Yields (bushels/acre) by N Rate (pounds/acre)					
	Corn			Grain Sorghum*		
	0	45/75†	90/190†	0	75	150
Hairy vetch	49	68	96	67	75	99
Bigflower vetch	44	71	99	72	79	97
Corn residue only	17	43	83	45	75	99

*Includes two years only, 1985-1986.
†45 and 90 pounds/acre in 1983; 75 and 150 pounds/acre in 1984, 1985, and 1986.

Cover crop samples were harvested before grain planting each year to determine dry matter yields and N content. Grain was harvested in early October each year and yields were adjusted to 15.5% moisture.

Results and discussion

Results for the first 4 years, 1980-1983, showed that dry matter production of hairy vetch was significantly higher than that of bigflower vetch (6). The total N content contained in the aboveground portion of hairy vetch was statistically higher than bigflower vetch for these years. During years when soil moisture was less limiting, 1981 and 1982, the additional N supplied by legume cover crops was more evident in terms of corn yield.

Corn grain yields (Table 1), averaged over 7 years, 1980-1986, were higher for both hairy vetch and bigflower vetch cover treatments at all N rates compared with corn residue only. Corn grain yields were similar for hairy vetch and bigflower vetch treatments, with slightly higher grain yields for bigflower vetch where fertilizer N was added. Corn grain yields in 1985 and 1986 under bigflower vetch cover treatments were considerably higher than corn yields after hairy vetch. The lower grain yield response for hairy vetch was due in part to winter kill that reduced the stands of hairy vetch.

Grain sorghum yields (Table 1) were comparable for all cover treatments at 75 and 150 pounds/acre N. With no fertilizer N applied, we observed significantly higher yield for the vetches compared with the corn residue cover treatment.

Based on the 1985-1986 corn grain yield data, we estimate that hairy vetch and bigflower vetch have the potential to provide N at a level equivalent to about 45 and 65 pounds/acre fertilizer N, respectively.

Both corn and grain sorghum responded to legume cover crops at the Princeton location. An advantage that grain sorghum offers in this climatic zone is that the optimum planting data for sorghum coincides better with the date that the legumes reach maturity. The legume maturity date is usually later than the optimum date for planting corn in western Kentucky.

REFERENCES
1. Buntley, G. J. 1986. *Tennessee no-tillage update.* In Proc. S. Reg. No-Till Conf. Reg. Series Bull. 319. Ky. Agr. Exp. Sta., Lexington.
2. Ebelhar, S. W., W. W. Frye, and R. L. Blevins. 1984. *Nitrogen from legume cover crops for no-tillage corn.* Agron. J. 76: 51-55.
3. Flannery, R. L. 1981. *Conventional vs. no-tillage corn silage production.* Better Crops 66: 3-6.
4. Hargrove, W. L. 1986. *Winter legumes as a nitrogen source for no-till grain sorghum.* Agron. J. 78: 70-74.
5. Mitchell, W. H., and M. R. Teel. 1977. *Winter annual cover crops for no-tillage corn production.* Agron. J. 69: 569-573.
6. Varco, J. J., W. W. Frye, J. H. Herbek, and R. L. Blevins. 1984.. *Effects of legume cover crops on yield of no-till corn.* In J. T. Touchton [ed.] *Proceedings of Seventh Annual Southeast No-Tillage Systems Conference.* Ala. Agr. Exp. Sta., Auburn.
7. Worhsam, A. D. 1986. *No-tillage research update-North Carolina.* In Proc. S. Reg. No-Till Conf. Reg. Series Bull. 319. Ky. Agr. Exp. Sta., Lexington.

Grow your own nitrogen

William H. Reichenbach

Farmers in southeastern Indiana are continually looking for ways to increase profits from their corn land. With stable or declining prices prevailing over the past several years, profits increasingly depend upon reducing production costs. Using legumes and no-till planting methods are two possible ways to cut production costs. Legumes in a rotation provide lower cost N than buying N in a bag. When the row-crop revolution hit southeastern Indiana several years ago, growing legumes in a rotation became a forgotten science. The result: more acres of corn and beans and, when prices were good, more profit. Under this system, however, production depended upon the purchase of commercial N to keep corn yields up. When corn prices fell, profits fell. Some believe it is time to reinvent the wheel and "grow" N.

Hairy vetch, a winter annual.

Long-time farmers remember a viney plant called hairy vetch. This is a winter annual legume that was seeded in the fall and matured in late spring or early summer. In the process it took N from the air and fixed it in the plant. As the plant deteriorates during the growing season, the nutrients, including N, are available for plant growth. Improved chemicals and planting equipment (no-till planters) made it possible to cut planting costs and reap the benefits of the cheaper N.

Soil conservationists have found that a winter cover crop, such as hairy vetch, can help control erosion on sloping fields and add needed organic matter to the soil. The result: a hairy vetch cover crop to grow N in the fall, winter, and spring, and to protect the soil from erosion and a no-till planting system to cut planting costs. Both would, hopefully, increase profit.

Hairy vetch, a summer annual

In 1983 the U.S. Department of Agriculture's payment-in-kind (PIK) program reduced corn production and the corn surplus. Farmers were required to seed PIK land to a summer cover to protect it from erosion. Innovative farmers in southeastern Indiana raised this question: If hairy vetch reduces soil erosion and fixes nitrogen in the fall, winter, and spring, what would happen if we plant it on PIK acres in the spring?

This was a good question, but there didn't appear to be good answers. The bottom line: they tried it and it worked. As a result, several farmers are regularly using hairy vetch, the proven winter annual legume, on annual set-aside or

William H. Reichenbach is an area conservationist with the Soil Conservation Service, U.S. Department of Agriculture, North Vernon, Indiana 47265.

diverted acres. I have not encountered anyone who does not like the results.

Crownvetch, a permanent cover

In the early 1980s some innovative farmers read about a permanent N-fixing legume called crownvetch. The overall idea was to establish a stand of crownvetch, which is perennial, and plant no-till corn into it each year. Chemicals applied at planting time would "put the vetch to sleep" until the corn was established and growing well. The crownvetch would then "wake up" and do its thing: take N from the air and put it in the soil. Again, research from several universities showed it might work. The bottom line: there are several trial fields established in southeastern Indiana from which we have learned a few things:

1. Crownvetch seed is expensive.
2. Crownvetch is a slow starter and may take 2 to 5 years to get established.
3. Crownvetch requires a well-drained soil.
4. Crownvetch requires a high phosphate and potash soil.
5. There are some tricks to the chemical treatment.
6. If a good stand is obtained, crownvetch provides excellent erosion control.

The jury is still out on whether or not crownvetch will provide erosion control, fix N, and not cut corn production/acre on a year-to-year basis.

In southeastern Indiana the search goes on for other ways to cut production costs. Farmers regularly ask USDA offices for help. Rotations, including legumes, remain a possibility. Many farmers use animal manure to cut fertilizer costs. Farmers are seeding wheat, rye, and annual ryegrass for winter cover crops to help control erosion, build organic matter, and aid in weed control in corn and soybean fields. Many farmers are sure that planting no-till corn or soybeans in a growing cover crop cuts the cost of chemicals needed to control weeds. Mixing hairy vetch and wheat-rye or ryegrass adds additional N for the corn crop.

One farmer used wheat as a cover crop in one-half of the field and wheat and hairy vetch in the other half. The farmer planted the whole field to green beans the following spring. He harvested and sold $50/acre more green beans on the wheat-hairy vetch half of the field. The only variable was the vetch with added N. He believes spendings $5/acre for hairy vetch seed (10 pounds at 50 cents/pound) was a sound investment.

Research from several universities indicates that hairy vetch, seeded in the fall, will produce enough N to produce 100 to 110 bushels of corn the following season without additional N. Much of the land in southeastern Indiana is rolling and subject to erosion. The row-crop revolution is here. Corn and bean prices are down so farmers must determine how to cut production costs. Legumes, used as a cover crop, are proving useful in lowering N costs, controlling erosion, and adding beneficial organic matter to the soil. All this is done when the soil is resting from its chief function: raising food for a hungry world!

Management of subterranean clover as a source of nitrogen for a subsequent rice crop

S. M. Dabney, G. A. Breitenbeck, B. J. Hoff, J. L. Griffin, and M. R. Milam

A legume cover crop can provide a substantial portion of the N required by a subsequent crop of flooded rice. However, little is known about the effects of various management practices on the availability of soil N in legume-rice rotations. To achieve the optimal benefit from a legume cover crop, the management system employed should promote the development of readily mineralizable forms of soil N but inhibit the oxidation of legume N to nitrate, thereby reducing N losses via denitrification and leaching when a permanent flood is established.

Work in California on the effects of timing and depth of placement of legume crop residue before water seeding rice indicated that deep incorporation of the legume residue by tillage immediately prior to flooding produced the most efficient recovery of legume N by rice plants (*4, 5*). Other studies, however, have shown that the availability of N from residues incorporated into the soil does not meet the demand of a young rice crop (*3*) and that small quantities of top-dressed inorganic N may be needed as a supplement to legume green manures (*2*).

The cost of planting a legume cover crop is another important consideration in legume-rice rotations. Because the cost of legume seed can approach the value of the N fertilizer replaced, management systems that promote natural reseeding of the legume are desirable. Recent studies in Louisiana showed that when sub clover was permitted to produce mature seed before no-till planting of rice, excellent stands of this legume were reestablished after the floodwater was removed (*1*).

Study methods

Sub clover (*Trifolium subterraneum* Woogenellup) was planted at a rate of 20 pounds/acre during November 1985 on plots located at Baton Rouge, Crowley, and Winnsboro, Louisiana. Soils at the three locations were Mhoon silty clay loam (Typic Fluvaquent), Crowley silt loam (Typic Albaqualf), and Gigger silt loam (Typic Fragiudalf), respectively.

One week prior to planting rice, we incorporated the sub clover cover into the soil on the tilled plots by rototilling and sprayed the clover on the no-till plots with glufosinate or glyphosate herbicide. Before performing these treatments,

S. M. Dabney is an assistant professor, G. A. Breitenbeck is an associate professor, and B. J. Hoff is an associate professor, Department of Agronomy, Louisiana Agricultural Experiment Station, Louisiana State University Agricultural Center, Baton Rouge, 70803. J. L. Griffin is an associate professor, Rice Research Station, Crowley, Louisiana 70527. M. R. Milam is an assistant professor, Northeast Research Station, St. Joseph, Louisiana 71366. Approved for publication by the Director of the Louisiana Agricultural Experiment Station as manuscript number 86-09-0373.

the aboveground clover biomass was 350, 380, and 650 pounds of dry matter/acre at Baton Rouge, Crowley, and Winnsboro, respectively. The stand of sub clover at Winnsboro was damaged severely by hail in April 1986.

Rice (*Oryza sativa* L. 'Labelle') was planted at a rate of 140 pounds/acre on May 12 and May 15, 1986, with a Moore Uni-drill at Crowley and Baton Rouge, respectively, and on June 18 with a Kincade seeder at Winnsboro. Thirty days after planting, urea was broadcast at rates of 0, 50, 100, and 150 pounds/acre N and a permanent flood was established. We replicated all treatments four times at each of the three locations. At maturity, we determined panicle density from duplicate 0.1-m² areas in each plot. After harvest, we determined weight/seed from the weight of duplicate lots of 500 combine-harvested seed. Grain yields were adjusted to 12% moisture.

To determine the effects of till and no-till management systems on the forms of soil N, we took composite soil samples from unfertilized plots at the Baton Rouge location before tillage or herbicide application and at approximately weekly intervals thereafter until imposition of a permanent flood on June 17. The field-moist samples were sieved (4-mm), mixed, and 10 g subsamples extracted with 2 N KC1. We determined the amount of exchangeable ammonium and nitrate in the extracts by steam distillation.

Results and discussion

With no urea N applied, rice yields at the Baton Rouge location were 290 pounds/acre greater in no-till plots than in plots where sub clover was tilled into the soil before planting (Table 1). The yields from the corresponding no-till plots at Crowley, however, were 400 pounds/acre less than tilled plots. No-till planting at Winnsboro resulted in complete

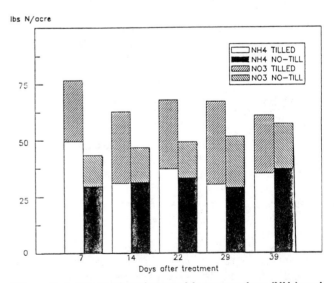

Figure 1. Amount of exchangeable ammonium (NH₄) and nitrate (NO₃) in soils at various times after tillage (tilled plots) or herbicide application (no-till plots). At the time these treatments were performed, the amount of exchangeable ammonium and nitrate in soil averaged 45.6 and 38.4 pounds/acre, respectively.

Table 1. Yield, panicle density, and weight per grain of rice planted with and without tillage and receiving four rates of fertilizer N at three locations, with a clover cover crop.

Urea N Applied (pounds/acre)	Till			No-Till		
	Grain Yield (pounds/acre)	Panicle Density (number/m²)	Weight per Grain (mg)	Grain Yield (pounds/acre)	Panicle Density (number/m²)	Weight per Grain (mg)
Baton Rouge						
0	5,140	340	21.1	5,430	364	21.4
50	5,180	380	21.8	5,310	375	21.1
100	5,460	384	20.9	4,780	363	20.1
150	4,180	404	20.2	4,810	387	20.0
Crowley						
0	4,280	383	21.9	3,840	385	22.1
50	4,640	393	22.0	5,420	410	21.7
100	4,840	325	21.3	5,750	392	20.9
150	5,120	353	20.4	5,050	392	20.5
Winnsboro						
0	2,590	228	19.9	ND*	ND	ND
50	3,500	265	18.9	ND	ND	ND
100	4,120	355	18.7	ND	ND	ND
150	3,980	317	18.6	ND	ND	ND

*No-till planting at Winnsboro resulted in stand failure.

stand failure due primarily to an unsatisfactory performance of the planting equipment used at that location.

Application of 50 or 100 pounds of urea N/acre to tilled plots increased rice yields at all three locations (Table 1). On the other hand, applications of urea at these rates to no-till plots led to yield increases only at the Crowley location. Application of 150 pounds of urea N/acre led to reduced rice yields at all three locations. The weight/rice grain decreased in response to increased rate of urea-N application at all three locations (Table 1). We attributed the yield increase in response to N fertilization at Winnsboro, where the amount of clover produced was small, solely to an increase in the number of panicles. In contrast, the yield increase observed at Crowley resulted from an increase in the number of grains produced/panicle. At Baton Rouge, a combination of inherent native fertility and clover-derived N in no-till plots appeared to provide sufficient N to produce maximum yields with no additions of fertilizer; addition of supplemental N led to increased lodging and reduced yields.

The management system used had a marked effect on the form and amount of inorganic soil N (Figure 1). The amount of inorganic soil N recovered from unfertilized plots where sub clover was incorporated by tillage was substantially greater than from the corresponding no-till plots where sub clover was treated with a herbicide. Although tillage appeared to increase mineralization of clover-derived N, it is unlikely that this increase led to a corresponding increase of plant-available N to the rice crop because most of the increase in inorganic soil N in the tilled plots could be attributed to an increase in the amount of nitrate (Figure 1). We assumed that most if not all of this nitrate was lost via denitrification and leaching when the permanent flood was established.

Our results indicate that when sub clover is permitted to produce mature seed prior to planting rice, desiccation of sub clover with a herbicide 40 days prior to flooding leads to greater availability of legume N than does incorporation of this legume by tillage. Clover stand ratings in December 1986 indicated that stands were better in no-till than in tilled areas although adequate sub clover stands reestablished in all plots.

REFERENCES

1. Eastman, J. S. 1986. *Potential for the use of legume cover crops, reduced tillage, and sprinkler irrigation in Louisiana rice production.* M. S. Thesis. Dept. Agron., La. State Univ., Baton Rouge. 165 pp.
2. Morris, R. A., R. E. Furoc, and M. A. Dizon. 1986. *Rice response to a short-duration green manure. II. N recovery and utilization.* Agron. J. 78: 413-416.
3. Westcott, M. P. 1982. *A comparison of leguminous green manure as sources of nitrogen for lowland rice.* Ph.D. Thesis. Univ. Calif., Davis.
4. Williams, W. A., and D. C. Finfrock. 1962. *Effect of placement and time of incorporation of vetch on rice yields.* Agron. J. 54: 547-549.
5. Williams, W. A., and J. H. Dawson. 1980. *Vetch is an economical source of nitrogen in rice.* Calif. Agr. (Aug.-Sept.): 15-16.

Legumes in crop rotations as an internal nitrogen source for corn

J. K. Radke, W. C. Liebhardt, R. R. Janke, and S. E. Peters

We studied three cropping systems, each with a 5-year cropping rotation, in a conversion experiment from 1981 to 1986. We designed the conversion experiment to determine the effects of converting from a farming system using normal levels of commercial fertilizers and pesticides to two other systems using no commercial chemicals.

The experimental design was a split-plot, randomized complete block with three farming systems (whole plots), three starting points (crops) into each of the 5-year crop rotations (sub-plots), and eight replications. The three cropping systems were a low-input animal system; a low-input, cash-grain system; and a conventional corn-soybean system. We added N to the two low-input systems by plowing under green legume crops. In addition, animal manure was applied prior to planting the corn crops in the low-input animal system. A basic rotation of small grain, red clover, corn, soybeans, silage corn was grown in the low-input animal system. The rotation in the low-input, cash-grain system was small grain/red clover, corn, small grain/red clover, corn, soy-

J. K. Radke is a soil scientist with the Agricultural Research Service, U.S. Department of Agriculture, stationed at the Rodale Research Center, Kutztown, Pennsylvania 19531. W. C. Liebhardt is director of science, R. R. Janke is farming systems coordinator, and S. E. Peters is a researcher with the Rodale Research Center.

beans (Table 1). The cropping sequence in the conventional system was corn, corn, soybeans, corn, soybeans. In 1984 we replaced the red clover crop with hairy vetch, which was plowed under in the spring of 1985 prior to planting corn in one of the low-input, cash-grain treatments. In 1986 we used a similar set of treatments in the continuation of this experiment, now best described as a farming system experiment.

In 1985 and 1986 we measured plant heights, crop biomass, leaf area, crop growth stages, yields, and other parameters several times during the growing seasons. Soil water was monitored weekly with a neutron probe. We took soil samples at 0 to 15 cm and 15 to 30 cm four times in 1985 and ten times in 1986 for nitrate and ammonium determinations. Weather parameters were measured hourly with an automated weather station.

Corn yields in the two low-input systems during the first 2 years (1981 and 1982) of the conversion process were 60% of the conventionally grown corn yields. In 1983 the low-input, cash-grain corn yielded 80% and the low-input-animal-system corn yielded 90% compared to the conventional system. By 1985 the corn yields in the two low-input systems increased to the point that yields were equal to or greater than the corn yields in the conventional system. In 1985 the corn grown following the plow-down of hairy vetch in the low-input, cash-grain system yielded 112% of the conventional corn (significantly different at P = 0.10). Corn yields in the low-input animal system and conventional system were not significantly different in 1986; however, the low-input, cash-grain corn yields were slightly lower. Leaf tissue N in the corn showed similar trends—lower for the low-input corn in the early years and equalling or exceeding the conventional corn levels in 1985 and 1986.

Table 1. Rotation sequences for the 5-year conversion project. Each of the three systems has three entry points into the rotations.

| Treatment | Conversion Project Rotations | | | | |
	1981	1982	1983	1984	1985
Low-input animal system					
1	Spring oats Red clover	Red clover	Manure Corn	Soybeans	Manure Corn silage
2	Corn	Soybeans	Manure Corn silage	Wheat Red clover*	Red clover
3	Manure Corn silage	Wheat Red clover*	Red clover	Manure Corn	Soybeans
Low-input cash-grain system					
1	Spring oats Red clover	Corn	Spring oats Red clover	Corn	Soybeans
2	Soybeans	Spring oats Red clover	Corn	Wheat Hairy vetch	Corn
3	Corn	Soybeans	Spring oats Red clover	Corn	Spring oats Annual legume*
Conventional cash grain					
1	Corn	Corn	Soybeans	Corn	Soybeans
2	Soybeans	Corn	Corn	Soybeans	Corn
3	Corn	Soybeans	Corn	Corn	Soybeans

*Overseeding

Soil nitrate levels were moderate in 1985. We observed the highest nitrate levels early in the season in the conventional tillage treatments receiving starter fertilizer and the low-input animal treatment receiving a manure application. Soil nitrates in the low-input, cash-grain system increased as the growing season progressed and were the highest at the last sampling date, August 13. Soil nitrates in both the conventional and low-input animal systems were low by this date even though N stress was not observed in any of the corn. Soil nitrate levels generally were higher in 1986. Heavier than normal applications of chicken manure to two low-input animal corn treatments resulted in soil nitrate levels exceeding 100 ppm in August. Obviously, excessive amounts of N can be applied with animal manure as well as with commercial fertilizer.

The two major problems that must be handled when using low-input systems are N and weeds. Some time is needed to establish new N equilibria when converting from commercial fertilizers to animal and green manures. For this reason farmers should grow crops other than corn for the first year or two. Starting the rotation with a small grain/legume or soybeans resulted in no yield losses. Sufficient N is available from green manures or green and animal manure applications to give corn yields comparable to conventionally grown corn. Weeds are best controlled by maintaining good groundcover and a vigorous crop.

We simulated the 1985 and 1986 corn treatments in a computer using the Nitrogen-Tillage-Residue Management Model. The model was calibrated to the 1985 conventional corn data. Subsequent runs for the low-input systems tended to give yield predictions that were two low, indicating that the model did not adequately simulate all of the biological processes important in systems making use of green and animal manures. Modifications to the model have made the predictions for all systems much more accurate.

Hairy vetch winter cover for continuous corn in Nebraska

P. T. Koerner and J. F. Power

Our purpose was to evaluate hairy vetch (*Vicia villosa* Roth) as a winter cover crop for continuous dryland corn in eastern Nebraska. The experiment was conducted for 3 crop years on a Sharpsburg silty clay soil at Lincoln, Nebraska.

We established 20-foot by 25-foot plots in a standing corn field in September 1981 by broadcast seeding half of the plots with 60 kg/ha of Madison hairy vetch.

We analyzed six different treatments on plots before corn planting: (1) no N and no vetch, (2) no vetch with 60 kg/ha N, (3) vetch left standing through growing season, (4) vetch double disked in before corn planting, (5) vetch sprayed with paraquat before corn planting, and (6) vetch sprayed with Banvel (decamba) before corn planting. We then planted all plots to corn with a no-till planter. Grain and stover were harvested at maturity. We measured soil moisture, soil nitrate, and vetch production during the year. The same procedure was repeated the next 2 years on the same plots. Weeds were controlled during the growing season by recommended no-till methods.

In all 3 years May and June precipitation were above the normal values of 3.84 and 4.09 inches, respectively. Precipitation in July and August also was above normal for 1982, but considerably below normal in 1983 and 1984. This resulted in ample water for corn establishment, but severe water deficiencies during the pollinating and grain development phases in 1983 and 1984.

Corn grain yield levels reflected water availability in July and August, with relatively high yields in 1982 and low yields in 1983 and 1984 (Table 1). Contrast statements showed that corn grain yields increased under three conditions: the use of vetch, killing the vetch at planting time, and disking the vetch in rather than leaving the dead residue on the soil surface. In all years near maximum corn yields occurred when the vetch was disked in. Average corn yields were essentially the same for the three treatments in which vetch growth was either unretarded, retarded with paraquat, or killed with Banvel.

Because N fertilization failed to stimulate corn yields, the high yield that resulted from disking in vetch residues cannot be explained by possible changes in N availability. We determined nitrate-N in the surface foot of soil periodically (Table 2). Contrasts indicate that differences were present for the fertilized versus unfertilized plots. We also found differences for the fertilized plots versus those that had vetch (treatments 3, 4, 5, and 6). These differences were present mainly for the fall sampling date. This may be an indication that time is required to change the inorganic N in the vetch to the nitrate form in the soil. We found no differences in

P. T. Koerner is a research technologist and J. F. Power is research leader, University of Nebraska, Agricultural Research Service, U.S. Department of Agriculture, Lincoln, Nebraska 68583.

to the nitrate form in the soil. We found no differences in nitrate levels between the four vetch treatments.

Growing vetch as a winter cover crop could deplete the available water supply and thereby reduce corn yields. This may have contributed to the lower corn yields from plots on which the vetch was not killed. If the vetch depleted soil water in the surface foot at corn planting time, corn stand could be affected. However, we observed no such problem in this experiment. Occasionally, however, measurements of soil water in the 0- to 3-inch, 3- to 6-inch, and 6- and 12-inch soil depths did show significant differences between treatments at planting time. The positive effect from disking vetch residue into the soil is hard to explain on the basis of soil water.

Having excluded N and water availability as being responsible for the favorable effect of disking in vetch residues, one must look to other explanations. Disking under the vetch residues may have created a more favorable soil environment for corn growth (temperature, porosity, etc). These yield results may be an expression of the rotation effect that is often observed and not explained. An explanation may require investigations into changes in activity of disease-producing organisms or other parameters of microbial activity in the soil. Because we made no such measurements in this experiment, no definitive information on these points can be presented.

Table 1. Corn grain yield as affected by N fertilizer and hairy vetch

	Corn Grain Yield (bushels/acre)			
Treatment*	1982	1983	1984	Average
1	93	23	16	44
2	94	28	15	46
3	55	20	7	27
4	101	24	31	52
5	60	7	22	30
6	69	5	18	31
LSD(.05)	18	8	8	

*See text for description of treatments.

Table 2. Soil nitrate-nitrogen in 1983 as affected by treatment.

	Soil Nitrate (ppm) by Depth							
	Spring Sampling				Fall Sampling			
Treatment*	0-3"	3-6"	6-12"	12-24"	0-3"	3-6"	6-12"	12-24"
1	7.2	6.4	5.8	5.2	7.3	6.0	6.2	5.1
2	10.9	7.8	8.8	5.7	15.0	10.0	10.0	8.3
3	9.7	7.5	7.0	6.1	7.3	6.0	6.8	5.8
4	6.4	6.1	5.5	6.0	5.3	5.2	5.3	4.6
5	9.2	7.6	7.7	6.6	5.4	5.2	5.0	5.4
6	7.3	5.8	5.6	4.5	5.0	4.5	5.5	4.4
LSD(.05)	3.6	1.9	2.7	1.1	4.7	2.6	2.6	1.7

*See text for description of treatments.

Recovery of nitrogen-15 from labeled alfalfa residue by a subsequent corn crop

G. H. Harris and O. B. Hesterman

One important role of forage legumes in conservation tillage systems is to provide N to subsequent nonlegume crops. To give proper credit to a legume for its N contribution, the quantity of legume N recovered by the subsequent crop must be known.

Several methods have been used in the past to quantify the N contribution of alfalfa (*Medicago sativa* L.) to a subsequent corn (*Zea mays* L.) crop. Nitrogen credits based on the N in alfalfa biomass range from 60 (7) to 170 (2) pounds/acre N. The fertilizer replacement value of alfalfa is the amount of inorganic N fertilizer required to produce a yield of the subsequent crop equivalent to that produced following alfalfa. Nitrogen credits based on this method range from 80 (6) to 120 (1) pounds/acre N.

A more accurate method of estimating legume N credits involves enriching legume residues with ^{15}N and tracing the label into the subsequent crop. Few field studies have been conducted using ^{15}N-labeled legumes. In Australia, Ladd and associates (4, 5) incorporated ^{15}N-labeled medic (*Medicago littoralis* L.) into field plots and traced the ^{15}N into a subsequent wheat (*Triticum aestivum* L.) crop. They found that wheat recovered 14% to 23% of the ^{15}N applied.

Alfalfa-corn cropping systems are common in the midwestern United States. Yet no field studies using ^{15}N-enriched alfalfa have been reported. Thus, we sought to quantify, using ^{15}N, the N contribution of alfalfa in an alfalfa-corn rotation. We also investigated the effect of alfalfa plant parts (shoots versus roots/crowns) and time of incorporation (fall versus spring) on alfalfa N recovery by corn.

Study methods

We conducted this experiment at two field locations in Michigan; on a Capac loam at East Lansing and on Oshtemo sandy loam at the Kellogg Biological Station. During the winter of 1985, we labeled Vernal alfalfa growing in a greenhouse sand bench by fertilizing it with 12% [15]N atom % excess K[15]NO$_3$. At harvest, we separated plants into shoots and roots/crowns and dried all material. We applied treatments of ^{15}N-enriched alfalfa shoots (8.1 % [15]N atom % excess) or roots/crowns (6.9 % [15]N atom % excess) to plots in either the fall of 1985 or the spring of 1986. We applied all treatments to plots at a rate equivalent to 100 pounds/acre N. Each plot consisted of a 2-foot-diamater by 2-foot-deep undisturbed soil column enclosed by a 22-gauge sheet metal cylinder. When applying the treatments, we first removed the top 6 inches of soil and handpicked and screened out all existing residue. We then mixed the ^{15}N-enriched

G. H. Harris is a graduate assistant and O. B. Hesterman is an assistant professor, Department of Crop and Soil Sciences, Michigan State University, East Lansing, 48824.

alfalfa with the screened soil and returned it to the plot.

After we applied the spring treatments, we sowed corn in all plots, eventually thinning the corn to 3 plants/plot. In the fall of 1986 we harvested the corn grain, stover (stalks plus cobs), and roots from each plot. We dried all harvested plant material and subsampled it for analysis of total N (%N) and ^{15}N-enrichment. We determined total N by the salicylate method using a Lachat flow-injector analyzer after a standard Kjeldahl digest and determined ^{15}N-enrichment on a Micromass 622 mass spectrometer after a standard Kjeldahl digest and steam distillation. With this information, we calculated the proportion of alfalfa ^{15}N applied that was recovered by the corn grain, stover, or roots. We analyzed ^{15}N recovery in whole corn plants and parts by analysis of variance for a randomized complete block (four replications).

Results and discussion

Whole corn plants recovered an average of 16.6% and 25.0% of ^{15}N from incorporated alfalfa residue at East Lansing and Kellogg, respectively (Table 1). At both locations we found about 62%, 35%, and 3% of the ^{15}N recovered in grain, stover, and roots, respectively.

At both locations whole corn plants recovered a higher percentage of incorporated ^{15}N from alfalfa shoots than from roots/crowns (Table 2). Whole corn plants recovered a higher percentage of ^{15}N when alfalfa was incorporated in the spring versus fall at East Lansing (Table 2).

To express our results in terms of conventional legume N credits, percent ^{15}N recovery must be converted to pounds/acre N. A typical quantity of alfalfa dry matter returned to the soil in an alfalfa-corn rotation would be 2.7 tons/acre—1.8 tons/acre and 0.9 ton/acre for roots/crowns and shoots, respectively (3). Thus, at an N concentration of 1.6% for roots/crowns and 2.4% for shoots, this would be equivalent to 57 and 43 pounds/acre N for roots/crowns and shoots, respectively. Using percentage recovery results from this experiment, 8 pounds/acre N would be recovered from both roots (57 x .14) and shoots (43 x .19) at East Lansing and 12 pounds/acre N would be recovered from both roots/crowns (57 x .21) and shoots (43 x .29) at Kellogg. This range of 16 to 24 pounds/acre N is well below the values reported for N credits from non-^{15}N methods and suggests that previous methods of calculating N credits may overestimate the N contribution of alfalfa.

Our results agree with those of Ladd and associates (5), who concluded that legume materials incorporated into the soil contribute little to soil-available N and to crop N uptake during the first year of decomposition. Instead, the main value of the legume appears to be long-term, that is, to maintain levels of soil N high enough to ensure adequate delivery to future crops.

REFERENCES

1. Baldock, J. O., and R. B. Musgrave. 1980. *Manure and mineral fertilizer effects in continuous and rotational crop sequences in central New York.* Agron. J. 72: 511-518.
2. Groya, F. L., and C. C. Sheaffer. 1985. *Nitrogen from forage legumes—harvest and tillage effects.* Agron. J. 77: 105-109.
3. Hesterman, O. B., C. C. Sheaffer, D. K. Barnes, W. E. Leuchen, and J. H. Ford. 1986. *Alfalfa dry matter and nitrogen production and fertilizer nitrogen response in legume-corn rotations.* Agron. J. 78: 19-23.
4. Ladd, J. N., J. M. Oades, and M. Amato. 1981. *Distribution and recovery of nitrogen from legume residues decomposing in soils sown to wheat in the field.* Soil. Biol. Biochem. 13: 251-256.
5. Ladd, J. N., M. Amato, R. B. Jackson, and J. H. A. Butler. 1983. *Utilization by wheat crops of nitrogen from legume residues decomposing in soils in the fields.* Soil. Biol. Biochem. 15: 231-283.
6. Schmid, A. R., A. C. Caldwell, and R. A. Briggs, 1959. *Effect of various meadow crops, soybeans, and grain on crops that follow.* Agron. J. 51: 160-162.
7. Smith, D. 1956. *Influence of fall cutting in the seeding year on the dry matter and nitrogen yields of legumes.* Agron. J. 48: 236-239.

Table 1. Recovery of nitrogen-15 from alfalfa residue (% of input) by a subsequent corn crop at two Michigan locations. Values are averaged over alfalfa plant parts (shoots and roots/crowns) and times of incorporation (fall or spring).

Plant Part	^{15}N Recovery (%) by Location	
	East Lansing	Kellogg
Grain	10.1 (0.9)*	15.9 (1.8)
Stover	6.1 (0.6)	8.3 (0.7)
Root	0.5 (0.0)	0.8 (0.1)
Whole plant	16.6 (1.4)	25.0 (2.2)

*Standard error shown in parenthesis.

Table 2. Effect of alfalfa plant part and time of incorporation on recovery of nitrogen-15 from alfalfa residue (% of input) by subsequent whole corn plants at two Michigan locations.

Factor	^{15}N Recovery (%) by Location	
	East Lansing	Kellogg
Plant part		
Shoots	19.3*	28.5
Roots/crowns	14.0	21.4
Time of incorporation		
Fall	13.9†	25.8
Spring	19.4	24.2

*Shoots greater than roots/crowns at P < 0.05 using a paired t test.
†Spring greater than fall at P < 0.05 using a paired t test.

Fate of nitrogen mineralized from nitrogen-15-labeled alfalfa in no-till, intercropped corn

D. Jordan, D. Harris, C. W. Rice, and J. M. Tiedje

The benefits of legumes in providing N to subsequent crops have long been recognized. However, one of the greatest challenges in producing corn (*Zea mays* L.) under no-till systems is N management (*1*). Quantifying the amount of legume N available to a subsequent corn crop has not been accurately assessed either in a no-till or an intercrop system. This has been hampered by a suitable method of assessing the quantity of N transferred in intercropping or crop rotation systems. The methods commonly used are measuring the total Kjeldahl N in legumes, calculating the fertilizer replacement value, and tracing ^{15}N from labelled legume residues. The ^{15}N tracer technique can be used to trace the release of N from the legume residue and uptake by the subsequent crop.

Typically, in a no-till system the N contribution from ^{15}N-labelled tops is measured, but root N contribution generally has not been assessed. If N from legume roots is assessed, then the soil and roots are disturbed. The roots are then mixed into the soil, typifying a plowed system. Therefore, our objective was to develop a system for measuring the N contribution from legume tops and intact roots in a undisturbed system, either no-till or intercropped. There are several possible fates of the legume N in intercropped soil following suppression of the legume crop. These include use for legume regrowth or corn growth or movement into the microbial biomass or soil mineral N pools.

Study methods

We measured the transfer of ^{15}N from labelled alfalfa tops and roots to a subsequent corn crop in a greenhouse experiment. We compared two levels of suppression: cutting versus herbicide followed by cutting.

We grew alfalfa in a Kalamazoo sandy loam (fine, loamy, mixed, mesic Typic Hapludalfs). The plants were foliar fed twice a week for 8 weeks with 99% ^{15}N urea using a hand sprayer. A protective apparatus prevented direct contamination of the soil with the spray solution.

We suppressed alfalfa growth by applying glyphosate followed by cutting 4 hours later (suppression 1) or by cutting the plant at 5 cm (suppression 2). We dried the vegetative parts at 60°C and placed 25 g dry matter containing 0.69 g N on the 0.035-m² soil surface. Corn was planted 5 days later and harvested at 2 and 4 months. Treatments were

Table 1. Nitrogen in corn tops and roots from nitrogen-15-labelled alfalfa at two different suppression levels.

Plant Part and Suppression	Total N (mg/kg)	Total N Uptake (g/plant)	^{15}N from Legume (g)	N Recovery (%)
Corn tops				
Herbicide and cutting	.258(.043)*	.185(.055)	.0065(.002)	15.55(4.23)
Cutting only	.228(.056)	.061(.030)	.002(.002)	4.91(1.99)
Corn roots				
Herbicide and cutting	.0804(.0459)	.028(.020)	.0005(.0006)	1.39(1.01)
Cutting only	.0608(.0298)	.007(.004)	.0002(.0001)	.58(.31)

*Standard deviation shown in parenthesis.

replicated six times. We determined total Kjeldahl N and percent ^{15}N in the vegtative and root material of corn and alfalfa.

Results and discussions

Nitrogen loss from the alfalfa residues was 76% with cutting alone and 71% with herbicide treatment. The corn tops and roots had greater N uptake and recovery in suppression treatment 1 (herbicide and cutting) as compared to that with suppression treatment 2 (cutting only) (Table 1). When alfalfa regrowth was suppressed by both cutting and herbicide treatment, the corn plants recovered 17% of the alfalfa N (24% of the N released). With cutting alone, the recovery of N by the corn fell to 6% of the total N (8% of the N released).

These differences were clearly reflected in the growth of the corn plant—plant heights of 80.6 cm and 62.5 cm and plant biomass of 11.2 g and 4.0 g for suppression 1 and 2, respectively. Total N was also greater for suppression 1 (Table 1).

Regrowth of alfalfa suppressed by cutting alone was vigorous while regrowth of alfalfa treated with herbicide and cutting was delayed and reduced. The additional suppression of alfalfa growth by the herbicide treatment thus reduced competition between the corn and alfalfa regrowth, allowing the corn plants to recover much more of the mineralized N. Our results suggest that the level at which legume plants are suppressed in an intercrop system will determine the uptake and percent recovery of mineralized N in subsequent crops.

REFERENCE
1. Ebelhar, S. A., W. W. Frye, and R. L. Blevins. 1984. *Nitrogen from legume cover crops for no-tillage corn.* Agron. J. 76: 51-55.

D. Jordan is a graduate assistant, D. Harris is a research specialist, C. W. Rice is an assistant professor, and J. M. Tiedje is a professor, Department of Crop and Soil Sciences, Michigan State University, East Lansing, 48824.

INSECTS AND DISEASES

Influence of legumes on insects and diseases in conservation tillage systems

R. A. Byers and E. L. Stromberg

Odum defined agriculture in the United States as agroecosystems that are domesticated ecosystems intermediate between natural ecosystems—forests and grasslands—and fabricated ecosystems—cities (59). Market and other economic and political forces, along with urbanization, have transformed agroecosystems from domesticated ecosystems, which 100 years ago were relatively harmonious with the environment, into increasingly fabricated ecosystems. Recently, energy subsidies and farm sizes have increased, and farmers have placed emphasis on continuous culture of grain and soybean cash crops, much for export (59). Unfortunately, this trend toward intensified and chemically dependent agriculture has led to serious problems, such as insects and diseases. Pimentel (60) claimed that pest problems often have an ecological basis. He listed 10 factors that could be responsible for pest problems in agriculture: monocultures, introduction of new crops into new biotic communities, accidental introduction of pest species, movement of crops into different climatic regions, limited diversity of plant species, changed plant spacings, continuous crop culture, increased soil nutrients, altered planting times, and pesticide alterations of crop physiology.

Obviously, some of these factors are difficult to control, especially introduction of foreign pests. However, most of the remaining factors are management variables. For example, farmers frequently try to solve pest problems by applying chemical pesticides, often without considering the density of the pest population. Foster and associates (29) recommended a routine application of a pesticide at planting time

R. A. Byers is a research entomologist with the U.S. Regional Pasture Research Laboratory, Agricultural Research Service, U.S. Department of Agriculture, University Park, Pennsylvania 16802, and E. L. Stromberg is an assistant professor, Department of Plant Pathology, Physiology, and Weed Science, Virginia Polytechnical Institute and State University, Blacksburg, Virginia 24061.

for control of corn rootworms (*Diabrotica* spp.) in continuous corn, based strictly on economic criteria. Such a policy not only leads to problems of environmental contamination and development of insecticide-resistant pest populations, but it also upsets the balance of natural enemies of pests and can alter crop physiology, making plants more susceptible to pests (60). Consequently, management decisions to control pests with insecticides often have backfired.

Ecologists have advocated modifying agroecosystems using new management methods to alleviate pest problems. One of the most popular schemes is to increase crop diversity, which should lower pest pressure either by increasing natural enemies or by lowering host plant availability in diversified systems (2). Hare (35), however, outlined major barriers to using vegetative diversity as a means of insect pest and disease management. First, there are difficulties in harvesting crops from diverse stands without incurring additional expenses. Second, farmers' incomes could fall because of dilution of valuable plant species per unit area of cropland. Dempster and Coaker (18) felt one way to keep insect populations at low levels would be to diversify both habitats surrounding the crop and the plant composition in the field. However, they felt the high mobility of insects would make changes in the habitat surrounding the cropland less effective than increasing diversity in the planted field.

One group of ecologists has recommended controlling pests with crop rotations and changing the spatial pattern of crops by stripfarming, intercropping, and crop mixtures (3). They also advocated changing management variables, such as timing of cultivation, sowing, plant population, amount of fertilizer, and/or irrigation. However, they conceded that even if management solutions to pest problems could be found, they would be difficult to sell to farmers if those options reduced yields.

Use of legumes in conservation tillage systems can lead

to an increase in crop diversity. Farmers can use legumes in crop rotations with grains, as green manure for grains, as winter cover, as an intercrop, as a trap crop, and as a renovator for unproductive pastures and range. Scientists are only beginning to understand how these diverse systems interact with pathogens and insects.

Crop rotations

In the 1940s researchers recommended crop rotations to reduce insect and disease problems (47). A 4-year rotation of corn (Zea mays L.), oats (Avena sativa L.), red clover (Trifolium pratense L.), and wheat (Triticum aestivum L.) with a green manure cover of sweetclover [Melilotus alba (L.) Desr.,] had no corn hills infested with corn rootworm. However, with shorter rotations of 2 and 3 years, with corn grown 1 out of 3 years or 2 out of 3 years and no legume, there were rootworm infestations of 15% and 32%, respectively (6). The failure of shorter rotations to control corn rootworms is now thought to be partly related to diapause in corn rootworm eggs (47). In 1981, 40% of corn rootworm eggs had extended dispause over two winters, which could negate short crop rotations (48). Also, short rotations of corn with soybeans [Glycine max (L.) Merr.], and oats did not effectively control northern corn rootworms (Diabrotica longicornis LeConte) because more larvae of this species occurred in corn following oats than in continuous corn (41). Therefore, short rotations are not effective for controlling rootworms.

Crop rotation is a proven method of reducing crop losses from disease. Many of the important pathogens of forages are associated with host debris. The populations of many of these pathogens can be reduced with time as the infested host debris decays. In a survey of alfalfa fields in western Virginia, VanScyoc and Stromberg (81) found that the incidence and severity of Sclerotinia crown and stem rot, caused by the fungus Sclerotinia trifoliorum Eriks., was significantly lower in fields that had not been associated with a known host 1 or more years prior to planting with conservation tillage.

Corn farmers have abandoned crop rotations for a variety of economic reasons (47). However, rotations of alfalfa (Medicago sativa L.) with corn are not less profitable than continuous corn. An alfalfa-corn sequence, with alfalfa subjected to a three-cut system, was the most economically optimum rotation in Minnesota (38). Two-year gross margins for continuous corn and soybean systems fell $15.60/acre and $80.00/acre, respectively (38). The researchers listed reduction of weed, insect, and pathogen infestations as one of several benefits from rotations, but presented no pest data to confirm this relationship.

Insects and diseases also cause problems in continuous alfalfa fields (78). Clover root curculio [Sitona hispidulus (F.)] reduced alfalfa seedling density 43.8% 22 days after planting alfalfa by conventional methods (30). In addition, soil-borne fungi reduced stand density by 40.5% in 1983 at Lexington, Kentucky (30). In that same study stresses from fungi and insects in 1984 significantly reduced dry weight yields by 49.5% and 19%, respectively.

Slugs and pathogens are often problems in continuous alfalfa established with conservation tillage methods. One study found that two slug species (Arion fasciatus Nilsson and Derocerus reticulatum Müller) and three leafhopper species [Empoasca fabae (Harris), Aceratagalia sanguinolenta (Provancher), and Macrosteles fascifrons (Stol)] were negatively associated with alfalfa plant density (7).

Several diseases are potentially more prevalent and severe under continuous alfalfa cropping. Crown wart [Physoderma alfalfae (Lagh.) Karling], previously limited to the western United States, has been reported in Pennsylvania (51). This chytrid fungus is associated with excessive soil moisture during early spring. Conservation tillage establishment methods in continuous crop production tend to maintain higher moisture levels in the upper soil profile that favor spring infection by the pathogen. The wider use of conservation tillage could increase the prevalence of crown wart throughout the mid-Atlantic and southeastern United States.

Violet root rot, caused by the fungus Rhizoctonia crocorum DC. ex FR., has been reported as a minor disease of alfalfa, but one study showed that rot severely reduced stand life and productivity of affected fields (32). Stromberg (76) identified five fields in one Virginia county infected by this disease within the past 3 years. This was the first confirmed report of this disease in Virginia, and all incidences of the pathogen have occurred in no-till seedings.

Phytophthora root rot, caused by the fungus Phytophthora megasperma (Drechs.), is a disease favored by cool (24° to 27°C), wet soils associated with inadequate vertical and horizontal drainage (32). This disease can devastate susceptible cultivars both as seedlings or older plants under favorable environmental conditions. Reestablishment of continuous alfalfa by conservation tillage provides a particularly favorable situation for the development of this disease for several reasons: (a) there is no host-free period; (b) water movement through the soil profile is retarded due to compaction from harvesting equipment; and (c) the accumulation of organic residues on the soil surface contributes to cooler, wetter soils. However, farmers can reduce the risk of this disease by planting resistant cultivars.

Sclerotinia crown and stem rot increases the risk of failure in continuous alfalfa. The resting structures or sclerotia are present at or near the soil surface, often in infested alfalfa debris. These sclerotia produce sporebearing structures called apothecia. Primary infection follows forcible discharge of ascospores. These ascospores germinate under favorable moisture conditions upon contacting susceptible host tissue and produce hyphae than directly penetrate leaf cuticle (61). An inconspicuous leaf spot forms within a few days after the infection. The pathogen grows from the leaf spot after a killing frost and colonizes the dying or dead leaf tissue (20). The pathogen invades the stem and crown tissue by late winter and early spring, frequently killing the young plant. It is during this secondary colonization of the stem and crown tissue that the fungus advances as mycelia—white, cottony vegetative growth—to adjacent healthy plants. Sclerotinia actually may spread down a drill slit or groove, killing all plants in its path (74, 75, 77). Young plants decay completely, leaving little evidence that plants ever existed. Colonized stems of larger plants wilt and become bleached in appearance as the tissue degrades and the plant eventually dies. In the spring, as the pathogen colonizes the crown, the mycelia form aggregations that mature into sclerotia. Under continued wet conditions, the fungus may remain active into early summer, causing wilting and bleaching of stems (*21, 62*). Severely infested fields often contain large patches of dead plants, or the entire field may be affected and weed-infested (*74, 75, 77, 81*). The grower must decide whether continuous alfalfa is worth the time, extra cost, and management effort in reestablishment because of the many stresses on new seedlings from slugs, insects, diseases, weeds, and even competition from old plants. In light of the many pest problems associated with continuous alfalfa production, perhaps a better management alternative would be to use the N increase in soil following alfalfa in a rotation with grain.

Stripcropping

Alfalfa has been used as a trap crop in California to control lygus bugs (*Lygus hesperus* Knight) when the alfalfa is planted in strips in cotton fields (*68*). Lygus bugs preferred alfalfa even with cotton growing within 3 feet. The bugs stayed on the alfalfa unless it was under water stress. Consequently, lygus damage to cotton was reduced (*69*). The stripcropping scheme also has worked in Mississippi (*67*). Because of the success in controlling lygus bugs, the use of stripcropping should be encouraged.

Intercropping

Drake proposed a unique cropping system with corn planted in rather wide row spacings within an alfalfa field, with the option to harvest both crops (*25*). He suggested other legumes that might be compatible with corn: birdsfoot trefoil (*Lotus corniculatus* L.], white clover (*Trifolium repens* L.) and crownvetch (*Coronilla varia* L.). Cardina and Hartwig (*14, 36*) have planted corn in crownvetch.

Presently, scientists do not know if insects and pathogens may become problems in such systems. Needless to say, there are so few experiments with intercropping systems using forage legumes that data on pest populations are lacking. In Colorado, Capinera and associates (*13*) studied intercropping pinto beans (*Phaseolus vulgaris* L.) and sweet corn to determine what level of diversity is necessary for a pest control benefit. They found that planting various combinations of alternating rows of corn and beans reduced the number of beetles (*Epilachna varivestris* Mulsant) on beans planted in an intercropped system. However, corn rootworms were more numerous in the most diverse systems. Risch and associates (*64*) tested whether a corn-bean-squash (*Curcubita pepo* L.) intercropped polyculture system reduced pest species. They found that populations of leaf-eating beetles were always lower on beans planted with corn than with beans alone. They thought that the corn exerted an inhibitory effect on beetle numbers by shading the bean plants and by posing a physical barrier to beetle movement. Beetle movement in a squash monoculture was different than movement in a polyculture. Beetles stayed longer on a single plant in a monoculture than in a polyculture because of the presence of nonhosts in the polyculture. They concluded that more studies on the effect of intercropping on insect control and the resulting yields in this systems are needed.

Winter cover

Use of legumes as winter cover crops and for grazing is important in areas of the United States with mild climates. Several annual clovers, including crimson (*Trifolium incarnatum* L.), arrowleaf (*T. vesiculosum* Savi), and sub (*T. subterraneum* L.), are important for winter grazing in the Southeast and in the dry areas of northern California and Oregon (*80*). However, growers have planted fewer acres of crimson clover in recent years because of attacks by the clover head weevil [*Hypera meles* (F.)] (*46*). Adults overwinter in trash near field borders and woodlands. They leave these sites and attack crimson clover in mid-February, and larvae feed on flower ovules and seeds in March (*37*). Adult feeding on stems results in lodging of flower heads (*79*). Weevils can take up to 80% of the seed (*73*). Plant resistance to clover head weevil is a possibility (*70, 71*).

Arrowleaf clover acreage has increased concommitant with reduced acreages of crimson clover. Initially, there were few pest problems with arrowleaf, introduced to the southern United States in the 1960s. But recently clover head weevils and lesser clover leaf weevils have been collected from arrowleaf in Mississippi (*57*). Other potential pests include the clover root curculio [*S. hispidulus* (F.)], clover stem borer (*Languria mozardi* Lat.), Fuller rose beetle (*Panntomorus cervinus* Koch), the pea aphid, [*Macrosiphon pisi* (Harris)], the potato aphid (*M. euphoreae* Thomas), and the green peach aphid (*Myzus persicae* Sulzer). Ellsbury and associates (*27*) reported pea aphids colonizing arrowleaf in the spring when plants are elongating. They found an interaction between plant diseases and aphids on arrowleaf clover. Aphids grew better on healthy versus diseased plants.

There are other legumes grown for winter cover. But the insect and disease situations on these species are not known or poorly understood. Because several of these crops are relatively new to these systems, insect and disease problems may become apparent when the amount of land planted to winter legumes increases.

Large insect populations on winter legumes are not always detrimental. Winter cover, such as clover, offers overwintering sites for insects, some of which are beneficial. Two Lepidopteran pests on *Brassica* sp., *Rachiplusia ou* Guenee and the bilobed looper [*Autographa biloba* (Stephens)], overwinter on clover with many of their parasites and predators (*56*). On the other hand, the clovers may act as a reservoir for such virus vectors as aphids (*34*). Clearly, scientists need to know more about insect populations on winter legumes and how they may damage ensuing crops.

Green manure

Sweetclover, alfalfa, ladino clover, red clover, and alsike clover (*Trifolium hybridum* L.) are used often as green manure crops (*72*). Biennial sweetclover generally is the best of these legumes because it produces the largest amount of dry matter and N. However, farmers have planted less sweetclover because of the sweetclover weevil [*Sitona cylin-*

dricolis (F.)) (*72*). Among 19 *Melilotus* species screened by Akeson and associates (*1*), *M. infesta* Guss. was resistant to feeding by adults, with a complex chemical basis for the resistance, involving both attractants and deterrents (*54*).

Sweetclover is also attacked by a root borer, *Walshia misecolorella* (Chambers), a native insect that attacks wild legumes (*55*).

Pasture and range renovation

No-till renovation of grasslands becomes economical when desirable grass and legume species decline to a point where both broadleaf and grassy weeds infest the pasture or range. Weed competition with alfalfa seedlings usually is not great in late summer, and the reduced weed pressure permits seedlings to become well established before spring. The general procedure is to graze or cut grass sods close in summer and apply herbicides to prepare a weed-free, killed sod for planting in August through October (*82*). Byers and associates (*9*) reviewed insect and slug pests of sod-seeded legumes in 1983. Since then, scientists have done considerable research on the damage caused by three species of slugs, *Derocerus laeve* Muller, *D. reticulatum* Muller, and *Arion fasciatus* Nilsson, to legumes planted into sods with conservation tillage methods (*7, 24, 33, 66*). In addition to these species, *D. agrestris* and *A. ater* (L.) are important pests of white clover in Washington (*44*). Legumes planted with no-till drills are susceptible to slug damage because the groove in the sod made by these drills is an ideal environment for slugs. Slugs are attracted to the moist and cool conditions in the groove and eat emerged seedlings. The reduction in plant density often is reflected in yield reductions up to 1 year later (*9, 24*).

Snails have been reported as pests of legumes in conservation tillage systems (*45*). The snail *Polygyra cereolus* Muhlfeld destroyed legume seedlings in a bahiagrass (*Paspalum notatum* Flugge) pasture. White clover was more susceptible than red clover or alfalfa. More than one snail per 20 seedlings constitutes a threat to legume establishment (*45*).

Researchers have indentified several insects as pests in pasture renovations with legumes. Crickets [*Allonemobius allardi* (Alexander and Thomas)] are limiting factors in establishing white clover and alfalfa into grass sods in late summer (*33, 43, 66*). Grasshoppers (*Melanoplus* sp.) also may have damaged legume seedings in North Carolina (*65*). However, there is a question of how important grasshoppers actually are in legume establishment. Mangan and associates (*53*) collected grasshoppers most often in plots of birdsfoot trefoil with open canopies and with high proportions of grass. Grasshoppers were less numerous on red clover, probably because of the lower density of the grass and the more closed canopies of red clover compared with trefoil. Furthermore, fewer than half of the Orthoptera species collected showed significant associations with the planted legumes (*53*).

Grasshoppers are important pests in rangeland in the western United States. In laboratory and field tests Hewitt and associates (*39*) evaluated seven legume cultivars as hosts for grasshoppers. The species all have potential for reseeding rangeland. Alfalfa had the greatest potential for reseeding

range sites because it was least preferred by grasshoppers (*39*). Later tests showed Spreador-2, Teton, and Travois alfalfas were least preferred by grasshoppers, although Roamer and Ranger were very susceptible (*40*). Further experiments are needed to determine if grasshoppers are pests of legume seedlings in conservation tillage systems.

Mangan and Byers (*52*) showed that the garden fleahopper [*Halticus bractatus* (Say)] is a pest of alfalfa and red clover but not birdsfoot trefoil. One to three insects confined on seedling plants in the greenhouse had negative effects on alfalfa and red clover. The insects failed to survive on trefoil. Conversion of sweep-net data to number of fleahoppers/plant indicated that field populations corresponded closely with numbers used in the greenhouse. Damage to seedlings (stippling, spotting, and death) was greater in fields with volunteer white clover and black medic (*Medicago lupulina* L.) than in grass fields without these legumes.

The clover root curculio is another important insect pest of legumes in pasture renovations. Larvae can be numerous enough to limit establishment of alfalfa due to destruction of root nodules and limited plant growth (*9*). Damage to tap roots by larvae accumulates over several years and increases with the age of the stand (*16*). Tap root damage is caused by the last two instars (*22*). Damage to the tap root by larvae has been associated with increased incidence of diseased roots (*19*). Yield reductions have not been determined for this species, but Goldson (*31*), working with a related species (*S. discoideus* Gyllenhal), determined the economic threshold in alfalfa to be 1,200 larvae/m² in dry years and 2,100 larvae/m² in wet years.

Researchers considered sowbugs (*Tracheniscus rathkei* Brandt) as pests in experiments with alfalfa in Illinois (*28*). However, Byers and Anderson (*10*) found that sowbugs should not be considered a limiting factor in establishing alfalfa based on free choice experiments in the laboratory. Sowbugs were confined to greenhouse flats containing alfalfa seedlings (no choice) or sod containing alfalfa seedlings, volunteer white clover, grass, and organic debris (free choice). Sowbugs ate less alfalfa in the free-choice situation (*11*). Therefore, sowbugs may be less detrimental to alfalfa establishment than was originally suspected.

The potato leafhopper is an important pest of spring-planted alfalfa in conventional tillage systems (*8*). Byers and Templeton (*11*) collected high numbers of potato leafhopper nymphs from alfalfa planted in orchardgrass sod. However,the impact of potato leafhopper in conservation tillage systems is unclear. Some researchers have reported fewer potato leafhoppers in alfalfa fields with grass than in weedy alfalfa (*50*). Weeds may serve as refuges for potato leafhopper adults during post-harvest periods until alfalfa regrowth occurs (*49*). Further studies are needed on populations of this important pest in conservation tillage systems.

Tillage methods influence beneficial insects as well as pests. Tillage affected 40 species of ground beetles in different ways in Kentucky. Some were more abundant in old alfalfa, some in no-till systems, and some in conventional tillage systems (*5*). Legumes need to be managed to increase these important predators in no-till systems.

Sclerotinia crown and stem rot is a disease that poses a significant threat to renovating degraded pastures and ranges

81). Such pastures and ranges contain some volunteer clovers and other small-seeded legumes or broadleaf weeds that are hosts for the pathogen. The probability of the pathogen being present as sclerotia on or within the killed sod is very high for grass fields in many areas of the United States. Reed and associates (*63*) showed that sclerotia are induced to produce apothecia 17 days after the mean soil temperature, at a 5-cm depth, falls to 15°C, and apothecia production continues until the mean temperature drops below 0°C. Rainfall seems to affect the number of apothecia produced, but the soil temperature is critical for initial and continued induction. Sclerotia that remain at or near the soil surface because of minimum soil disturbance with conservation tillage are well positioned to infect susceptible young plants. Plant debris associated with the killed sod creates a rich, moist medium upon which the pathogen can exist as a saprophyte. A typical seeding rate for alfalfa of 15 pounds/acre produced seedling densities of 22 to 26 plants/foot of drill slit (*23*). Excessive plants in narrow drill slits, along with elevated moisture and relative humidity provided by the groove, makes conditions extremely favorable for secondary spread of the pathogen by mycelia.

Controlling pests in legumes

There are no recommended control procedures for many of the pests of no-till forages. Methiocarb baits are effective against slugs (*12, 24*), but baits are too expensive and are not labeled for field crops. Carbofuran controls clover root curculious (*58*), but this insecticide is not labeled for this purpose. Clements (*15*) recommended insecticides at planting time to control pests in the short term. Incorporation of granules at planting time is effective for some pests but detrimental to earthworms and ground beetles (*42*). Scientists also should seek cultivars that are resistent to insects and pathogens. Barker (*4*) recommended development of cultural control methods and resistant cultivars to control slugs. Plant resistance is needed for potato leafhoppers, crickets, grasshoppers, clover head weevils, and slugs.

Biological control of legume pests has not been exploited fully. However, the recent success of parasite programs for the alfalfa weevil (*Hypera postica* Gyllenhal) (*17*) and the alfalfa blotch leafminer [*Agromyza frontella* (Rondani)] (*26*) offers promise for such control methods.

Past research on tillage and its effect on plant diseases has concentrated primarily on operations that bury plant residues in single-crop production systems (*78*). Burial of plant debris by tillage to destroy pathogens is a principle of plant disease control. However, there are no specific disease control recommendations for conservation tillage systems. Disease control methods for forage legumes has involved the use of locally adapted cultivars and, for alfalfa, those cultivars resistant to one or more of three diseases—Bacterial wilt, caused by *Clavibacter michiganeuse* subsp. *insidiosum* (McCulloch) Davis et a. 1984, anthracnose [*Colletotrichum trifolii* [Bain]], and Phytophthora root rot. Shallow planting of high-quality seed in a firm, weed-free seedbed that is well drained, fertile, and at suitable pH (6.5-8.0) is essential to obtain a thick and vigorous stand (*32*). Recently, the fungicide metalaxyl has been registered for use as a seed treatment for alfalfa

and other forage legumes. This treatment should be beneficial in conservation tillage systems for controlling preemergence and postemergence damping-off caused by *Pythium* spp. and seedling injury by Phytophthora root rot. However, attention to site selection, site preparation, crop rotation, and cultivar selection currently provides the best means for reducing losses from diseases in forage legumes produced with conservation tillage methods.

Summary

Ecologists have advocated changing agroecosystems to alleviate pest problems. They have advocated increasing crop diversity by using legumes in crop rotations, stripcropping, winter cover crops, green manure crops, and grassland renovation with legumes. However, pests of legume crops interfere with legume establishment and use. The biological consequences of crop residues on pests need to be studied to develop methods of pest control, such as cultural management, plant resistance, and biological control agents, before farmers can reap the benefits of using legumes in these systems.

REFERENCES

1. Akeson, W. R., G. R. Manglitz, H. J. Gorz, and F. A. Haskins. 1967. *A bioassay for detecting compounds which stimulate or deter feeding by the sweetclover weevil.* J. Econ. Entoml. 60: 1082-1084.
2. Andow, D. 1983. *Effect of agricultural diversity on insect populations.* In W. Lockeretz [ed.] *Environmentally Sound Agriculture.* Praeger Sci., New York, N.Y. pp. 91-115.
3. Anonymous. 1976. *Ways of preventing problems arising by ecosystem management.* In Cherrett, J. M. and G. R. Sagar [eds.] *Origins of Pest, Parasite, Disease and Weed Problems.* 18th Symp., British Ecol. Soc., Blackwell Sci., London, Eng. pp. 367-369.
4. Barker, G. M. 1987. *Biology of pest slugs and their significance in conservation-tillage systems.* In R. R. Hill, Jr., R. O. Clements, A. A. Hower, Jr., T. A. Jordan, and K. E. Zeiders [eds.] *Proceeding, International Symposium on Establishment of Forage Crops by Conservation-Tillage: Pest Management.* U. S. Regional Pasture Res. Lab., Univ. Park, Pa. pp. 83-106.
5. Barney, R. J., and B. C. Pass. 1986. *Ground beetle (Coleoptera: Carabidae) populations in Kentucky alfalfa and the influence of tillage.* J. Econ. Entomol. 79: 511-517.
6. Bigger, J. H. 1932. *Short rotation fails to prevent attack of Diabrotica longicornis Say.* J. Econ. Entomol. 25: 196-199.
7. Byers, R. A., and D. L. Bierlein. 1984. *Continuous alfalfa: Invertebrate pests during establishment.* J. Econ. Entomol. 77: 1,500-1,503.
8. Byers, R. A., J. W. Neal, Jr., J. H. Elgin, Jr., K. R. Hill, J. E. McMurtrey III, and J. Feldmesser. 1977. *Systemic insecticides with spring-seeded alfalfa for control of potato leafhopper.* J. Econ. Entomol. 70: 337-340.
9. Byers, R. A., R. L. Mangan, and W. C. Templeton, Jr. 1983. *Insect and slug pests in forage legume seedlings.* J. Soil and Water Cons. 38: 224-226.
10. Byers, R. A., and S. G. Anderson. 1985. *Are sowbugs pests when establishing alfalfa in conservation-tillage systems?* Melsheimer Entomol. Ser. 35: 7-12.
11. Byers, R. A., and W. C. Templeton, Jr. 1987. *Effects of planting date, placement of seed, vegetation suppression, slugs, and insects upon establishment of no-till alfalfa in orchardgrass sod.* Grass and Forage Sci. (In review).

12. Byers, R. A., W. C. Templeton, Jr., R. L. Mangan, D. L. Bierlein, W. F. Campbell, and H. J. Donley. 1985. *Establishment of legumes in grass swards: Effects of pesticides on slugs, insects, legume seedling numbers and forage yield and quality.* Grass and Forage Sci. 40: 41-48.

13. Capinera, J. L., T. J. Weissling, and E. E. Schweizer. 1985. *Compatibility of intercropping with mechanized agriculture: Effects of strip intercropping of pinto beans and sweet corn on insect abundance in Colorado.* J. Econ. Entomol. 78: 354-357.

14. Cardina, J., and N. L. Hartwig. 1981. *Influence of nitrogen and corn population on no-tillage corn yield with and without crownvetch.* Proc. Northeast Weed Sci. Soc. 35: 27-31.

15. Clements, R. O. 1987. *The impact of insect pests during the establishment of forage crops and some possible solutions.* In R. R. Hill, Jr., R. O. Clements, A. A. Hower, Jr., T. A. Jordan, and K. E. Zeiders [eds.] *Proceedings, International Symposium on Establishment of Forage Crops by Conservation-tillage: Pest Management.* U. S. Regional Pasture Res. Lab., Univ. Park, Pa. pp. 7-23.

16. Cranshaw, W. S. 1985. *Clover root curculio injury and abundance in Minnesota alfalfa of different stand age.* Great Lakes Entomol. 18: 93-95.

17. Day, W. H. 1981. *Biological control of the alfalfa weevil in the Northeastern United States.* In G. C. Papavizas [ed.] *Biological Control in Crop Production* (BARC symp. no. 5). Allanheld, Osmun, Totowa, N. J. pp. 361-374.

18. Dempster, J. P., and T. H. Coaker. 1971. *Diversification of crop ecosystems.* In D. P. Jones and M. E. Solomon [eds.] *Biology in Pest and Disease Control.* Blackwell Sci., Oxford, Eng. pp. 106-115.

19. Dickason, E. A., C. M. Leach, and A. E. Gross. 1968. *Clover root curculio injury and vascular decay of alfalfa roots.* J. Econ. Entomol. 61: 1163-1168.

20. Dijkstra, J. 1964. *Inoculation with ascospores of Sclerotinia trifoliorum for detection of clover rot resistance in red clover.* Euphytica 13: 314-329.

21. Dillon-Weston, W. A. R., A. R. Loveless, and R. L. Taylor. 1946. *Clover rot.* J. Agr. Sci. 36: 18-28.

22. Dintenfass, J. P., and G. C. Brown. 1986. *Feeding rate of larval clover root curculio, Sitona hispidulus (Coleoptera: Curculionidea), on alfalfa tap roots.* J. Econ. Entomol. 79: 506-510.

23. Donohue, S. J., R. L. Harrison, and H. E. White. 1984. *A Handbook of Agronomy.* Pub. 424-100. Va. Coop. Ext. Serv., Blacksburg, Va. 88 pp.

24. Dowling, P. M., and D. L. Linscott. 1983. *Use of pesticides to determine relative importance of pest and disease factors limiting establishment of sod-seeded lucerne.* Grass and Forage Sci. 38: 179-185.

25. Drake, K. 1976. *Prairie models for agricultural systems.* In Proc., 5th Midwest Prairie Conf. Iowa State Univ., Ames. pp. 226-230.

26. Drea, J. J., and R. M. Hendrickson, Jr. 1986. *Analysis of a successful classical biological control project: The alfalfa blotch leafminer (Diptera: Agromyzidae) in the Northeastern United States.* Environ. Entomol. 15: 448-455.

27. Ellsbury, M. M., G. R. Pratt, and W. E. Knight. 1985. *Effects of single and combined infection of arrowleaf clover with bean yellow mosaic virus and Phytophthora sp. on reproduction and colonization by pea aphids (Homoptera: Aphididae).* Environ. Entomol. 14: 356-359.

28. Faix, J. J., R. S. Hooten, and C. J. Kaiser. 1981. *Pest identification and efficacy of carbofuran in improving no-till alfalfa in sod.* Ill. Agr. Exp. Sta. DSAC 9: 81-86.

29. Foster, R. E., J. J. Tollefson, J. P. Nyrop, and G. L. Hein. 1986., *Value of adult corn rootworm (Coleoptera: Chrysomelidae) population estimates in pest management decision making.* J. Econ. Entomol. 79: 303-310.

30. Godfrey, L. D., D. E. Legg, and K. V. Yeargan. 1986. *Effects of soil-borne organisms on spring alfalfa establishment in an alfalfa rotation system.* J. Econ. Entomol. 79: 1,055-1,063.

31. Goldson, S. L., C. B. Dyson, and R. J. Proffitt. *The effect of Sitona discoideus Gyllenhal (Coleoptera: Curculionidae) on lucerne yields in New Zealand.* Bull. Entomol. Res. 75: 429-442.

32. Graham, J. I., F. I. Frosheiser, D. L. Stuteville, and D. C. Erwin. 1979. *A compendium of alfalfa diseases.* Am. Phytopathol. Soc., St. Paul, Minn. 65 pp.

33. Grant, J. F., K. V. Yeargan, B. C. Pass, and J. C. Parr. 1982. *Invertebrate organisms associated with alfalfa seedling loss in complete-tillage and no-tillage plantings.* J. Econ. Entomol. 75: 822-826.

34. Hagel, G. T., and R. O. Hampton. 1970. *Dispersal of aphids and leafhoppers from red clover to red Mexican beans, and the spread of bean yellow mosaic by aphids.* J. Econ. Entomol. 63: 1057-1060.

35. Hare, J. D. 1983. *Manipulation of host suitability for herbivore pest management.* In R. F. Denno and M. S. McClure [eds] *Variable Plants and Herbivores in Natural and Managed Systems.* Academic Press, New York, N.Y. pp. 655-680.

36. Hartwig, N. L. 1974. *Crownvetch makes a good sod for no-till corn.* Crops and Soils 27: 16-17.

37. Hays, S. B. 1965. *Insecticidal control of the clover head weevil and the lesser clover leaf weevil on crimson clover and effects of control measures on honey bees.* J. Econ. Entomol. 58: 481-484.

38. Hesterman, O. B., C. C. Sheaffer, and E. I. Fuller. 1986. *Economic comparisons of crop rotations including alfalfa, soybean, and corn.* Agron. J. 78: 24-28.

39. Hewitt, G. B., A. C. Wilton, R. J. Lorenz. 1982. *The suitability of legumes for rangeland interseeding and as grasshopper food plants.* J. Range Manage. 35: 653-656.

40. Hewitt, G. B., and J. D. Berdahl. 1984. *Grasshopper food preferences among alfalfa cultivars and experimental strains adopted for rangeland interseeding.* Environ. Entomol. 13: 828-831.

41. Hill, R. E., and Z. B. Mayo. 1980. *Distribution and abundance of corn rootworm species as influenced by topography and crop rotations in eastern Nebraska.* Environ. Entomol. 9: 122-127.

42. House, G. J., and R. W. Parmlee. 1985. *Comparison of soil arthropods and earthworms from conventional and no-tillage agroecosystems.* Soil Tillage Res. 5: 351-360.

43. Hoveland, C. S., M. W. Alison, Jr., R. F. McCormick, Jr., W. B. Webster, V. H. Calvert, II, J. T. Eason, M. E. Ruf, W. A. Griffey, H. E. Burgess, L. A. Smith, and H. W. Grimes, Jr. 1981. *Seeding legumes into tall fescue.* Bull. 531. Ala. Agr. Exp. Sta., Auburn. 15 pp.

44. Howitt, A. J. 1961. *Chemical control of slugs in orchardgrass-ladino white clover pastures in the Pacific-Northwest.* J. Econ. Entomol. 54: 778-781.

45. Kalmbacher, R. S., D. R. Minnick, and F. G. Martin. 1979. *Destruction of sod-seeded legume seedlings by the snail, Polygyra cereolus.* Agron. J. 71: 365-368.

46. Knight, W. E., and C. S. Hoveland. 1973. *Crimson clover and arrowleaf clover.* In M. E. Heath, D. S. Metcalf, and R. F. Barnes [eds] *Forages: The Science of Grassland Agriculture.* Iowa State Univ. Press, Ames. pp. 199-207.

47. Krysan, J. L., D. E. Foster, T. F. Branson, K. R. Ostlie, and W. S. Cranshaw. 1986. *Two years before the hatch: Rootworms adapt to crop rotation.* Bull. Entomol. Soc. Am. 32: 250-253.

48. Krysan, J. L., J. J. Jackson, and A. C. Lew. 1984. *Field termination of egg diapause in Diabrotica with new evidence of extended diapause in D. barberi (Coleoptera: Chrysomelidae).* Environ. Entomol. 13: 1237-1240.

49. Lamp, W. O., M. J. Morris, and E. J. Armbrust. 1984. *Suitability of common weed species as host plants for the potato leafhopper,* Empoasca fabae. Ent exp. appl. 36: 125-131.

50. Lamp, W. O., R. J. Barney, E. J. Armbrust, and G. Kapusta. 1984. *Selective weed control in spring-planted alfalfa: Effect on leafhoppers and planthoppers (Homoptera: Auchenorrhyncha) with emphasis on potato leafhoppers.* Environ. Entomol. 13: 207-213.

51. Leath, K. T. 1978. *Crown wart of alfalfa in Pennsylvania.* Plant Dis. Rep. 62: 621-623.

52. Mangan, R. L., and R. A. Byers. 1982. *Evaluation of* Halticus bractatus *as a probable pest of minimum-tillage legume establishment.* Melsheimer Entomol. Ser. 32: 25-31.

53. Mangan, R. L., R. A. Byers, A. Wutz, and W. C. Templeton, Jr. 1982. *Host plant associations of insects collected in swards with and without legumes seeded by minimum tillage.* Environ. Entomol. 11: 255-260.

54. Manglitz, G. R., H. J. Gorz, F. A. Haskins, W. R. Akeson, and G. L. Beland. 1976. *Interactions between insects and chemical components of sweet clover.* J. Environ. Quality 5: 347-352.

55. Manglitz, G. R., H. J. Gorz, and H. J. Stevens, Jr. 1971. *Biology of the sweetclover root borer.* J. Econ. Entomol. 64: 1154-1158.

56. Martin, P. B., P. D. Lungren, G. I. Greene, and A. H. Baumhover. 1981. *Seasonal occurrence of* Rachiplusia ou *and* Autographa biloba, *and associated entomophages in clover.* J. Ga. Entomol. Soc. 16: 288-295.

57. Miller, J. D. and H. D. Wells. 1985. *Arrowleaf clover.* In N. L. Taylor [ed.] *Clover Science and Technology.* Mono. 25. Am. Soc. Agron., Madison, Wisc. pp. 503-514.

58. Neal, Jr., J. W., and R. H. Ratcliffe. 1975. *Clover root curculio: Control with granular carbofuran as measured by alfalfa regrowth, yield, and root damage.* J. Econ. Entomol. 68: 829-831.

59. Odum, E. P. 1984. *Properties of agroecosystems.* In R. Lowrance, B. R. Stinner, and G. J. House [eds.] *Agricultural Ecosystems: Unifying Concepts.* Wiley-Interscience, New York, N.Y. pp. 5-11.

60. Pimental, D. 1977. *The ecological basis of insect pest, pathogen, and weed problems.* In J. M. Cherrett and G. R. Sagar [eds.] *Origins of Pest, Parasite, Disease and Weed Problems.* Blackwell Sci., London, Eng. pp. 1-33.

61. Prior, G. D., and J. H. Owen. 1964. *Pathological anatomy of* Sclerotinia trifoliorum *on clover and alfalfa.* Phytopathol. 54: 784-787.

62. Reed, H. E., and J. H. Felts. 1965. *Crown rot of alfalfa.* Tenn. Farm and Home Sci. Progress Rpt. 54 (Apr-Jun): 14-15.

63. Reed, K. L., E. L. Stromberg, and S. W. VanScoyoc. 1987. *The influence of field micrometeorological events on the development of* Sclerotinia trifoliorum Eriks., *casual agent of crown and stem rot of alfalfa.* In R. R. Hill, Jr., R. O. Clements, A. A. Hower, Jr., T. A. Jordan, and K. E. Zeiders [eds.] *Proceedings, International Symposium on Establishment of Forage Crops by Conservation-Tillage Methods: Pest Management.* U. S. Regional Pasture Res. Lab., Univ. Park, Pa. p. 76.

64. Risch, S. J., D. Andow, and M. A. Altieri. 1983. *Agroecosystem diversity and pest control: Data, tentative conclusions, and new research directions.* Environ. Entomol. 12: 625-629.

65. Rogers, D. D., D. S. Chamblee, J. P. Mueller, and W. V. Campbell. 1983. *Fall sod-seeding of ladino clover into tall fescue as influenced by time of seeding, and grass and insect suppression.* Agron. J. 75: 1,041-1,046.

66. Rogers, D. D., D.S. Chamblee, J. P. Mueller, and W. V. Campbell. 1985. *Conventional and no-till establishment of ladino clover as influenced by time of seeding and insect and grass suppression.* Agron. J. 77: 531-538.

67. Schuster, M. F. 1980. *Cotton ecosystem diversification and plant bug trapping with interplanted alfalfa in the Delta of Mississippi.* Miss. State, The Station Tech. Bull. 98: 16 p.

68. Sevarcherian, V., and V. M. Stern. 1974. *Host plant preferences of lygus bugs in alfalfa-interplanted cotton fields.* Environ. Entomol. 3: 761-766.

69. Sevarcherian, V., and V. M. Stern. 1975. *Movement of lygus bugs between alfalfa and cotton.* Environ. Entomol. 4: 163-165.

70. Smith, C. M., J. L. Frazier, and H. N. Pitre. 1975. *Feeding preference of the clover head weevil on clovers on the genus* Trifolium. J. Econ. Entomol. 68: 165-166.

71. Smith, C. M., W. E. Knight, and J. L. Frazier. 1977. *Oviposition and larval survival of the clover head weevil on crimson clover.* Crop Sci. 17: 162-164.

72. Smith, D. 1981. *Forage management in the North.* Kendal Hunt Pub. Co., Dubuque, Iowa. pp. 133-134.

73. Stanley, R. L., N. M. Randolph, and G. L. Teets. 1970. *Control of the clover head weevil on crimson clover.* J. Econ. Entomol. 63: 256-258.

74. Stromberg, E. L. 1983. *Death of fall seeded alfalfa stands.* Va. Coop. Ext. Serv. Plant Protection Newsletter 2(Feb): 7.

75. Stromberg, E. L. 1984. *Stand failures in fall planted alfalfa.* Va. Coop. Ext. Serv. Plant Protection Newsletter 3(Apr): 8.

76. Stromberg, E. L. 1985. *Violet root rot of alfalfa.* Va. Coop. Ext. Serv. Plant Protection Newsletter 4(Feb): 3.

77. Stromberg, E. L. 1986. *Sclerotinia crown and stem rot: Three case histories.* Va. Coop. Ext. Serv. Plant Protection Newsletter 5(Apr): 1-2.

78. Sumner, D. R., B. Doupnik, Jr., and M. G. Boosalis. 1981. *Effects of reduced tillage and multiple cropping on plant disease.* Ann. Rev. Phytopathol. 19: 167-187.

79. Tippins, H. H. 1958. *Granulated insecticides for the control of seed weevils on crimson clover.* J. Econ. Entomol. 51: 459-460.

80. Van Keuren, R. W., and C. S. Hoveland. 1985. *Clover management and utilization.* In N. L. Taylor, [ed.] *Clover Science and Technology.* Mono. 25. Am. Soc. Agron., Madisoin, Wisc. pp. 325-354.

81. VanScoyoc, S. E., and E. L. Stromberg. 1984. *Results of the 1984 Sclerotinia crown and stem rot survey.* Va. Coop. Ext. Serv. Plant Protection Newsletter 2(Jul): 7-13.

82. White, H. E., E. S. Hagood, D. D. Wolf, and J. M. Luna. 1985. *No-till seeding of forage grasses and legumes.* Pub. 418-007. Va. Coop. Ext. Serv. Blacksburg. 6 pp.

Legume cover cropping, no-till practices, and soil arthropods: Ecological interactions and agronomic signficance

G. J. House and D. A. Crossley, Jr.

No-till practices generate a favorable habitat for soil- and surface-dwelling arthropods by reducing moisture loss, ameliorating temperature extremes, and providing food resources (4). Although crop damage by some pest insects, such as corn billbugs [Sphenophorus callosus (Olivier)], increase under no-till (1), damage from other pests, such as lesser cornstalk borers [Elasmopalpus lignosellus (Zeller)], decrease due to the lack of tillage (3). Also, such positive interactions as biological control of lepidopteran pests by ground beetles apparently occur more frequently with no-till than with conventional tillage conditions (2).

The objectives of our study were twofold. First, we sought to quantify and compare the effects of conventional and no-till cropping practices, including different winter legume and grain cover crops, on soil arthropod population dynamics and community structure. Second, because large numbers of secondary composers, such as mycophagous microarthropods, occur in no-till systems (5), we also investigated the influence these fauna may have on decomposition and N release from surface crop residues using litterbag techniques.

We collected soil arthropods from conventional and no-till agroecosystems in Guilford and Johnston Counties, North Carolina, sites representative of the southern Piedmont and Coastal Plain, respectively. Both areas are relatively level (2% to 6% slope), with prime agricultural land of moderate fertility. Corn was planted in the spring using conventional or no-till methods into fall-planted hairy vetch (Vicia villosa Roth), crimson clover (Trifolium incarnatum L.), rye (Secale cereale L.), and wheat (Triticum aestivum L.). Soil arthropods were elutriated from 10 cm x 15 cm soil cores and extracted from surface-placed mesh bags containing cover crop residues.

Results and discussion

Legume cover crops. Cover crop species effects on the number of soil arthropods and trophic diversity were most pronounced early in the season (Table 1). Initially, vetch supported higher numbers of herbivores, predators, and decomposers than clover or wheat. Thereafter, differences in numbers of arthropods among cover crops and trophic diversity were less discernable. Predator and decomposer numbers, especially in wheat, steadily increased throughout

G. J. House is an assistant professor of soil entomology and ecology, Department of Entomology, North Carolina State University, Raleigh, 27695, and D. A. Crossley, Jr. is a professor of entomology and ecology, Department of Entomology, University of Georgia, Athens, 30602.

the season in contrast to herbivore numbers, which paralleled crop phenology and biomass. The destruction of cover crops in late April substantially reduced herbivore populations in May (Table 1). Similarity of taxa among cover crops was lowest in April (55% shared taxa) and increased to 90% by July, indicating a continuous dissipation of cover crop influence.

Comparison of soil arthropod trophic composition among cover crops did not show a clear pattern for plant type (Table 2). We suggest these varying soil arthropod trophic compositions reflect differences in the decomposition rates of each crop residue and, hence, the availability and subsequent nutrient uptake by crop plants (corn). The slower decomposition of wheat and to some extent clover relative to vetch promoted a robust decomposer soil arthropod fauna. Decomposers comprised almost one-half of the total fauna from wheat and clover but only about one-fourth of the total vetch fauna (Table 2).

Tillage effects. Tillage method had a significant effect on specific soil arthropod taxa as well as on trophic groups. The seedcorn maggot [Delia platura (Meigen)] was the dominant herbivore and occurred in large numbers ($P < 0.05$) in con-

Table 1. Seasonal dynamics of soil arthropod herbivores, predators, and decomposers in crimson clover, hairy vetch, and wheat.

	Mean Number Per Sample Unit*		
Categories	Early (April)	Mid-season (May)	Late (July)
Herbivores			
Crimson clover	4.25	0.67	3.17
Hairy vetch	9.58	0.71	3.67
Wheat	0.75	0.58	5.75
Predators			
Crimson clover	0.17	1.25	2.25
Hairy vetch	0.67	1.58	3.17
Wheat	0.21	0.42	2.92
Decomposers			
Crimson clover	0.10	4.13	4.75
Hairy vetch	0.44	3.83	3.42
Wheat	0.13	3.00	6.33

*Sample unit, 3 cores (ca. 500 cm³ of soil).

Table 2. Soil arthropod trophic composition for crimson clover, hairy vetch, and wheat

	Mean Seasonal Percentage		
Categories	Crimson Clover	Hairy Vetch	Wheat
Herbivores	34.6	52.0	35.5
Predators	15.8	20.0	17.5
Decomposers	49.6	28.0	47.0

Table 3. Mean number of seed corn maggots collected per 500 cm³ of soil in conventional and no-till systems.

	Number of Seed Corn Maggots	
Date	Conventional Tillage	No-till
April	7.4	1.7
July	2.6	1.8
Seasonal mean	10.1	3.6

Table 4. Soil arthropod trophic composition occurring in conventional and no-till systems.

| | Soil Arthropod Trophic Composition (%) | | | | | |
| | Conventional Tillage | | | No-till | | |
Date	Herbivores	Predators	Decomposers	Herbivores	Predators	Decomposers
April	94	1	5	83	15	2
May	10	6	84	11	27	62
July	37	23	40	34	24	42
Seasonal mean	48	13	39	32	24	44

ventional tillage plots compared with no-till plots (Table 3). Both Elateridae (wireworms) and Staphylinidae (rove beetles) larval numbers were higher in no-till than conventional tillage systems (P < 0.05). We found Chrysomelidae (leaf beetles) larvae almost exclusively in conventional tillage plots, while Carabidae (ground beetles) larvae and adults and Curculionidae (weevils) were present predominantly in no-till treatments.

Trophic composition analysis (Table 4) showed that during the critical early season months of April and May, no-till systems supported a higher number of predators than conventional tillage systems. In contrast, herbivorous soil arthropods were higher on conventional tillage plots compared to no-till plots during these same periods (Table 4). However, by July herbivore, predator, and decomposer percentages were similar in both tillage systems.

Nevertheless, the elevated number of soil arthropod predators occurring in no-till plots during early and mid-season apparently provided timely and, thus, effective biological control of pest insects. For example, southern corn rootworm egg predation by beneficial arthropods occurred earlier and was consistently higher (P < 0.01) in the no-till plots compared to conventional tillage conditions (Brust and House, unpublished data). The higher early and mid-season values of the Simpson and Shannon-Wiener indicies (Table 5) calculated for no-till versus conventional tillage systems indicated an additional positive influence of tillage elimination, that is, in enhancing soil arthropod species diversity.

Decomposition study. Hairy vetch decomposed faster than crimson clover, which in turn decomposed faster than rye straw. Differences in decay rates among litterbag mesh sizes within a crop residue were not significant (P > 0.05), suggesting that microarthropod comminution of crop residue was minimal in these systems. However, percent N of crop residue was significantly lower (P < 0.05) in larger mesh bags versus smaller mesh bags, suggesting a soil faunal influence on N release and cycling.

Summary

Conventional tillage methods, as a consequence of soil disturbance, had a greater impact on the structure of the soil arthropod community than no-till systems. Importantly, no-till promoted and maintained a more trophically balanced soil arthropod community structure than the conventional tillage system during early and mid-season when crops were sensitive to insect damage. Finally, although the influence of soil arthropods on decomposition of crop residue in no-till systems remains unclear, any involvement of soil arthropods

Table 5. Comparison of soil arthropod diversity occurring in conventional (CT) and no-till (NT) systems.

| | Soil Arthropod Diversity | | | | | |
| | April | | May | | July | |
Index	CT	NT	CT	NT	CT	NT
Simpson's index*	1.38	3.03	4.06	5.59	11.02	10.76
Shannon-Wiener index†	1.18	2.16	2.48	3.51	8.54	8.15

*Simpson's index of dominance: $1/C = 1/\text{sum } p_i^2$.
†Shannon-Wiener index of general diversity: $H^1 = -\text{Sum } p_i \log p_i$.

in nutrient release from crop residues would have broad agronomic significance.

REFERENCES

1. All, J. N., R. S. Hussey, and D.G. Cummins. 1984. *Southern corn billbug (Coleoptera: Curculionidae) and plant parasitic nematodes: Influence of no-tillage, coulter-in-row-chiseling, and insecticides on severity of damage to corn.* J. Econ. Entomol. 77: 178-182.
2. Brust, G. E., B. R. Stinner, and D. A. McCartney. 1985. *Tillage and soil insecticide effects on predator - black cutworm (Lepidoptera: Noctuidae) interactions in corn agroecosystems.* J. Econ. Entomol. 78: 1,389-1,392.
3. Cheshire, J. M., and J. N. All. 1979. *Feeding behavior of lesser cornstalk borer larvae in simulations of no-tillage, mulched conventional-tillage and conventional-tillage corn cropping systems.* Environ. Entomol. 8: 261-264.
4. Crossley, Jr., D. A., G. J. House, R. M. Snider, R. J. Snider, and B. R. Stinner. 1984. *The positive interactions in agroecosystems.* In R. R. Lowrance, B. R. Stinner, and G. J. House [eds.], *Agricultural Ecosystems: Unifying Concepts.* John Wiley, New York, N.Y. pp. 73-81.
5. House, G. J., and R. W. Parmelee. 1985. *Comparison of soil arthropods and earthworms from conventional and no-tillage agroecosystems.* Soil Tillage Res. 5: 351-360.

Influence of legume cover crops and conservation tillage on soil-borne plant pathogen populations

C. S. Rothrock and W. L. Hargrove

We designed a study to examine the influence of winter cover crops, especially legumes, on the populations of soil-borne plant pathogens during the subsequent crop. Cover crop treatments exerted the most influence on Pythium populations in soil. Thus, herein we present data for this group of pathogens.

Study methods

We conducted the experiment on the Bledsoe Research Farm near Griffin, Georgia. In 1983 the experiment included rye, crimson clover, hairy vetch, and no cover crop (2). In 1984 and 1985 we used a randomized, complete-block, split-strip plot design with rye, crimson clover, and no cover crop as the main plots. Subplots were no-till and conventional tillage (moldboard plowing and disking). Sub-subplots were with and without cover crop residue. Cover crops were planted in the fall and killed with paraquat prior to planting sorghum.

We sampled soils at sorghum growth stage GS_1, growing point differentiation, and GS_3, physiological maturity (1). In 1984 and 1985 we also took samples at the time the cover crop was killed and when the sorghum crop was planted. In 1985 these dates were the same and only one sample was taken. We sampled 10 soil cores from the top 6 inches of soil from each plot. Samples were stored at $36°F$, then the soil was screened, suspended in water, blended, and plated by dilution plating onto a selective medium for *Pythium* spp. (3). Counts were made after 24- to 48-hour incubation.

Results

In 1983 Pythium populations increased following crimson clover, compared to no cover crop or rye, at maturity of the sorghum crop (growth stage GS_3) (Table 1). Populations following hairy vetch increased at both sampling dates compared to no cover or rye. Pythium populations at both sampling dates were greater following hairy vetch than following crimson clover.

In 1984 and 1985 Pythium populations following crimson clover also were significantly greater compared to no cover crop or rye, at the time the cover crops were killed, at sorghum planting, and at sorghum growth stage GS_1 and GS_3 (Table 2). The only exception was the population at GS_1 in 1985. Tillage did not have a large influence on Pythium

populations, with the exception of the GS_1 sample in 1984 and the GS_3 sample in 1985, when the populations under the no-till treatment were significantly lower.

Removal of aboveground residue did not influence the Pythium population in the initial samples in 1984 and 1985. Removal of clover residue did reduce Pythium populations from 122,000 propagules/pound of soil to 68,000 propagules/pound of soil and from 110,000 to 59,000 propagules/pound of soil for the GS_1 and GS_3 samples in 1984, respectively, and from 109,000 to 75,000 propagules/pound of soil for the GS_3 sample in 1985.

Legume cover crops elevated the population of *Pythium* spp. in the subsequent crop compared to a nonlegume cover crop or no cover crop. The large populations were not a result of residue decomposition alone, although removal of the aboveground residue did reduce populations. Legumes may differ in their ability to influence soil Pythium populations.

We sampled both experiments after several consecutive years of growth of the cover crops. We are uncertain if this effect is found during the first growing season of the legume cover crop. Our results indicate that cover crops will influence certain soil-borne plant pathogen populations, and the importance of *Pythium* spp. needs to be examined in terms of damage to specific crops used in conjunction with legume cover crops.

REFERENCES

1. Eastin, J. D. 1972. *Efficiency of grain dry matter accumulation in grain sorghum.* In D. Wilkinson [ed.] *Proceedings of*

C. S. Rothrock is an assistant professor and W. L. Hargrove is an associate professor, Department of Plant Pathology and Agronomy, University of Georgia, Georgia Experiment Station, Experiment, 30212.

Table 1. Influence of cover crop on soil Pythium populations, 1983.

Cover crop	Pythium Population (1,000 propagules/pound of soil)	
	GS_1*	GS_3*
None	44.5 A†	29.0 A
Rye	31.3 A	41.7 A
Crimson clover	77.6 A	138.8 B
Hairy vetch	202.3 B	179.6 C

*Sorghum growth stages: GS_1, growing point differentiation; GS_3, physiological maturity.
†Means in a column followed by the same letter are not significantly different, LDS (P = 0.05)

Table 2. Influence of cover crop and tillage on soil Pythium populations, 1984 and 1985.

Treatment	Pythium Population (1,000 propagules/pound of soil)						
	Death*	Planting		GS_1†		GS_3†	
	1984	1984	1985	1984	1985	1984	1985
Tillage							
Conventional	66.7A‡	84.8A	73.5A	101.2A	106.6A	68.5A	83.0A
None	75.8A	59.9A	65.3A	48.5B	93.9A	47.6A	60.6B
Cover crop							
None	47.6A	46.3A	51.7A	52.2A	99.8A	31.3A	54.4A
Rye	50.3A	59.4A	43.5A	51.3A	74.4A	33.1A	49.0A
Crimson clover	115.2B	111.1B	113.4B	121.6B	126.6B	110.2B	109.3B

*Killing of cover crop.
†Sorghum growth stages: GS_1, growing point differentiation, and GS_3, physiological maturity.
‡Means within a column and main effect followed by the same letter are not significantly different, LSD (P = 0.05).

the 27th Annual Corn and Sorghum Research Conference. Am. Seed Trade Assoc., Washington, D.C. pp. 7-17.

2. Hargrove, W. L. 1986. *Winter legumes as a nitrogen source for no-till grain sorghum.* Agron. J. 78: 70-74.

3. Tsao, P. H., and G. Ocana. 1969. *Selective isolation of species of Phytophthora from natural soils on an improved antibiotic medium.* Nature 223: 636-638.

Potential disease problems in winter annual legumes used in conservation tillage systems in the Southeast

R. G. Pratt, G. L. Windham, and M. R. McLaughlin

A wide range of annual legumes, including species of *Trifolium, Vicia, Pisum, Lupinus, Medicago,* and other genera, have potential for use as winter cover crops in conservation tillage systems in the southeastern United States. These species can be fall-seeded and grown as winter cover crops in rotation with summer annual row crops, such as corn, sorghum, and cotton. In this role, they can provide a groundcover or mulch to prevent soil erosion during the winter and supply part or all of the N requirements for the summer row crops. Some early maturing annual legumes may also be chosen and managed to favor seed production in the spring for volunteer stand establishment each fall.

Many diverse diseases may afflict cover crops of winter annual legumes in the Southeast. Damage or losses caused by diseases may range from trace levels to almost total losses of productive stands and elimination of natural reseeding. By killing plants or reducing their size and vigor, diseases may reduce or eliminate all of the desirable soil conservation and residual-N fertilization effects that are sought from use of winter annual legumes in conservation tillage systems.

Despite their potential severity, the nature or impact of diseases on these crops seldom can be predicted. All stands of legume cover crops may not be seriously affected by diseases. Even in the same fields, disease severity may vary greatly from year to year under different environmental conditions. Generally, with repeated growth of the same cover crop in any location increased disease severity and losses over years would be expected as new diseases appear and inoculum levels of pathogens gradually increase.

Pathogens that may damage or limit winter annual legumes include fungi, nematodes, and viruses. Some diseases caused by these pathogens are well known and appreciated as important limiting factors, while others have received little study. Most or all known diseases, however, should be considered potentially important for production of winter annual legume cover crops.

Fungal pathogens that are known or considered likely to cause serious disease problems in winter annual legumes in the Southeast include *Sclerotinia trifoliorum, Phytophthora* spp., *Fusarium oxysporum* (wilt-inducing), *Sclerotium rolfsii, Cercospora,* and certain other foliage-infecting fungi. *Sclerotinia* is likely to be the most widespread and damaging because of its broad host range, its nearly universal occurrence, and the fact that it is capable of pathogenesis at low temperatures. Root rot and vascular wilt diseases caused

R. G. Pratt, G. L. Windham, and M. R. McLaughlin are research plant pathologists, Crop Science Research Laboratory, Forage Research Unit, Agricultural Research Service, U.S. Department of Agriculture, Mississippi State, Mississippi 39762.

by *Phytophthora* spp. and *Fusarium oxysporum*, respectively, may be very damaging on susceptible crops in some fields. These diseases are partly or entirely host-specific; thus, farmers might avoid losses by choosing alternate legume species for cover crops if these diseases appear. Diseases caused by warm-season fungal pathogens, such as *Sclerotium rolfsii*, *Rhizoctonia*-like organisms, *Cercospora*, and *Macrophomina,* seem likely to be damaging mainly on crops sown too early in the fall or on late-maturing annual legumes that may not be the most preferred winter cover crops.

Plant parasitic nematodes may be important in conservation tillage systems as pathogens of both the legume cover crops and the subsequent summer row crops. Root-knot nematodes are likely the most important. *Meloidogyne incognita, M. javanica, M. arenaria*, and *M. hapla* may parasitize winter legume cover crops in the Southeast. Collectively, these species have a broad host range with little strong host specialization, and they occur widely throughout the region. Their broad pathogenicity could allow them to parasitize and maintain high populations in winter legumes and then infect the major summer crops. Most species and cultivars of winter annual legumes are susceptible to one or more root knot nematode species.

However, actual losses that may be caused by the nematodes in winter cover crops are uncertain, except in warm, sandy soils. In general, nematodes are warm-temperature pathogens. The amount of damage that they cause depends upon the number of nematodes present at planting and the number of life cycles of the nematodes that develop on a crop. Root-knot nematodes seem likely to be most damaging to winter legume cover crops in the more southerly or coastal regions of the Southeast. Warmer temperatures would enable pathogenesis for longer periods during the winter, and nematodes are generally more damaging to crops sown in sandy soils. *M. hapla,* however, is adapted to cooler temperatures and may limit cover crop production in the upper regions of the Southeast.

Several cultivars of winter legumes are resistant to one or more root-knot nematodes, and researchers are conducting breeding programs for nematode resistance at several locations. The anticipated future availability of nematode-resistant cultivars of winter annual legumes should enhance their value and adaptability as cover crops in conservation tillage systems.

Scientists have identified at least six aphid-transmitted viruses that infect winter annual legumes in the Southeast: alfalfa mosaic, bean yellow mosaic, clover yellow vein, cucumber mosaic, peanut stunt, and red clover vein mosaic. Introduction of viruses into a stand and subsequent disease development within it depend upon the feeding and movement of aphids.

Viruses must be considered potentially important limiting factors for annual legume cover crops because nearly all such crops are susceptible to one or more common viruses. Although some instances of host specificity among these pathogens are known, most have wide and overlapping host ranges. Although the natural spread of viruses by aphids has been documented in the field at Mississippi State during every month of the year, peak periods for virus movement occur in mid- to late spring and occasionally in early fall. Late spring infections in winter annual legumes may be of little consequence unless seed production is affected. Fall infections, however, may weaken plants and severely limit winter survival or subsequent spring growth. Most virus diseases reduce the vigor, root and top growth, and seed production of susceptible hosts, but they usually do not kill plants alone or entirely prevent reseeding. However, viruses may also predispose plants to damage from fungal diseases.

Because of the great diversity of diseases that can limit winter annual legumes in the Southeast, only a few general recommendations can be made for avoiding or reducing losses to disease. One broad recommendation is to choose legume species or varieties that are well-adapted for growth in each area. Well-adapted varieties are most likely to withstand damage from diseases and to continue growth and seed production in spite of infection. If farmers suspect diseases are causing poor performance or failure of annual legume cover crops, then proper diagnosis is the essential first step in initiating specific corrective actions.

Although cultivars resistant to specific diseases for the most part are not yet available among winter annual legumes, the substitution of different genera or species of legume cover crops in problem fields may enable a comparable form of control if diseases are accurately diagnosed.

Predatory arthropod ecology in conservation tillage systems using legumes

Benjamin R. Stinner, Gerald E. Brust, and David A. McCartney

Multiple cropping with legumes in conservation tillage agriculture can be accomplished by rotating legume and non-legume crops or by interplanting two or more crops on the same land during the same growing season. Major agronomic reasons for multiple cropping with legumes include a reduction in soil erosion and an increase in N input (8, 9). Studies comparing different tillage practices have established that predatory insects and other arthropods, such as ground beetles (Carabidae) and spiders, increase in numbers under conservation tillage, especially no-till (1, 6). Although an increase of predatory fauna under conservation tillage conditions should suppress populations of insect pests, quantitative documentation of this phenomenon are lacking.

Compared to conventional plowing, conservation tillage results in a more diverse assembly of organisms, including bacteria, fungi, nematodes, earthworms, and arthropods (2, 4, 7). A gradient of decomposition products ranging from fresh surface crop residues to humified organic matter within the upper soil strata occurs under continuous, reduced, and especially no-till management. These residues create environmental conditions conducive to the proliferation of an abundant and diverse soil fauna and floral community (3, 6).

We hypothesized that multiple cropping with conservation tillage systems should further increase predatory arthropods and their activity against pest populations. To test this hypothesis, we evaluated predatory arthropod activity and predation on pest species in a range of conservation tillage systems that included legumes in rotation and interplanted systems.

Study methods

We conducted our experiments at the Ohio Agricultural Research and Development Center, Wayne County, Ohio, over a 2-year period. The study was conducted on 0.5-acre plots (four replications) in the following cropping systems: (a) conventional (moldboard) tillage corn (*Zea mays*) following corn, (b) no-till corn following corn, (c) no-till corn following soybeans (*Glycine max*), and (d) corn interplanted with alfalfa (*Medicago sativa*). For the first three treatments we used standard recommendations provided by the Ohio State Cooperative Extension Service. Our interplanted system included corn planted in rows into a previously established

Benjamin R. Stinner is an assistant professor, Department of Entomology, Ohio Agricultural Research and Development Center, Ohio State University, Wooster, 44691; Gerald E. Brust is a graduate research assistant, Department of Entomology, North Carolina State University, Raleigh, 27695; and David A. McCartney is a research associate, Department of Entomology, Ohio Agricultural Research and Development Center, Ohio State University, Wooster.

stand of alfalfa that had been mowed before planting the corn to reduce competition between the legume and the germinating corn plants. Cyanazine and alachlor herbicides were applied to all treatments. These herbicides suppressed, but did not kill, the alfalfa in the interplanted system so that a living mulch (5) system resulted.

We used Lepidopterous larvae as bait to evaluate predator activity. Prey species examined were the black cutworm (*Agrotis ipsilon*), armyworm (*Psuedaletia unipuncta*), European corn borer (*Ostrinia nubilalis*), and stalk borer (*Papaipema nebris*). We measured predator activity in the different cropping systems by observing the number of times predatory arthropods attacked the pest larvae. We also measured absolute density estimates for predators in each cropping system at the same time predator activity was quantified. Arthropod density was evaluated by excavating litter and soil within metal quadrats. We evaluated both predator activity and predator density every 2 weeks throughout the growing season (1).

Results and discussion

We found that about 30 taxa of arthropods were predators in the four different agroecosystems. Predators included 17 species of ground beetles (Coleoptera: Carabidae) as well as ants (Hymenoptera:Formicidae), slugs (Mollusca), and spiders (Lycosidae). There was significantly more predator activity and predation in the no-till cropping systems compared with the conventional tillage crops on all sampling dates and significantly more predation in the no-till-following-soybeans cropping system compared with the no-till-following-corn system. Absolute density estimates showed that total predator densities were significantly greater in the no-till systems than in the conventional tillage treatment after planting. Carabid beetle density usually was greater in the no-till-following-soybeans systems than in the no-till-following-corn systems. Thus, we concluded that the legume in the rotation system better maintained the food chains necessary to sustain generalist predators for long periods of time, particularly when pest populations were at low densities.

We found significantly (P = 0.05) greater densities of arthropod predators and predatory activity in the interplanted no-till corn-alfalfa system compared to the monoculture corn system during the 2-year period. We attributed this difference, in part, to more favorable microhabit for the predators provided by the living mulch forage legume.

In the most recent phase of this research we are growing corn and alfalfa in biculture using narrow strips or ribbons—planting six 30-inch rows of no-till corn adjacent to an equivalent width of alfalfa, then repeating this pattern across an entire field. Agronomically, such a system has advantages in that the two crops can be rotated so that the corn receives a N benefit and the presence of the legume reduces soil erosion, especially if the ribbons are oriented perpendicular to slopes. Also, existing equipment can be used without modifications. We are experimenting with this system because it is a method for achieving biculture without having severe competition between crops, a problem encounted with living mulch, alfalfa-corn systems. In terms of predatory arthropod ecology, we hypothesize that the presence of the legume will generate sufficient predator populations that will migrate into the corn rows and contribute significantly to pest control.

REFERENCES

1. Brust, G. E., B. R. Stinner, and D. A. McCartney. 1986. *Predator activity and predation in corn agroecosystems.* Environ. Entomol. 15: 1,017-1,021.
2. Dick, W. A. 1984. *Influence of long-term tillage and crop rotation combinations on soil enzyme activities.* Soil Sci. Soc. Am. J. 48: 569-574.
3. Doran, J. W. 1980. *Soil microbial and biochemical changes associated with reduced tillage.* Soil Sci. Soc. Am. J. 44: 765-771.
4. Edwards, C. A. 1984. *Changes in agricultural practice and their impact on soil organisms.* In Proc., Inst. Terrestrial Ecol. Symp. No. 13. Inst. Terrestrial Ecol., Edinburgh, Scotland. pp. 26-29.
5. Elkins, D., D. Fredesking, R. Marashi, and Byron McVay. 1983. *Living mulch for no-till corn and soybeans.* J. Soil and Water Cons. 38: 431.
6. House, G. J., and B. R. Stinner. 1983. *Arthropods in no-tillage soybean agroecosystems: Community composition and ecosystem interactions.* J. Environ. Manage. 7: 23-28.
7. House, G. J., B. R. Stinner, D. A. Crossley, Jr., and E. P. Odum. 1984. *Nitrogen cycling in conventional and no-tillage agroecosystems: Analysis of pathways and processes.* J. Appl. Ecol. 21: 991-1,012.
8. Power, J. F., R. F. Follett, and G. E. Carlson. 1983. *Legumes in conservation tillage systems: A research perspective.* J. Soil and Water Cons. 38: 217-218.
9. Triplett, Jr., G. B., F. Haghiri, and D. M. Van Doren. 1979. *Plowing effect on corn yield response to N following alfalfa.* Agron. J. 71: 801-803.

Root diseases in crops following legumes in conservation tillage systems

Donald R. Sumner

Root disease severity and populations of soilborne pathogenic fungi have been studied in numerous agronomic and vegetable crops following legumes at the Coastal Plain Experiment Station from 1978 to 1985 (*3, 7, 10, 11, 12*). Soils at the site included Tifton loamy sand (fine loamy, siliceous, thermic Plinthic Paleudult) or Bonifay sand (loamy, siliceous, thermic grossarenic Plinthic Paleudult). Tillage practices compared included no-till, chisel plowing, disk harrowing, subsoiling under the row and planting, subsoiling under the row and ridging, and conventional tillage with a moldboard plow. Legumes planted in mono-, double-, or triple-cropping systems were soybeans (*Glycine max* L.), snap beans (*Phaseolus vulgaris* L.), lima beans (*P. lunatus* L.), cowpeas [*Vigna unguiculata* (L.) Walp], and peanuts (*Arachis hypogaea* L.).

Roots of plants in the seedling or juvenile stage in each crop were dug, washed, and rated for root and hypocotyl disease severity on a 1-5 scale, where 1 = <2, 2= 2-10, 3 = 11-50, 4 = >50% discoloration and decay and 5 = dead plant. Soil cores 1 inch in diameter and 6 inches deep were taken from each plot after tillage practices were established. Soil was assayed on various selective media for populations of *Rhizoctonia solani* Kuhn, *Pythium* spp., *Fusarium* spp., and numerous other fungi (*6*).

Rhizoctonia solani, *Pythium* spp., and *Fusarium* spp. are pathogens on roots of legumes and other crops (*4, 5*). In experiments comparing moldboard plowing, disk harrowing, and subsoiling under the row, plowing consistently reduced populations of *R. solani* compared with conservation tillage (Table 1). Tillage had less influence on populations of *Pythium* spp. (Table 1), *Fusarium* spp., and numerous saprophytic fungi, but populations of total fungi in soil were usually lower with plowing than with conservation tillage.

Rhizoctonia solani and related pathogens survive in surface soil in colonized plant debris, such as roots, stems, peanut shells, and seeds (*1, 2*), and do not survive well in the subsoil below 6 to 8 inches because of low O_2 and high CO_2. Plowing buries surface debris and propagules of *R. solani* and brings soil to the surface with low populations of *R. solani*. In contrast, *Pythium* spp. and *Fusarium* spp. survive many months as dormant propagules in soil to a 12-inch depth, and tillage practices have less influence on the distribution of these fungi in the root zone.

In most experiments root disease severity was slight to moderate in corn following legumes, and tillage practices did not affect root disease severity. Many of the fungi pathogenic on legumes are not pathogenic on corn. But *R.*

Donald R. Sumner is a professor in the Plant Pathology Department, University of Georgia Coastal Plain Experiment Station, Tifton, 31793-0748.

Table 1. Populations of *Rhizoctonia solani* and *Pythium* spp. in soil following legumes with various tillages practices, 1978-1983.

Previous Crop and Pathogen	Pathogen Populations by Tillage Practice		
	Moldboard Plow	Subsoil Plant	Disk-harrow
	Colony-forming units/100 g oven-dried soil*		
Peanuts (3)†			
Rhizoctonia solani	2	5	5
Pythium spp.	76	67	76
Soybeans (4)			
R. solani	4	11	14
Pythium spp.	43	94	98
Snap beans or lima beans (5)			
R. solani	4	17	14
Pythium spp.	23	53	68

*In *Pythium* spp. units are per gram of oven-dried soil.
†Number in parenthesis indicates the number of experiments conducted.

solani AG-2 type 2 and some *Pythium* spp. are pathogenic on both corn and legumes. Soil fumigation consistently increased corn yields an average of 7.5% following cowpeas-turnips, indicating that soil-borne pathogens do cause a significant yield loss in corn (*9*).

Root diseases of vegetables have been moderate following legumes, and yields are frequently lower with conservation tillage than with plowing. *Rhizoctonia* root and hypocotyl rot decreased, plant stands increased, and plants were more uniform following plowing than with conservation tillage in collard, spinach, and snap beans. In contrast, tillage practices had no influence on root disease severity in tomato transplants, turnips, or squash. Fewer root diseases occurred in cucumbers in multiple-cropping systems following turnips than following peanuts (*8*). Vegetables in most multiple-cropping rotations were planted after corn, small grains, or other nonlegumes to reduce the risk of premergence and postemergence damping-off and stunted plants.

REFERENCE

1. Bell, D. K., and D. R. Sumner. 1984. *Ecology of a sterile white basidiomycete in corn, peanut, soybean, and snap bean microplots.* Plant Dis. 68: 18-22.
2. Bell, D. K., and D. R. Sumner. 1984. *Unharvested peanut pods as a potential source of inoculum of soilborne plant pathogens.* Plant Dis. 68: 1,039-1,042.
3. Parker, M. B., N. A. Minton, E. D. Threadgill, C. C. Dowler, and D. R. Sumner. 1985. *Tillage, nitrogen, herbicide, and nematicide effects on irrigated double-cropped corn and soybean in the coastal plain.* Res. Bull. 326. Ga. Agr. Exp. Sta., Athens. 35 pp.
4. Sumner, D. R. 1982. *Crop rotation and plant productivity.* In Rechcigl, Jr., Miloslave [ed.] *Handbook of Agricultural Productivity. Vol. 1.* CRC Press, Boca Raton, Fla. pp. 273-313.
5. Sumner, D. R. 1984. *Cropping practices and root diseases.* In C. A. Parker, A. D. Rovira, J. J. Moore, P. T. W. Wong, and J. F. Kollmogen [eds.] *Ecology and Management of Soilborne Plant Pathogens.* Am. Phytopathol. Soc., St. Paul, Minn. pp. 267-270.
6. Sumner, D. R., and A. S. Csinos. 1986. *Assaying populations and evaluating fungicides for control of soilborne pathogenic fungi.* In K. D. Hickey [ed.] *Methods for Evaluating Pesticides for Control of Plant Pathogens.* Am. Phytopathol. Soc. Press, St. Paul, Minn. pp. 277-280.
7. Sumner, D. R., B. Doupnik, Jr., and M. G. Boosalis. 1981. *Effects of reduced tillage and multiple cropping on plant diseases.* Ann. Rev. Phytopathol. 19: 167-187.
8. Sumner, D. R., C. C. Dowler, A. W. Johnson, N. C. Glaze, S. C. Phatak, R. B. Chalfant, and J. E. Epperson. 1983. *Root disease of cucumber in irrigated, multiple-cropping system with pest management.* Plant Dis. 67: 1,071-1,075.
9. Sumner, D. R., C. C. Dowler, A. W. Johnson, R. B. Chalfant, N. C. Glaze, S. C. Phatak, and J. E. Epperson. 1985. *Effect of root diseases and nematodes on yield of corn in an irrigated multiple-cropping system with pest management.* Plant Dis. 69: 382-387.
10. Sumner, D. R., D. A. Smittle, E. D. Threadgill, A. W. Johnson, and R. B. Chalfant. 1986. *Interactions of tillage and soil fertility with root diseases in snap bean and lima bean in irrigated multiple-cropping systems.* Plant Dis. 70: 730-735.
11. Sumner, D. R., and D. K. Bell. 1986. *Influence of crop rotation on severity of crown and brace root rot caused in corn by* Rhizoctonia solani. Phytopathology 76: 248-252.
12. Sumner, D. R., E. D. Threadgill, D. A. Smittle, S. C. Phatak, and A. W. Johnson. 1986. *Conservation tillage and vegetable diseases.* Plant Dis. 70: 906-911.

The effect of weed and invertebrate pest management on alfalfa establishment in oat stubble

C. C. Bahler, R. A. Byers, and W. L. Stout

Conservation tillage systems reduce soil erosion and water loss compared to conventional tillage systems (5). Farmers have used conservation tillage methods with some success to plant alfalfa directly into a grass sward. However, competition from weeds and invertebrate pests increase in no-till environments because of the suitable habitats provided.

Some of the major invertebrate pests of alfalfa in no-till situations are slugs [*Arion fasciatus* (Nilsson), *Deroceras laeve* (Muller), and *D. reticulatum* (Muller)] (1, 2). Slugs can be controlled by chemicals, methiocarb bait, or environmental manipulation—plowing. Plowing acts as a slug control by disrupting the soil environment (6). Other crop residues also influence no-till alfalfa establishment (3, 4, 7). Dowling and Linscott (4) found that slug feeding occurred when alfalfa was no-tilled into small grain stubble.

Our objectives were to study the effect of weed and invertebrate pest management on alfalfa establishment with either conventional or conservation tillage methods.

Study methods

We planted alfalfa into a conventionally prepared seedbed or with a no-till drill into oat stubble. Weed management consisted of a control or an application of 0.4 pounds active ingredient/acre of sethoxydim (Poast). Invertebrate pest management consisted of a control; an application of 2 pounds active ingredient/acre of methiocarb bait (Mesurol), 2 pounds active ingredient/acre of carbofuran (Furadan), or both methiocarb and carbofuran to no-till-seeded alfalfa; or plowing. We conducted the study in 1985 and 1986, with a new planting each year. The experimental design was a randomized complete block with a factorial treatment design and six replications. Plots were 16 feet x 30 feet.

Results

In 1985 methiocarb bait increased the percent alfalfa plant frequency when used alone or with carbofuran (Figure 1). The percent frequency of alfalfa plants was lower for the carbofuran alone or the control compared to methiocarb bait. In addition, there were no significant differences in percent frequency of alfalfa plants between plowing and methiocarb treatments (Figure 2).

In 1986 the use of methiocarb, carbofuran, and both methiocarb and carbofuran did not significantly affect the

C. C. Bahler is a research associate, Pennsylvania State University, University Park, 16802, and R. A. Byers is a research entomologist and W. L. Stout is a research soil scientist, U.S. Regional Pasture Research Laboratory, Agricultural Research Service, U.S. Department of Agriculture, University Park, Pennsylvania 16802.

percent frequency of alfalfa plants. Plant frequency declined significantly with no chemical treatment (Figure 1). Comparison of the plowed and no-till plots was not possible in 1986 because of a seeding failure in the plowed plots (Figure 2).

Sethoxydim application in 1985 significantly increased the percent frequency of the alfalfa plants compared to the treated plots. The major weed species was volunteer oats. In 1986 sethoxydim application did not significantly affect percent alfalfa plant frequency mainly because the competition by the volunteer oats was significantly less than in the previous year (Figure 1).

Invertebrate pest and weed management affected shoot weight in both 1985 and 1986 (Figure 3). Application of

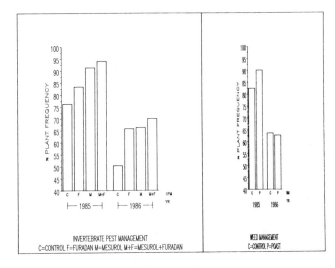

Figure 1. The effect of invertebrate pest and weed management on the percent alfalfa plant frequency, 1985 and 1986.

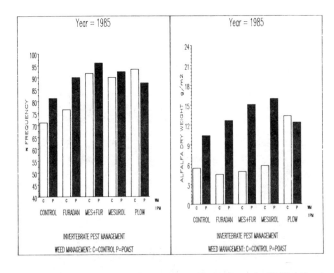

Figure 2. The effect of plowing on the percent alfalfa plant frequency and shoot weight compared to the other invertebrate management methods and their interaction with weed management, 1985.

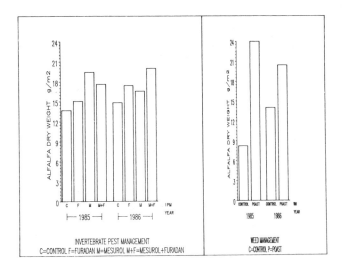

Figure 3. The effect of invertebrate pest and weed management on shoot weight, 1985 and 1986.

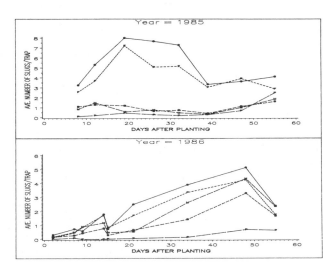

Figure 4. Slug population dynamics during the alfalfa establishment period.

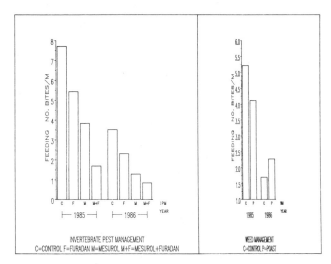

Figure 5. The effect of weed and invertebrate pest on invertebrate feeding, 1985 and 1986.

methiocarb bait increased the shoot weight of the alfalfa plants 6 weeks after planting compared to the control. In 1985 shoot weight in the plowed plots was not significantly different compared to the methiocarb-treated plots (Figure 2). Elimination of volunteer oats with sethoxydim in both 1985 and 1986 increased the shoot weight of alfalfa plants (Figure 3).

We monitored soil moisture, soil temperature, and slug populations during the establishment period in both years. In 1985 the slug population peaked during the second and third weeks of establishment. During 1986 there was a gradual increase in the slug population, reaching a peak at 5 to 6 weeks after planting (Figure 4). However, in both cases the number of slugs per trap never reached the threshold for economic damage, estimated at 3 slugs/square yard or 33 slugs/trap/plot (Byers, unpublished data). The amount of invertebrate feeding on alfalfa plants/3.3 feet of row declined in both years with the use of carbofuran and methiocarb. The largest amount of feeding reduction occurred when methiocarb and carbofuran were used in combination (Figure 5).

Weed management was the major factor affecting the establishment of alfalfa when no-tilled into oat stubble. However, invertebrate pest management treatments also contributed to successful alfalfa establishment. Applying methiocarb and carbofuran showed the highest percent frequency of plants and shoot weight. However, the number of slugs/trap did not correlate with the greater plant frequency when methiocarb and carbofuran were used.

We cannot explain these results based on slug management alone, but insect management may account for the increased frequency. The insect data has not been analyzed yet. Our results show that to establish no-till alfalfa successfully in oat stubble farmers must control both weeds and invertebrate pests.

REFERENCES

1. Byers, R. A., R. I. Mangan, and W. C. Templeton, Jr. 1983. *Insect and slug pests in forage legume seedings.* J. Soil and Water Cons. 38: 224-226.
2. Byers, R. A., W. C. Templeton, Jr., D. L. Bierlein, W. F. Campbell, and H. J. Donley. 1985. *Establishment of legumes in grass swards: Effect of pesticides on slugs, insects, legume seeding numbers, and forage yield and quality.* Grass Forage Sci. 40: 41-48.
3. Fornay, D. R., C. L. Foy, and D. D. Wolf. 1985. *Weed suppression in no-till alfalfa* (Medicago sativa) *by prior cropping summer annual forage grasses.* Weed Sci. 33: 490-497.
4. Dowling, P. M., and D. L. Linscott. 1983. *Use of pesticides to determine relative importance of pest and disease factors limiting establishment of sod-seeded lucerne.* Grass Forage Sci. 38: 179-185.
5. Gebhardt, M. R., T. C. Daniels, E. E. Schweizer, and R. R. Allmaras. 1985. *Conservation tillage.* Science 230: 625-630.
6. Godan, D. 1979. *Pest slugs and snails: Biology and control.* Springer-Verlag. New York, N.Y.
7. Wolf, D. D., E. S. Hagood, Jr., and M. Lentee. 1985. *No-till alfalfa establishment as influenced by previous cover crops.* Can. J. Plant Sci. 65: 609-613.

Legume mulch effects on no-till cotton survival and development

Diane H. Rickerl, W. B. Gordon, and J. T. Touchton

Winter annual legumes improve soil characteristics and provide N for subsequent crops. Legume mulches in reduced tillage systems conserve soil and water while reducing fertilizer N requirements. Acres in conservation tillage in Alabama in 1984 averaged 22% for all crops. However, the percentage of cotton acres in conservation tillage was only 1.5%. Adequate cotton stand establishment has been a major problem when using legume mulches for no-till cotton production. We initiated a study to determine the cause of cotton seedling losses and to identify management practices that would improve stands and yields.

Study methods

We conducted a field study in 1984 and 1985 on a Norfolk sandy loam soil in the Coastal Plains of southern Alabama. Areas of winter management were split between crimson clover and fallow. Prior to cotton planting in May, the crimson clover was killed with Paraquat, leaving a dense mulch on the soil surface.

Treatments at cotton planting (Table 1) consisted of tillage, N source, P starter fertilizer, and fungicides. Preplant tillage comparisons were disk, chisel, and Ro-terra about 1 month prior to planting. We planted all other plots with a Ro-till planter with in-row subsoilers. Nitrogen fertilizer sources included urea, ammonium nitrate, or calcium nitrate banded beside the row at planting at a rate of 90 pounds/acre N. Phosphorus starter, triple superphosphate, was banded beside the row at 18 pounds/acre P for the starter treatment. Fungicides compared were Terachlor, a fungicide commonly used for cotton in this area, and Tilt, a product labeled for grass seed production in the Northwest. The standard treatment consisted of Ro-till planting, ammonium nitrate N, Terachlor fungicide, and no starter fertilizer.

We took samples at the time of cotton planting to determine soil Collembola populations and to assess *Rhizoctonia solani* infestation. Cotton was mechanically picked when 70% to 80% of the bolls were open. A second picking occurred about 2 weeks later if necessary.

Results

Winter management and the planting treatment greatly affected cotton populations. In 1984 populations without fungicide application were 9% greater in the fallow plots than

Diane H. Rickerl is an assistant professor and W. B. Gordon is a Ph.D. candidate, Plant Science Department, South Dakota State University, Brookings, 57007; and J. T. Touchton is a professor, Agronomy and Soils Department, Auburn University, Auburn, Alabama 36849.

in the clover mulch plots. In 1985 the populations without fungicide were 267% greater in the fallow than in the clover mulch. Terachlor, the standard fungicide, improved stands in both the fallow and clover mulch plots, but populations in the clover mulch plots were still less than those in the fallow plots. Tilled and bedded treatments also improved populations in the clover compared to the standard Ro-Till treatment. However, because Terachlor was the common factor among the treatments that improved populations, it appears that the stand losses were due to seed and seedling disease. Terachlor does not control *Pythium;* thus, it is likely that the causal agent was *Rhizoctonia solani.*

Table 2 shows levels of Rhizoctonia infestation and

Table 1. Treatments established at time of cotton planting.

Treatment	Tillage	N Source	Fungicide	P Starter
1	Till	NH$_4$NO$_3$	Terachlor	no
2	Bed	NH$_4$NO$_3$	Terachlor	no
3	Ro-till	Urea	Terachlor	no
4	Ro-till	Urea + DCD	Terachlor	no
5	Ro-till	Ca(NO$_3$)$_2$	Terachlor	no
6	Ro-till	None	Terachlor	no
7	Ro-till	NH$_4$NO$_3$	Terachlor	yes
8	Ro-till	NH$_4$NO$_3$	Tilt	no
9	Ro-till	NH$_4$NO$_3$	None	no
10(Standard)	Ro-till	NH$_4$NO$_3$	Terachlor	no

Table 2. Rhizoctonia infestation and Collembola populations at cotton planting.

Winter Management	Severe	Moderate	Slight	None	Collembola (number/6.6 pounds of cotton)
Clover	1	84	15	0	12
Fallow	0	34	51	14	3

Table 3. Seed cotton yields as affected by winter management and planting treatment, 1985.

Year	Treatment	Clover	Fallow
1984	1	773	1,002
	2	857	831
	3	711	639
	4	642	925
	5	617	932
	6	638	573
	7	842	1,051
	8	904	755
	9	552	817
	10	562	838
1985	1	2,730	1,885
	2	2,146	1,913
	3	1,679	1,795
	4	1,036	1,741
	5	1,687	1,906
	6	1,871	1,266
	7	2,478	1,866
	8	2,042	1,829
	9	1,390	1,906
	10	1,615	2,521

numbers of Collembola, a natural predator of *Rhizoctonia.* Moderate infestation levels were greater in soils with clover mulch than in fallow soil. This corresponded to cotton population data where clover systems had lower plant numbers than fallow systems. Collembola populations also were greater in clover mulch plots where their food supplies were more abundant than in fallow soils. As many as 90 Colembola were found in 6.6 pounds of soil, but their average populations were inadequate for disease control.

In 1984 early season drought limited seed cotton yield; we found no differences in yield. Growing conditions in 1985 were excellent (Table 3), and seed cotton yields in the clover mulch plots were maximized by use of tillage or application of starter P. In the fallow soil no N (treatment 6) resulted in the lowest yields, and the standard treatment had the highest yields. All other treatments averaged 1,855 pounds/acre seed cotton.

Influence of winter cover crops on disease development and grain sorghum yield in the Southeast

R. R. Duncan, W. L. Hargrove, and R. E. Dominy

No-till and minimum tillage production practices increased from about 13% of total grain sorghum [*Sorghum bicolor* (L.) Moench] acreage nationally in 1972 to almost 40% in 1985 (*5*). In Georgia 44% of the total sorghum (grain/forage/silage/sweet sorghum) acreage in 1985 was planted using conservation tillage practices. By the year 2000, 65% of the major grain crops in the United States may be planted with no-till methods.

The need to conserve soil and water resources, the necessity to reduce production costs, and the adoption of various double-cropping practices have contributed to this increase in sorghum production by conservation tillage. The incidence of diseases, particularly foliar/stalk rot diseases, and their potential interaction with tillage systems and/or cover crops regarding grain yield stability are critical factors in sorghum production under conservation tillage practices. Hybrid-tillage system interactions have been found and disease resistance may be a contributing factor to increased yields (*3, 4, 5*).

Sorghum can be grown in several different double-cropping systems in the Southeast (*5, 7*). One of the most promising systems involves a winter legume, such as crimson clover (*Trifolium incarnatum*), that provides protection against soil erosion during the high-rainfall winter and early spring months and also provides sufficient N for optimum grain sorghum yield (*6, 8, 9*). Winter legumes have produced 70 to 100 kg N/ha/year for the subsequent sorghum crop. Grain sorghum yields in a winter legume double-cropping system have ranged from 4,000 to 6,500 kg/ha with no additional fertilizer N.

Dominy (*1*) investigated the influence of tillage and cover crop on sorghum grain yield and anthracnose (*Colletotrichum graminicola*) development (Table 1). He found that cover crop treatment had no effect on yield or anthracnose development. However, yields were 23% greater on no-till plots than on conventional tillage plots. Tillage method did not affect anthracnose development on sorghum leaves, stems, or panicles. We attributed the higher disease readings in 1984 on a previously resistant hybrid to the emergence of a new pathotype of the causal organism, but the spread of the disease was uniform across cover crops and tillage systems. We monitored *Fusiarum*-induced stalk rot development but found no significant tillage or cover crop effect (data not shown). Doupnik and Boosalis (*2*) concluded that plant

R. R. Duncan and W. L. Hargrove are associate professors and R. E. Dominy is a former research associate, Department of Agronomy, University of Georgia, Georgia Experiment Station, Griffin 30212-5099.

Table 1. Influence of tillage and cover crop on grain yield and anthracnose development for DeKalb DK-64 sorghum.

| | 1983 | | | | 1984 | | | |
| | Yield* | Disease Rating | | | Yield | Disease Rating | | |
	(pounds/acre)	Leaves	Stems	Panicles†	(pounds/acre)	Leaves	Stems	Panicles†
Tillage								
No-till	3,740	2.0	1.4	1.4	3,960	3.8	3.3	3.7
Conventional	3,050	2.1	1.4	1.6	3,200	4.4	3.9	4.5
LSD (0.05)	360	NS	NS	NS	530	0.3	0.3	0.2
Cover crop								
Wheat‡	3,410	2.0	1.4	1.3	3,700	4.2	3.5	4.1
Stubble§	3,515	2.0	1.6	1.4	3,240	4.2	3.8	4.0
Clover‖	3,505	2.1	1.4	1.8	3,790	4.0	3.4	4.1
Fallow	3,140	2.0	1.3	1.5	3,605	3.9	3.7	4.2
LSD (0.05)	NS	NS	NS	NS	NS	NS	NS	NS

*Oven dry weights; n = 8.
†Panicle at 112 and 114 days from planting in 1983 and 1984, respectively; 1 = no disease; 5 = dead plant part.
‡Soft red winter wheat = *Triticum aestivum* L. em Thell.
§Sorghum stubble from previous year.
‖Crimson clover = *Trifolium incarnatum* cv. Tibbee.

Table 2. Hybrid-tillage yield and disease interaction for selected grain sorghum hybrids in Georgia, averaged over 1984 and 1985.

| Sorghum Hybrid | No-till | | Minimum Tillage* | | Conventional Tillage† | | Yield LSD (0.05) | Disease LSD (0.05) |
	Yield (pounds/acre)	Disease Rating‡	Yield (pounds/acre)	Disease Rating‡	Yield (pounds/acre)	Disease Rating‡		
DeKalb DK-64	5,440	2.4	5,360	2.1	5,150	2.4	NS	NS
Paymaster 1022	6,250	2.4	5,930	2.5	4,915	3.1	784	0.4
Pioneer 8333	5,870	1.9	5,770	1.8	4,770	2.5	935	0.4
DeKalb DK-59	4,720	3.0	4,825	3.8	3,825	4.3	790	0.5
Pioneer 8311	4,320	4.6	4,890	4.8	3,890	4.9	NS	NS
LSD (0.05)	780	0.7	740	0.6	625	0.6	—	—

*Tillage included fall deep plowing prior to planting winter small grain. Sorghum was planted no-till.
†Deep plowing and disking during fall and early summer.
‡Anthracnose: *Colletotrichum graminicola* (Ces.) Wils.; 1 = no disease. 5 = dead plants.

diseases did not increase under reduced tillage systems, and the incidence of stalk rot in grain sorghum actually decreased. They suggested that crop rotation coupled with reduced tillage were important factors in preventing the buildup of plant diseases.

Another study (Hargrove, unpublished data) investigated hybrid-tillage system interactions on grain yield and disease development (Table 2). Yields were significantly different among hybrids for each tillage system. The variable disease ratings may have contributed to some of that variability. Consequently, commercial hybrids need to be evaluated for their yield responses under various conservation tillage systems.

Environmental stress factors, such as moisture and temperature, can predispose sorghum plants to infection by a number of pathogens. A plant that is well buffered against environmental stresses and that remains metabolically and physiologically active during late reproductive stages of development should better withstand attack by root and stalk-rotting organisms. Improvements in genetic resistance/tolerance mechanisms, especially on the multiple-gene level, should help to stabilize plant performance. Reduced tillage and surface mulches will minimize problems with high soil temperatures and erratic moisture availability that could result in predisposition to various pathogens.

REFERENCES

1. Dominy, R. E. 1984. *Influence of cropping systems on the anthracnose/fusarium complex of grain sorghum.* M.S. Thesis, Univ. Ga., Athens. 77 pp.

2. Doupnik, Jr., Ben, and M. G. Boosalis. 1980. *Ecofallow—a reduced tillage system—and plant diseases.* Plant Dis. 64: 31-35.

3. Duncan, R. R. 1985. *Disease factors which affect grain sorghum production in the Southeast.* In Dolores Wilkinson [ed.] *Proceedings, 40th Annual Corn and Sorghum Research Conference,* Am. Seed Trade Assoc., Washington, D. C. pp. 28-41.

4. Duncan, R. R. 1986. *The influence of* Colletotrichum *and* Fusarium *diseases on grain sorghum in the southeastern USA.* In M. A. Foale and R. G. Henzell [eds.] *Proceedings, 1st Australian Sorghum Conference,* Queensland Agr. Col., Gatton, Australia. pp. 3.56-3.62.

5. Duncan, R. R., and W. L. Hargrove. 1986. *Sorghum grain yields in conventional and conservation tillage systems in the S.E. USA.* In M. A. Foale and R. G. Henzell [eds.] *Proceedings, 1st Australian Sorghum Conference,* Queensland Agr. Col., Gatton, Australia. pp. 5.12-5.19.

6. Hargrove, W. L. 1986. *Winter legumes as a nitrogen source for no-till grain sorghum.* Agron. J. 78: 70-74.

7. Hargrove, W. L., and C. C. Dowler. 1985. *Sorghum in multiple-cropping systems.* In R. R. Duncan [ed.] *Proceedings, Grain Sorghum Shortcourse.* Spec. Publ. 29. Univ. Ga., Athens. pp. 38-45.

8. Hargrove, W. L., and G. W. Langdale. 1985. *Sorghum in no-tillage production systems.* In R. R. Duncan [ed.] *Proceedings, Grain Sorghum Shortcourse.* Spec. Publ. 29. Univ. Ga., Athens. pp. 46-53.

9. Touchton, J. T., W. A. Gardner, W. L. Hargrove, and R. R. Duncan. 1982. *Reseeding crimson clover as an N source for no-tillage grain sorghum production.* Agron. J. 74: 283-286.

CROPPING PRACTICES

Cropping practices using legumes with conservation tillage and soil benefits

L. F. Elliott, R. I. Papendick, and D. F. Bezdicek

Farmers have been able to maintain high crop production levels with farming practices that use intensive tillage, monoculture or 2-year rotations, and high fertilizer and pesticide inputs. However, such farming practices result in high-priced food and in many cases significant soil erosion, reduced soil organic matter, deteriorating soil structure, reduced water infiltration, increased compaction, weed infestations, and severe plant pathogen problems. Usually, these cropping systems do not lend themselves to the use of legumes in rotation for forage or green manure, intercropping of legumes or grasses, and grass-legume mixtures in the rotation—all of which improve soil structure, soil organic matter, and soil N and can decrease crop pest problems.

Conservation tillage is defined as any tillage sequence that reduces loss of soil and water relative to conventional tillage. Crops are important also because in most situations a legume, grass, or legume-grass mix used as a cover crop or green manure in the rotation reduces erosion, and the legume provides residual N for the following crop. These benefits are not realized always when a legume is harvested for seed.

Legumes in conservation tillage

Several factors govern conservation cropping practices that successfully employ legumes in rotation: the type of tillage;

L. F. Elliott is a microbiologist and R. I. Papendick is a soil scientist with the Agriculture Research Service, U.S. Department of Agriculture, Washington State University, Pullman, 97164-6421. D. F. Bezdicek is a professor of soils, Washington State University. This paper is a contribution from USDA-ARS in cooperation with the College of Agriculture and Home Economics Research Center, Washington State University, Pullman; manuscript number 7717. Trade and company names are included for the benefit of the reader and do not imply endorsement or preferential treatment by the U.S. Department of Agriculture.

the amount and timing of rainfall; availability of and need for supplemental water; whether spring or fall seeding options are available; length and temperature of growing season; and whether the legume is used for green manure or as a cover, strip, forage, or seed crop. When these factors are considered, economic constraints must be included also.

Conservation tillage normally includes practices that increase soil surface roughness and maintain significant crop residues on the soil surface. Conservation tillage usually involves reduced tillage, which means less field traffic, less fuel consumed, lower machinery investment, and sometimes, but not always, more chemicals for pest control. The extreme case of reduced tillage is no-till, where the crop is planted with just enough tillage to place and cover the seed in the soil.

Success with reduced tillage depends upon soil type, drainage, climate, and management practices. The presence of surface residues and less disturbance of the soil with reduced tillage systems may change drastically the soil-plant environment compared with that of conventional tillage. With reduced tillage the soil is wetter and cooler in the spring, which may delay germination and slow early growth of some crops. All of these factors resulting from the soil-residue management system must be considered when a farmer introduces a legume into the rotation. For example, if the soil surface is rough, it will be difficult to establish a small-seeded legume, such as alfalfa (*Medicago sativa*), without additional tillage or an increase in seeding rate. Additional tillage promotes erosion unless the farmer avoids critical periods. Increased seeding rates raise farming expenses. Seeding a legume into heavy surface residues may fail if the environment is cold and wet for a prolonged period, which can cause seed rotting and reduced vigor, a problem with many legumes.

The amount and timing of rainfall is critical to successfully establishing and growing legumes. Poor moisture during establishment of such legumes as alfalfa can result in a poor stand and plants that never root deeply. Drought stress during bloom, particularly with such legumes as peas (*Pisum sativum*) or lentils (*Lens culinaris*), usually results in a greatly reduced yield. These crops do not tolerate prolonged wet conditions either. Where irrigation is available, farmers can manage drought problems. Unfortunately, in many areas irrigation has become too expensive.

The time in which the legume can be established in the rotation depends upon many factors, including the times of the year when seeding is possible, whether the legume is interseeded or used as a cover crop and/or a strip crop, and whether it is to be used for green manure, seed harvest, or forage. For example, only spring seeding may be possible in cold winter climates unless adequate moisture is available in the fall to initiate legume growth and no-till or conservation seeding leaves enough stubble or residue for adequate snow catchment for overwinter freeze protection. In other cases the season may be long enough and rainfall well-distributed enough so a farmer could establish the legume in late spring or after another crop had been harvested in the summer. Harvesting legumes for seed usually creates a negative N budget (*2*), although benefits are still realized from the rotational effect. Farmers will realize the maximum benefit from a legume when it is used as a green manure. A legume's beneficial effects in the rotation may be more valuable than the cash crop it replaces.

Harvested for forage, legumes provide a beneficial rotational effect, and if farmers allow the crop to regrow before it is plowed or killed, N and sanitizing effects are gained, as from green manuring. In many cases the benefit is greater than using a green manure crop in the rotation because the stand has remained long enough for an enhanced positive effect on soil N, soil organic matter, soil structure, soil compaction, and reduced erodibility.

There are situations with legumes in the rotation in which conservation tillage, by itself, may not control erosion adequately. In the Pacific Northwest erosion on no-till fields following peas or lentils can be relatively high, especially on steeper land. These crops produce small amounts of residues, and the residues decompose more rapidly than small grain cereal straw. Hence, the residues lose their effectiveness to control erosion much earlier. Secondly, these grain legumes tend to produce a loose, mellow soil surface, similar to the well-known condition for soybeans (*Glycine max*), that is highly vulnerable to erosion (*25*). Erosion control following these crops may require supporting practices, such as divided-slope or cross-slope farming and early fall planting of winter wheat (*Triticum aestivum*) in combination with no-till seeding. However, it is difficult to evaluate what the soil erodibility would be if the rotation contained a meadow or green manure crop. The benefit of these practices for erosion control and improved soil structure is well documented (*30*).

The agroclimatic region and economic factors dictate to the grower the choice of legumes and conservation tillage practice and the intended use of the legume. The benefits derived from legumes and grasses in the rotation in a conservation tillage system are such that legumes can be no longer ignored. Benefits include the N value; disease control; improved soil structure, compaction, and organic matter; and beneficial effects on microbiological relations in the rhizosphere.

Soil structure

Soil structure and organic matter content are related closely. Loss of organic matter results in loss of soil tilth and productivity. Although the amount of organic matter in most soils is relatively small, its influence on soil properties can be large. Organic matter serves as the chief granulating agent in soils and provides structural stability and optimum air and water relations. The organic matter content of virgin soil depends upon the climatic zone under which the soil is cultivated. The equilibrium organic matter content in the soil depends upon the amount of tillage, the amount and type of organic material added, the soil environment, and the cropping system. Organic matter losses about 50% below virgin levels appear inevitable for many soils after 50 to 60 years of cultivation. This loss, however, varies considerably, depending upon the crop rotation and tillage management. Organic matter losses are greatest with intensive row cropping and least with meadow in the rotation.

The burnout of soil organic matter by cultivation also may increase soil organic matter losses by erosion. Soil erosion is a selective or an enrichment process (*31*). The eroded soil fraction generally is higher in organic matter and clays than the original soil. Moreover, because of their colloidal nature, organic matter and clays most likely remain in runoff water and are carried off the land.

Erosion from a specific area can accelerate with time as soil organic matter, sediments, and associated nutrients are lost, resulting in a degradation of soil physical, chemical, and biological properties (*16*). In turn, plant growth is affected adversely because of reduced water intake and storage, reduced nutrient levels, a poorer rooting medium, and, in many cases, greater soil compaction (*27*). While additional fertilizer can help maintain a certain level of soil productivity, it is difficult to prevent the decline in other soil properties with conventional, high-intensity cropping practices.

Reduced soil productivity indicates that something is wrong with the management system. If the problem is due to deterioration of a physical property of the soil, further investigation is warranted to determine if the cause is from excessive soil erosion, excess water, salt buildup, water stress, low organic matter content, or other factors. However, these factors are only symptoms that something is wrong with the soil system. Rather than treat the symptom, it is best to treat the cause. For example, where soil compaction has become a serious problem, it is better in the long term to treat the cause of compaction rather than devise mechanical ways to break up the hard soil because that solution is only temporary. To accomplish this, it may be necessary to design a different cropping or tillage system, apply organic manures, grow legumes in the rotation, or manage crop residues differently (*27*).

Conservation tillage usually results in increased soil organic matter in soil that has been conventionally tilled for extended time periods. Soil organic matter increases near the soil surface because the crop residues are maintained on and near the surface. Doran, in a survey of plots across the United States (9), found that microbial numbers and soil organic matter were greater near the surface in no-till plots than in conventionally tilled plots. Soil physical properties benefit most from soil organic matter if the organic matter is concentrated in the upper few inches of soil. This is most beneficial for such factors as infiltration; stabilizing the soil against erosion; and improving the structure and tilth for seedbed preparation or planting, seedling establishment, and any subsequent cultivation.

Experiments generally show that proper crop rotations and tillage, residue management, and regular applications of animal manures or other organic materials can maintain soil organic matter levels. Legumes or other forage crops in the rotation effectively slow the decline of organic matter or increase its equilibrium level in the soil. Organic materials added to the soil at regular intervals as green manures, such as rye, oats, sweetclover, vetch, or alfalfa; barnyard manure; sewage sludge; or other organic materials also help to maintain soil organic matter content. These practices, combined with reduced tillage, will increase organic matter content and improve soil properties fairly rapidly (28).

Dollette (8) described an ideal cropping system to improve soil organic matter and structure in semiarid regions. Australian research has shown that wheat yields increased with a bare fallow-wheat rotation using intensive tillage. The yield increase appeared to be due to increased moisture storage and increased mining of the soil N. However, the yield increase was only temporary. There was a decline in soil fertility paralleling a deterioration of soil structure and an increase in soil erosion. In a red brown earth, soil aggregates greater than 0.2 mm declined from 52% to 24% and water infiltration decreased from 200 mm to less than 20 mm in 3 hours. With reduced tillage and a legume in the rotation, the Australians reported soil N increases. One would also expect eventual increases in soil organic matter and soil aggregation (21). In another Australian study Loch and Coughlan (24) found that retention of surface residues increased the organic C content of the 0- to 10-cm layer after 5 years and the dispersion ratio and the dry aggregate size in the 0- to 4-cm layer of the no-till treatments.

Bakhtri (1) pointed out the benefits of *Medic* in a rotation. In Algeria and Tunisia soil N increased from 18 to 54 pounds/acre after one year of a clean *Medic* pasture. He quoted Mouffak's work in Tunisia that showed the same wheat yields with fewer cultivations when *Medic* was in the rotation (Table 1). He also pointed out that with *Medic* in

the rotation the soil was easier to work and that the same yields might have been obtained with less N added.

In climates where they can be used, winter cover crops increase soil organic C and total N compared to fields without cover crops (23, 32). Including a winter legume as a cover crop also results in higher soil pH. Hargrove (18) found that soil organic matter and organic N increased in a no-till clover-sorghum system using clover as the winter cover. In a 6-year study Buntley (5) reported increased yields for both tilled and no-till corn following a wheat-vetch cover crop compared to conventionally seeded corn. These results relate primarily to the southeastern United States.

Lal and associates (22) addressed the rapid deterioration of soil physical properties in a tropical soil following deforestation. They found that rice (*Oryza sativa*) straw mulches rapidly increased total porosity and that there were positive effects on relative proportions of macropores, saturated hydraulic conductivities, field infiltration rates, and moisture retention at -0.1 and -0.3 bars. The mulches also reduced runoff from the plots. From previous studies one could postulate that if legumes were used in the mulch the positive effect would be greater (17).

Soil biological properties

Tillage, residue management, and crop rotation influence soil biological properties. In most cases soil biological properties relate to soil microbial biomass and soil organic matter. Carter (6) studied cultivation and cropping effects on microbial biomass in semiarid and per-humid regions of Canada. He found that permanent crops, such as grassland, generally enhanced levels of microbial biomass, while arable crops and soil cultivation reduced biomass. However, Carter cited several studies showing that this decline can be tempered by additions of fertilizers, crop residues, or other sources of organic matter. Under intensive tillage and a wheat-fallow system, there was signficantly less microbial biomass C and N in the Ap horizon compared with a continuous wheat or cereal-grass system (6). The cereal-grass system had the highest microbial biomass C and N. One would expect similar differences to occur in soil organic matter.

Legumes in a rotation positively influence soil biological properties. Bolton and associates (4) compared soils on a farm that had used leguminous crops since 1909 and more recently Austrian winter peas (*Pisum sativum* spp., *arvense* L., Poir) as a green manure plus native soil fertility for N

Table 1. Effect of *Medicago* on wheat yield, Tunisia, 1974. (1).

			Wheat/Medic		Wheat/Fallow	
Region	Variety	N (pounds/acre)	Number of Cultivations	Wheat Yield (pounds/acre)	Number of Cultivations	Wheat Yield (pounds/acre)
Tunis	INRAT-69	59	3	1,249	4	1,071
Fahs	Mhamoudi	30	4	892	5	803
Fahs	Ariana-66	45	3	2,855	5	2,320
Gaafour	Fx Aurore	30	3	1,517	5	1,428
Gaafour	INRAT 69	30	2	1,874	4	1,874
El Aroussainia		32	4	178	5	1,963

with soils on an adjacent farm that had received regular applications of anhydrous ammonia, P, and S at recommended application rates for the past 30 years. They found that urease, phosphatase, and dehydrogenase were significantly higher at all samplings and that soil microbial biomass was significantly higher on two of three sampling dates in the soil that had not received fertilizer. Soil organic matter also was greater, and indications were that erobibility was less. Both operators used intensive tillage. Presumably the legume effect would have been greater under minimum tillage.

Effects on rhizosphere biology

Herein, we primarily discuss winter wheat, but many of the principles apply to other crops. However, climatic zone and whether the crop is a winter or spring crop may have a dominant effect.

Crop growth problems are still prevalent in many conservation tillage situations, and they may occur with grain and legume crops. The problems relate to pathogens, cold soils, saturated soils, crop residue decomposition products, and harmful bacteria on the root surface. In the Pacific North-west, for example, no-till-seeded winter wheat can appear to grow poorly and suffer overwinter stand losses, especially in cold, wet springs. As the weather warms, the crop appears to recover and grow normally. These differences usually are obvious only if compared to adjacent areas without residues. The problem cannot be attributed to root pathogens, decomposition products from the previous crop residues, or soil fertility. The problem seems to be related to changes in root colonization by microorganisms.

Initial studies indicated that pseudomonads and total bacteria colonized roots of no-till-seeded winter wheat more so in fields where residues were left in the seedling row compared to conventionally tilled and seeded areas (*10*). Later, several pseudomonads isolated from the roots of plants growing poorly in residues inhibited winter wheat seedling root growth. Testing selected isolates against several wheat cultivars, Elliott and Lynch (*11*) found that the plants varied in susceptibility to the organisms.

Inhibitory pseudomonads grow well at low temperatures and are nonfluorescent. The inhibitory effect is due to toxin production, and in the majority of the organisms toxin production is unstable. Toxin-producing pseudomonads are difficult to detect on wheat roots in the fall but are abundant in the spring, indicating that overwinter cold stress may be necessary for toxin expression. The organisms do not appear to penetrate the plant root (*11*). Inhibitory pseudomonads are aggressive root colonizers, and their severity appears affected by soil type (*13, 14*). The organisms appear less severe as soil organic matter increases, and they seem to be somewhat host-specific. The pseudomonads inhibited spring wheat and winter barley (*Hordeum vulgare*) less than winter wheat. Oats (*Avena sativa*), lentils, and peas were unaffected (*15*); however, the bacteria did colonize the roots of the unaffected crops in high numbers. Reasons for

Figure 1. Inhibition of winter wheat seedlings in soil by an inhibitory psuedomonad (inoculated seedlings on right.)

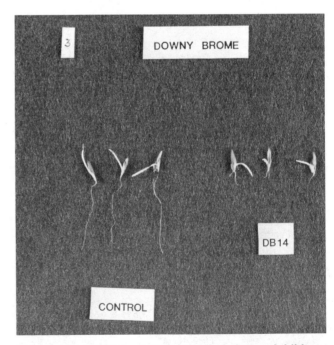

Figure 2. Inhibition of downy brome by an inhibitory psuedomonad (inoculated plants on right.)

this apparent host specificity are not clear. The degree of winter wheat root growth inhibition that can occur in soil is shown by plants grown in the greenhouse (Figure 1).

Inhibitory pseudomonads isolated from plant roots from the field will strongly colonize winter wheat roots when reintroduced in the field (H. Bolton, Jr., H. F. Stroo, J. K. Frederickson, and L. F. Elliott, unpublished results). There can be significant overwinter stand reduction and yield reductions because of the reintroduced organisms.

Cherrington and Elliott (7) conducted a survey to determine the incidence in the field of inhibitory pseudomonads on the rhizoplane of and inhibitory to crops other than winter wheat. Those tests included peas, lentils, and winter barley. Pea and lentil roots harbored low levels of inhibitory pseudomonads. Whether the pseudomonad presence was due to the fact these plants were legumes or that they were spring crops is unknown. Several isolates inhibitory to the weed downy brome (*Bromus tectorum*) but not winter wheat also were found during this study (7). Laboratory trials showed severe retardation of the weed root growth, but field testing remains (Figure 2).

The inhibitory pseudomonads appear to constrain winter wheat yields, especially in conservation tillage systems. If legumes were a significant part of a rotation and were used as green manure, problems with these organisms would be less likely.

Bhattacharyya and associates (3) studied the effect of manures, including green manure, on selected microbes and processes in the rhizosphere of rice and the residual effect in the rhizosphere of a following wheat crop. While the results and implications were unclear, there was no question that green manure had a large effect on microorganisms in the rhizosphere. There is a need for more such studies to measure the harmful or beneficial microbial-rhizosphere relationships and to devise the most beneficial system. The beneficial effects of green manuring may relate in large part to the rhizosphere effect. Plant cultural systems and residue and soil management are still the most viable and economical approaches for achieving optimum plant health while avoiding soil productivity losses.

Legumes for conservation tillage

The suitability of a legume for conservation tillage depends upon its adaptability to conservation tillage and how well it grows in a particular agroclimatic region. Hoyt and Hargrove (20) summarized the characteristics of legumes, adapted from Heath and associates (19), primarily as cover crops in the South, including some soil and climatic considerations (Table 2). Their summary will allow researchers to select legumes for conservation tillage trials in the region. There is an abundance of older literature on winter legumes; however, cultivars and cultural practices are much different today. Such summaries, expanded to include seed legumes, are needed for all regions of the United States.

Research in the Pacific Northwest

Methods. We conducted a tillage, rotation, and fertilizer study to determine the effect of legumes in the rotation on

Table 2. Commonly planted legumes and their characteristics (*19, 20*).

Common Name	Scientific Name	Characteristics
Arrowleaf clover	*Trifolium vesiculosum* Savi	Adapted to lower South; late maturity; good reseeding; good insect resistance; grows best on well-drained soils with near neutral pH.
Ball clover	*T. nigrescens* Viv.	Adapted to lower South; reseeds well; matures later than crimson clover but earlier than arrowleaf clover.
Berseem clover	*T. alexandrinum* L.	Generally not winter hardy but produces abundant fall and winter growth where it survives.
Crimson clover	*T. incarnatum* L.	Adapted to all of the South; moderately tolerant of soil acidity; does not grow on calcareous soils or poorly drained soils; produces abundant growth in fall and winter months.
Persian clover	*T. resupinatum* L.	Adapted to clayey, wet, or poorly drained soils of the lower South.
Red clover	*T. pratense* L.	Grown as a winter annual only in the lower South; best production of fertile, well-drained soils.
Rose clover	*T. hirtum* All.	Adapted to lower South on well-drained soils; has low fertility requirement.
Subterranean clover	*T. subterraneum* L.	Excellent reseeding; moderately tolerant of soil acidity; adapted to most of the South.
White clover	*T. repens* L.	Grown as a winter annual only in the lower South; adapted to well-drained, medium-textured soils of near neutral pH.
Blue lupine	*Lupinus angustifolius* L.	Adapted to neutral or slightly acid soils of the Coastal Plain; not very cold-tolerant.
White lupine	*L. albus* L.	Adapted to neutral fertile soils, especially in the lower Mississippi Delta; more winter hardy than blue lupine.
Bigflower vetch	*Vicia grandiflora* Scop.	Intermediate in cold tolerance between hairy vetch and common vetch; adapted to most of the South.
Common vetch	*V. sativa* L.	Less winter hardy than hairy or bigflower vetch; adapted to well-drained soils, particularly in the Coastal Plain.
Hairy vetch	*V. villosa* Roth	Most winter hardy of cultivated vetches; well adapted to all of the South.
Field (winter) peas	*Pisum sativum arvense* (L.) Poir	Winter hardy; well adapted to most of the South.
Rough peas	*Lathyrus hirsutus* L.	Adapted to heavy soils in the lower South; reseeds well.

wheat yield. In the spring of 1981 we established 8-foot by 30-foot plots at the Palouse Conservation Field Station 2 miles northwest of Pullman, Washington. We established four rotations: spring wheat-winter wheat (SW-WW), field peas-winter wheat-spring wheat-field peas-winter wheat (P-WW-SW-P-WW), spring wheat-winter wheat-spring wheat-spring wheat-winter wheat (SW-WW-SW-SW-WW), and Austrian winter peas-winter wheat-spring wheat-Austrian winter peas-winter wheat (AWP-WW-SW-AWP-WW).

We used three N rates: no added N; one-half the recommended N rate (35 and 45 pounds/acre for spring and winter wheat, respectively); and the full recommended rate (70 and 90 pounds/acre, respectively). Fertilizer was applied at seeding as NH_4NO_3. On the conventionally tilled plots NH_4NO_3 was surface broadcast. The plots were deep-tilled to 8 inches with sweeps (two passes), double-disked to 6 inches, harrowed, and seeded (double-disk drill, 7-inch rows.) No-till plots were sprayed with 32 ounces of active ingredient/acre glyphosate 3 to 5 days before seeding.

We seeded the no-till plots with the USDA III no-till drill using a paired row configuration, with the NH_4NO_3 deep-banded between the paired rows (2, 6). Waverly spring wheat was seeded at 90 pounds/acre and Dawes winter wheat at 80 pounds/acre. Tracer field peas and Austrian winter peas were seeded in the same manner at 150 pounds/acre with no fertilizer; we used both as green manures. The pea vines were flail-chopped in both the tilled and no-till plots. We incorporated the green manure into the tilled plots with a 6-inch deep double disking and into the no-till plots with a single, 2-inch deep disking. The plot design was four blocks, each representing a different rotation, in each replication. Each block was split three times for the three fertilizer rates. Each of the fertilizer rates was split again for the two tillage treatments. The main blocks were randomized within three replications. We harvested 20-foot by 5-foot yield strips

from each plot with a plot combine. Yields were reported as clean dry seed. Table 3 shows the rotations. Winter wheat was grown 50% of the time in rotations A and C and 33% of the time in rotations B and C. We established the Austrian winter peas for 1981 in the spring that year but in the fall of 1983 for 1984.

While yields were not high for this area, there was a strong rotational effect for the winter wheat between rotations A and C with B and D (Table 4). For winter wheat tilled treatments, there appeared to be little advantage to adding N, except in rotation C. For winter wheat, no-till treatment rotations B and D, the first increment of N appeared to increase yield. Spring wheat did not respond to the rotation like winter wheat, possibly because it followed winter wheat and not a legume. Also, spring wheat appeared to respond more to fertilizer N. While these results are not clear cut, they do establish the value of a legume in the rotation for conservation tillage.

We established other plots in the same area in the fall of 1983 to evaluate several tillage options for legumes prior to planting winter wheat. Our objectives were threefold: (1) to evaluate and compare the effect of moldboard plowing, shallow tillage, or chemically killing the legume on winter wheat yields; (2) to evaluate residue from Austrian winter peas, red clover, hairy vetch, and spring peas on winter wheat yields; and (3) to assess overseeding of a forage legume into a spring cereal as a means of establishing legumes.

We established replicated, 12- by 150-foot plots in spring barley residue in the fall of 1983 using four rotations:

1. SB-SW/RC-RC-WW. We planted spring wheat in the spring of 1984 and immediately overseeded it with red clover. Three tillage options were imposed on the red clover in summer 1985, followed by the planting of winter wheat.

2. SB/WW-WW/HV-HV-WW. We established winter wheat in the fall of 1983, followed by no-till seeding of hairy vetch in fall 1984. Hairy vetch residue was chemical killed in the summer of 1985, followed by the no-till planting of winter wheat.

3. SB/WW-WW/AWP-AWP-WW. We established winter wheat in the fall of 1983, followed by fall seeding of Austrian winter peas in 1984. Three tillage options were imposed on the pea residue in 1985, followed by the planting of winter wheat.

4. SB-WW-SP-WW. This rotation served as a conventional rotation comparison. We established winter wheat in the fall

Table 3. Rotations employed in tillage, nitrogen, and legume study.

Rotation	Rotation 1981	1982	1983	1984	1985
A	SW	WW	SW	WW	SW
B	P	WW	SW	P	WW
C	SW	WW	SW	SW	WW
D	AWP	WW	SW	AWP	WW

Table 4. Effect of tillage, nitrogen rate, and rotation on wheat yield, 1981-1985.

Crop and Tillage	Wheat Yields by Rotation and N Rate											
	Rotation A*			Rotation B*			Rotation C*			Rotation D*		
	0 N†	½ N	N	0 N†	½ N	N	0 N†	½ N	N	0 N†	½ N	N
	bushels/acre											
Spring wheat‡												
Tilled	32.6	38.7	38.9	31.1	41.3	46.2	28.2	35.5	33.7	37.0	41.1	45.7
No-till	28.5	37.8	35.7	23.6	32.2	38.4	24.6	30.2	30.2	27.8	34.6	39.4
Winter wheat§												
Tilled	28.8	26.0	32.7	55.2	56.7	50.8	34.5	40.3	44.8	51.6	53.9	52.5
No-till	24.8	27.3	29.3	42.7	51.2	51.6	28.2	30.1	33.9	50.7	57.6	58.5

*See table 3 for rotations.
†0 N = no N; ½ N = one-half the recommended rate, 35 and 45/pounds acre for spring wheat and winter wheat, respectively; N = full recommended rate, 70 and 90 pounds/acre N for spring wheat and winter wheat, respectively.
‡LSD for spring wheat = 9.0.
§LSD for winter wheat = 16.2

of 1983, followed by planting of spring peas in the spring of 1985. Spring peas were seed-harvested in 1985, followed by the no-till planting of winter wheat.

The three tillage options imposed on the red clover and winter pea residue in summer and fall 1985 included:

1. Moldboard plow/conventional plant. Legume residue plots were flail-chopped in September, moldboard-plowed, shallow-disked, and planted to winter wheat in October 1985.

2. Shallow tillage/conventional plant. Legume residue plots were flail-chopped in September, disked twice, and planted to winter wheat in October 1985.

3. Chemical kill/no-till plant. In mid-July 1985 we sprayed appropriate plots of red clover, Austrian winter peas, and hairy vetch with 32 ounces active ingredient/acre of Dinosoeb (2-secbutyl-4-6, -dinitrophenol, 55%). Only the red clover residue necessitated a second spraying 2 weeks later with 32 ounces active ingredient/acre of glyphosate to completely kill the plants. Plots were flail-chopped prior to no-till planting of winter wheat in October 1985. After seed harvest spring pea plots were no-till planted to winter wheat as well.

We seeded Dawes winter wheat on October 1 at 80 pounds/acre with the USDA III no-till drill. Pressure was taken off the drill when we seeded the conventionally tilled plots. We deep-band applied 0, 60, or 120 pounds/acre of N as ammonium nitrate sulfate (30-0-0-06) with the no-till drill within each existing plot. We applied glyphosate at 32 ounces active ingredient/acre to all plots prior to winter wheat planting. We applied buctril in mid-April at 12 ounces active ingredient/acre to winter wheat for broadleaf weed control. We determined soil gravimetric water and available N at 1-foot increments to a depth of 6 feet in each of the plots in mid-July 1985 before residue plowdown. We also measured legume residue biomass and N at this time.

Results. Growth of clover in spring wheat was slow during the 1984 growing season but improved after fall rains. Legume heights usually did not exceed 4 inches in late fall of 1984. In 1985 growth of all legumes increased dramatically during May and June and was sufficient to control all weeds during the growing season. Table 5 shows total aboveground dry matter and other relevant data for the four legumes.

Dry matter and aboveground plant N of Austrian winter peas were higher than for any other legume. The higher total plant N of Austrian winter peas than spring peas may have reflected the earlier establishment of Austrian winter peas. In Washington, Smith and associates (*29*) showed similar

values for these two legumes when they were both spring planted. Nitrogen fixation was highest for Austrian winter peas and lowest for red clover. However, N_2-fixation values for the forage legumes may be questioned because their root systems are known to be more extensive than annual-seeded legumes and would, therefore, accumulate more soil N and more biologically fixed N in their root systems.

Residual soil NH_4^+-N and NO_3^--N prior to tillage in 1985 was lowest for red clover and highest for spring peas. These differences probably reflected the differences in water usage between the forage and seed legumes rather than any contribution of the legume N to the soil-available N pool. Forage legumes that extracted the greatest quantities of soil water also depleted the soil N to a greater extent than did the seed legumes, which removed the least amount of available water. The correlation of available soil N (NH_4^+ + NO_3^-) with soil water (r = 0.873) suggests that legumes with their deeper root systems (presumably forage legumes) depleted the available soil N to a greater extent than did the seed legumes.

Residual soil moisture was the lowest for winter wheat, intermediate for the forage legumes, and highest for the seed legumes (Figure 3). Below 3 feet, peas extracted considerably less water than did all other crops. In cereal-pea rotations the residual soil moisture and available N remaining after peas may benefit a following wheat crop and may contribute to the overall rotational effect as compared to rotations of continuous cereals. This effect would be more evident in drier years when the soil profile is not completely recharged with water during the winter months. An estimate for recharge of soil profile water was not made in the spring of 1986. However, full recharge usually occurs in the Pullman area during the winter.

The highest winter wheat yield response, exceeding 5,000 pounds/acre, occurred at all N levels following clover residue that was either moldboard-plowed or shallow-tilled (Figure 4). Yield response to added fertilizer N was minimal. Wheat seed yield without supplemental N following clover residue either plowed down or shallow-tilled was equal to or greater than yield of no-till wheat following spring peas at all levels of supplemental N. These results suggest that, in addition to potential residual N from the clover residue, there were rotational benefits as well. Winter wheat yield following chemically killed red clover residue was substantially below (33%) that of clover residue plowed down. Although there was a substantial increase in wheat yield up to the application of 60 pounds/acre N, wheat yield decreased when 120

Table 5. Total dry matter production, total plant N, fixed N, soil profile inorganic N, and profile water prior to tillage, 1985.

Legume	Rotation	Dry Matter	Total Plant N	N Fixed*	Soil Profile Inorganic N to 3 Feet	Soil Profile Water (inches/3 feet)
		pounds/acre				
Red clover	SB-SW/RC-RC-WW	3,624	73	53	71	6.4
Hairy vetch	SB/WW/HV-WW	4,081	116	96	73	6.4
Austrian winter pea	SB/WW/AWP-WW	5,742	122	101	82	7.0
Spring pea	SB-WW-SP-WW	3,450	88	67	89	7.7

*Based on the modified ^{15}N different method (*29*).

pounds/acre N were applied to wheat following chemically killed residue. Wheat yield was slightly higher following red clover residue that was shallow-tilled as compared to plowed residue.

Wheat yield following Austrian winter peas, which was 20% lower than following red clover residue, was maximized at 60 pounds/acre supplemental N and agreed closely with yield following spring peas at all supplemental fertilizer rates. As was the case for red clover residue, wheat yield declined substantially when Austrian winter peas were chemically killed and was lower than yield obtained for chemically killed red clover residue. We found little response to supplemental fertilizer N following chemical kill. This suggests that factors other than N were responsible for the relatively poor wheat yield. The poorest wheat yield occurred following chemically killed vetch. Generally, these yields were 50% or less of the yields obtained following other optimum tillage options for clover residue. An infestation of cheat grass was a problem in all of the plots in early summer. However, these weeds were much less of a problem in wheat following red clover than in any of the other plots. We obtained excellent weed control in the fall-seeded legumes and red clover.

Discussion. We observed excellent wheat yields without supplemental N following red clover residue that was either moldboard-plowed or shallow-tilled. These yields were higher than the wheat yield following spring peas with supplemental N up to 120 pounds/acre N applied to the wheat. Wheat yield was 10% to 20% lower following residues of Austrian winter peas.

Winter wheat yield following chemically killed residue declined drastically following red clover and Austrian winter peas compared to when these residues were tilled in. Wheat yield following chemically killed hairy vetch residue was the poorest—50% or less of the yields following clover residue that was tilled in. Except for a positive wheat yield response to 60 pounds/acre N in red clover residue, wheat yield did not generally respond to supplemental N after residues were chemically killed. This would suggest that N was not limiting

and that other factors were responsible for the poor yields. Such factors as allelopathy, pathogens, and toxins may have contributed to the lower yields. Residual herbicide on the legume residues also could have been a factor. However, the relatively long period of time from the spraying to wheat emergence (3 months) should have reduced this effect considerably.

Summary

Legumes in a rotation, especially legumes used for green manure, do benefit conservation tillage systems. Legumes reduce the N demand from external sources, provide a large rotational effect, reduce pest problems, and improve soil structure and soil organic matter. Many grasses or grass-legume mixtures in the rotation will provide these benefits, except that grasses do not provide the N benefit.

Many legumes are better than others in rotations for several reasons. Some legumes do not produce the biomass or fixed N that others do. The rotational effect may not be as marked from some as others, and there may be alleopathy problems with certain legume residues. Some rotations may promote better microbial-rhizosphere relations than others. There are many questions about the adaptability of some legumes to conservation tillage, and some legumes are more suited to the prevailing climatic conditions than others. Much research remains to be done in these areas.

Research needs

Researchers need to (a) determine which legumes are suitable for each agroclimatic region; (b) establish which forage and seed legumes grow well in conservation tillage systems; (c) determine cultural conditions under which various legumes are most suited, for example, interseeding, cover crop, or green manure; (d) study the legume effect on beneficial and deleterious microbial relationships in the rhizosphere of the following crops; and (e) provide economic assessments of legumes in conservation tillage

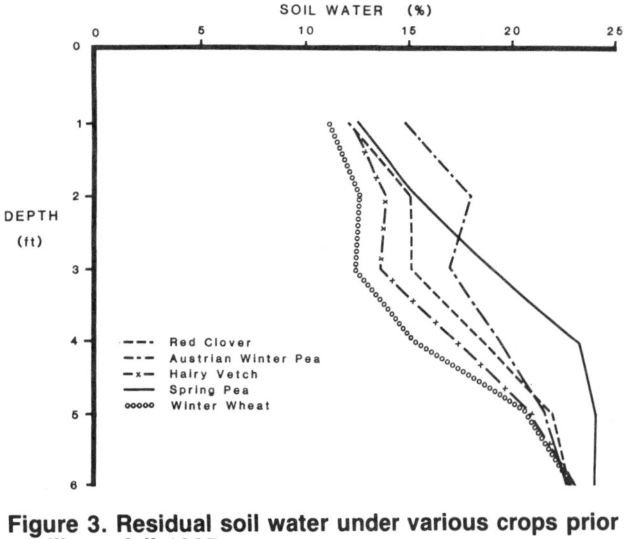

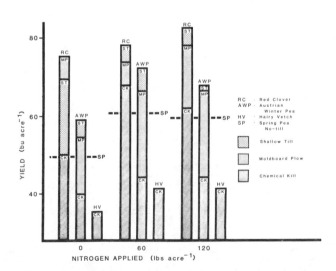

Figure 3. Residual soil water under various crops prior to tillage, fall 1985.

Figure 4. Winter wheat yield as affected by type of green manure residues, tillage, and nitrogen fertilizer, 1986.

REFERENCES

1. Bakhtri, M. N. 1978. *Wheat/forage legume rotation and integration of crop and sheep husbandry in the near East and North Africa.* In Glen H. Cannell [ed.] Proc., Int. Symp. Rainfed Agr. in Semi-Arid Reg. Univ. Calif., Riverside, and Oregon State Univ., Corvallis. pp. 520-538.

2. Bezdicek, D. F., D. W. Evans, B. Abebe, and R. E. Witters. 1978. *Evaluation of peat and granular inoculum for soybean yield and N fixation under irrigation.* Agron. J. 78: 865-868.

3. Bhattacharyya, P. B., K. Dey, S. Nath, and S. Banik. 1986. *Organic manures in relation to rhizosphere effect. III. Effect of organic manures on population of ammonifying bacteria and mineralization of nitrogen in rice and succeeding wheat rhizosphere soils.* Lentralbl. Mikrobiol. 141: 267-277.

4. Bolton, Jr., H., L. F. Elliott, R. I. Papendick, and D. F. Bezdicek. 1985. *Soil microbial biomass and selected soil enzyme activities: Effect of fertilization and cropping practices.* Soil Biol. Biochem. 17: 297-302.

5. Buntley, George J. 1986. *Tennessee no-tillage update.* In Proc., S. Reg. No-till Conf. Series Bull. No. 319. Ky. Agr. Exp. Sta., Lexington. pp. 100-102.

6. Carter, M. R. 1986. *Microbial biomass as an index for tillage-induced changes in soil biological properties.* Soil and Tillage Res. 7: 29-40.

7. Cherrington, C. A., and L. F. Elliott. 1987. *Incidence of inhibitory pseudomonads in the Pacific Northwest.* Plant and Soil (in press).

8. Doolette, John G. 1978. *The application of the Austrian farming system in North Africa.* In Glen H. Cannell [ed.] Proc., Int. Symp. Rainfed Agr. in Semi-Arid Reg. Univ. Calif., Riverside, and Oregon State Univ., Corvallis. pp. 389-408.

9. Doran, J. W. 1980. *Microbial changes associated with residue management with reduced tillage.* Soil Sci. Soc. Am. J. 44: 518-524.

10. Elliott, L. F., C. M. Gilmour, V. L. Cochran, C. Coley, and D. Bennett 1980. *Influence of tillage and residues on wheat root microflora and root colonization by nitrogen-fixing bacteria.* In J. L. Harley and R. S. Russell [eds.] The Soil Root Interface. Academic Press, London, Eng.

11. Elliott, L. F., and J. M. Lynch. 1984. *Pseudomonads as a factor in the growth of winter wheat* (Triticum aestivum L.). Soil Biol. Biochem. 16: 69-71.

12. Elliott, L. F., and J. M. Lynch. 1985. *Plant growth-inhibitory pseudomonads colonizing winter wheat* (Triticum aestivum L.) *roots.* Plant and Soil 84: 57-65.

13. Fredrickson, J. K., and L. F. Elliott. 1985. *Colonization of winter wheat roots by inhibitory rhizobacteria.* Soil Sci. Soc. Am. J. 49: 1,172-1,117.

14. Fredrickson, J. K., and L. F. Elliott. 1985. *Effects on winter wheat seedling growth by toxin-producing rhizobacteria.* Plant and Soil 83: 399-409.

15. Fredrickson, J. K., L. F. Elliott, and J. C. Engibous. 1987. *Crop residues as substrate for host-specific inhibitory psuedomonads.* Soil Biol. Biochem. (in press).

16. Frye, W. W., O. L. Bennett, and G. J. Buntley. 1985. *Restoration of crop productivity on eroded or degraded soils.* In R. F. Follett and B. A. Stewart [eds.] Soil Erosion and Crop Productivity. Am. Soc. Agron., Madison, Wisc. pp. 335-356.

17. Gilmour, C. M., O. N. Allen, and E. Truog. 1948. *Soil aggregation as influenced by the growth of mold species, kind of soil and organic matter.* Proc. Soil Sci. Soc. Am. 13: 292-296.

18. Hargrove, W. L. ed. 1982. *Proceedings of the minisymposium on legume cover crops for conservation tillage production systems.* Spec. Publ. 19. Univ. Ga., Athens.

19. Heath, M. E., R. F. Barnes, and D. S. Metcalfe eds. 1985. *Forages: The Science of Grassland Agriculture.* Iowa State Univ. Press, Ames.

20. Hoyt, Greg D., and William L. Hargrove. 1986. *Legume cover crops for improving crop and soil management in the southern United States.* Hort. Sci. 21: 397-402.

21. Lal, R. 1976. *No-tillage effects on soil properties under different crops in Nigeria.* Soil Sci. Soc. Am. J. 40: 762-768.

22. Lal, R., D. De Vleeschauwer, and R. Malafa Nganje. 1980. *Changes in properties of a newly cleared tropical Alfisol as affected by mulching.* Soil Sci. Soc. Am. J. 44: 827-833.

23. Lewis, R. D., and J. H. Hunter. 1940. *The nitrogen, organic carbon, and pH of some southeastern coastal plain soils as influenced by green-manure crops.* J. Am. Soc. Agron. 32: 586-601.

24. Loch, R. J., and K. J. Coughlan. 1984. *Effects of zero tillage and stubble retention on some properties of a cracking clay.* Aust. J. Soil Res. 22: 91-98.

25. Moldenhauer, W. C., G. W. Langdale, W. Frye, D. K. McCool, R. I. Papendick, D. E. Smika, and D. W. Fryear. 1983. *Conservation tillage for erosion control.* J. Soil and Water Cons. 38: 144-151.

26. Papendick, R. I. 1984. *Changing tillage and cropping systems: Impacts, recent developments, and emerging research needs.* In Glen Hass [ed.] The Optimum Tillage Challenge. Univ. Sask. Printing Serv., Saskatoon.

27. Papendick, R. I., and L. F. Elliott. 1984. *Soil physical factors that affect plant health.* In Thor Kommendahl and Paul H. Williams [eds.] Challenging Problems in Plant Health. Am. Phytopathol. Soc., St. Paul, Minn. pp. 168-180.

28. Papendick, R. I., and L. F. Elliott. 1984. *Tillage and cropping systems for erosion control and efficient nutrient utilization.* In D. F. Bezdicek and J. F. Power [ed.] Organic Farming: Current Technology and Its Role in a Sustainable Agriculture. Am. Soc. Agron., Madison, Wisc. pp. 69-81.

29. Smith, S. C., D. F. Bezdicek, H. H. Cheng, and R. F. Turco. 1987. *Seasonal N_2 fixation by cool-season pulses based on several ^{15}N methods.* Plant and Soil 97: 3-13.

30. Stewart, B. A., D. A. Woolhiser, W. H. Wischmeier, J. H. Caro, and M. H. Frere. 1976. *Control of water pollution from cropland, Vol. II.* U.S. Dept. Agr. and Environ. Protection Agency, Washington, D.C. 187 pp.

31. Stoltenberg, N. L., and J. L. White. 1953. *Selective losses of plant nutrients by erosion.* Soil Sci. Soc. Am. Proc. 17: 406-410.

32. Wilson, G. F., R. Lal, and B. N. Okigbo. 1982. *Effects of cover crops on soil structure and on yield of subsequent arable crops grown under strip tillage on an eroded Alfisol.* Soil Tillage Res. 2: 233-250.

Legume interseeding cropping systems research at the Rodale Research Center

R. R. Janke, R. Hofstetter, B. Volak, and J. K. Radke

Researchers at the Rodale Research Center have conducted studies on interseeded cropping systems since 1978. This research has focused primarily on four cropping systems: (1) broadcasting legumes and grasses in mid-summer into field corn at the final cultivation, (2) broadcasting legumes and grasses into soybeans in the fall at soybean leaf-yellowing, (3) broadcasting and drilling legumes in the spring into tillering small grains, and (4) drilling soybeans into tillering small grains.

Research objectives are to incorporate legumes and grasses into a cash-grain cropping system (a) to fix and/or conserve N and other nutrients that would otherwise be lost; (b) to hold the soil and reduce erosion over the winter or during other fallow, noncrop months, and (c) in some systems to provide an additional cash crop or efficient method of establishing a hay or grain.

There are two constraints in such a system: the interseeded species must not compete with the cash-grain crop and must be able to be established under somewhat adverse conditions, for example, low light levels or dry soil surfaces. To be a good N source for the following crop, the interseeded species must be able to accumulate signficant biomass either in the late summer, fall, or winter months or begin growth early in the spring for mid-spring plow-down. Much of the Rodale research has been directed toward screening potential grass and legume species for vigor and biomass production in an interseeding system and toward refining these cropping systems with respect to defining optimal planting dates and seeding rates.

Corn interseeding

A 1980 experiment compared five legumes interseeded into field corn following cultivation, either June 23 or July 7. The first seeding date was optimal in this experiment, partly due to lower-than-average rainfall in July and August of 1980. By the fall Arlington red clover and white clover achieved 80% ground cover. Medium red clover, crimson clover, and hairy vetch achieved 60% cover from the June 23 seeding. The ground cover from species interseeded on July 7 ranged from 10% to 30%, an unacceptable level.

A 1982 trial compared the performance of medium red clover, hairy vetch, Austrian winter peas, and annual ryegrass interseeded on June 22 or July 6 following cultivation; corn was planted on May 11 in 30-inch rows. Medium

R. R. Janke is agronomy coordinator and R. Hofstetter and B. Volak are agronomy researchers, Rodale Research Center, Kutztown, Pennsylvania 19530. J. K. Radke is a soil scientist, Agricultural Research Service, U.S. Department of Agriculture, Rodale Research Center, Kutztown, Pennsylvania.

red clover performed well in this trial, producing 52% and 60% ground cover for the first and second seeding dates, respectively. Hairy vetch seeded on June 22 established consistently well, but at the second date establishment was erratic. Biomass and ground cover were good by the fall, however—60% and 61% cover for the first and second seeding dates, respectively. Austrian winter peas seeded on June 22 germinated well, but did not withstand the light reduction (83% interception at silking) within the corn canopy in August. No peas survived into October. Germination was erratic for peas seeded on July 6, and no pea plants survived in this treatment either. Annual ryegrass performed well at the first seeding date, producing 41% ground cover by October. Germination at the second date was erratic due to dry soil, but there was 34% ground cover by mid-October. Follow-up studies performed in 1983 showed that corn leaf tissue N was higher for hairy vetch and red clover plots compared to ryegrass and control plots.

Soybean interseeding

In a 1982 trial soybeans were interseeded at full-leaf-yellowing (October 6) with either medium red clover (c.v. Arlington), hairy vetch, or annual rye grass. All species germinated well. But by early November plant height was only 1.5 inches for red clover and 3 inches for the vetch and ryegrass.

In 1983 a slightly earlier interseeding date was used (September 28) for interseeding 14 species of grasses, cereals, or legumes into two soybean varieties. Spring oats and spring barley showed excellent germination, biomass production, and ground cover going into the winter months. Winter rye grain also provided winter ground cover and, in addition, put on growth in the early spring. Annual ryegrass, perennial ryegrass, and Kentucky bluegrass germinated rapidly. But due to slow growth and their fine leaves, these species provided insufficient ground cover during the winter months. Hairy vetch germinated well, but a combination of slow growth and damage from combine wheel traffic reduced the cover provided by this species. Germination was poor for interseeded Austrian winter peas, although the plants that survived the winter showed excellent spring regrowth. Three varieties of alfalfa germinated well. But slow growth in the fall resulted in no protective ground cover over the winter.

A soybean interseeding trial was established in the fall of 1986 at two seeding dates—pre-leaf-yellowing (September 10) and full-leaf-yellowing (September 22)—into Asgrow A3127 soybeans. Species in this trial include crimson clover, medium red clover, hairy vetch, annual ryegrass, Aroostook rye grain, and winter wheat, all planted as monocultures and in grass-legume combinations. We think the grass-legume mixture will produce rapid growth and good ground cover in the fall and that the legume component will increase the N content of the biomass for plowdown.

Small grain interseeding

Legume interseeding into oats and wheat has been practiced in the farming systems experiment as a regular part of the rotations for the low-input treatments since 1981. Arlington red clover is seeded in mid-March to early April, either broadcast into standing winter wheat or planted simultaneously with spring oats. The wheat and oats are harvested for grain in mid to late July. The red clover is then either plowed down the following spring prior to corn planting in the low-input, cash-grain treatment or harvested as a hay crop the second year in the low-input, animal treatment.

In 1986 reseachers initiated a separate experiment to look at the potential weed control benefits of interseeding crimson clover and hairy vetch into winter wheat. A broadcast seeding on March 27 was compared to drilling the legumes into live wheat on May 5. Better stands resulted from drilling at the later date than broadcast seeding, and the drill did not reduce wheat yields. Hairy vetch broadcast in March grew tall and interferred with the wheat harvest. Crimson clover seeded at this time remained short, matured early, and did not produce much biomass. Both May-seeded legumes produced a large amount of biomass in 1986: 5,000 pounds/acre for crimson clover and 3,400 pounds/acre for hairy vetch by August 15.

Interseeding soybeans into small grains

Observations on this fourth interseeding cropping system were initiated in 1985 at Rodale by drilling soybeans into two fields of wheat on May 6. Yields were 62 and 52 bushels/acre for the wheat and 52 and 42 bushels/acre for the soybeans from these fields, respectively. Weed control was excellent. In 1986 experiments were initiated to compare a grain drill to a no-till planter for bean establishment and to compare bean growth and yield in winter barley, winter wheat, Aroostook rye, and spring oats. There were no signficant differences in bean yields with the type of planter used. Yields were highest for beans seeded into barley, followed by beans in wheat and rye. Soybeans seeded into spring oats did poorly, probably due to the fact that oats come off later than the other grains.

Soybeans also were drilled into spring barley in 1986 in the farming systems experiment. Soybeans yielded 48 bushels/acre, significantly higher than the 42 bushels/acre yield obtained from conventionally grown monoculture beans in this experiment. In addition, barley yield was 30 bushels/acre from the interseeded plots.

In summary, interseeding legumes and grasses into corn and soybeans can provide winter ground cover and high-N biomass for spring plowdown. Interseeding legumes into small grains in the spring is a way to establish either a hay crop or provide a source of plowdown N the following year. Interseeding soybeans into a small grain will yield two crops out of a single field under growing season conditions in Pennsylvania.

Intercropping corn and forage legumes in Michigan

M. A. Schultz, A. E. Erickson, and J. A. Bronson

The objectives of corn and forage legumes growing together are often antagonistic, although the concept has many attractions. A mat of vegetation should be retained at all times to obtain maximum erosion protection. This should be a living mat to be self-sustaining. The potential for continued N fixation exists if this living mat is a legume. Appreciable N transfer from the legume to corn will occur only if the legume is killed or severely suppressed. Legume growth and resource requirements also must be arrested to achieve economical corn yields. If pests are to be controlled, consideration should be given to removing the legume cover, at least during the initial stages of corn growth.

We conducted experiments at the Kellogg Biological Station in southwestern Michigan during 1984 and 1985 to determine the feasibility of intercropping corn with established forage legumes (1). The four legumes investigated at corn (*Zea mays*) planting in early May exhibited different growth stages. Vernal alfalfa (*Medicago sativa*) and mammoth red clover (*Trifolium pratense*) had produced active growth, Empire birdsfoot trefoil (*Lotus corniculatus*) had only begun to grow, and Penngift crownvetch (*Coronilla varia*) was still dormant. We investigated herbicides and mowing as means of temporarily suppressing this early spring growth. Initial broadcast herbicide treatments, applied at corn planting or when legume growth was deemed excessive, included glyphosate at 0.25-1.0 pounds/acre active ingredient; paraquat at 0.25-0.5 pounds/acre active ingredient plus 0.25% X77 surfactant; and 2,4-D amine at 0.12-0.25 pounds/acre active ingredient. The rates varied among legumes. Mowing was done prior to corn emergence. We also experimented with planting corn in 6- or 12-inch-wide, chemically killed strips and combinations of broadcast herbicides, mowing, and banding. We side-dressed ammonium nitrate, 100 pounds/acre N, as a treatment in 1985. No other fertilizer was applied in either year. We applied simazine, 2 pounds/acre active ingredient, plus alachlor, 1 pound/acre active ingredient, for annual grass control in 1985.

Efforts to produce corn in living legume swards resulted in large yield reductions for one or both components. With more than 50% of the ground cover retained, maximum corn silage yield was 74% of the yield in chemically killed sod. This occurred in clover plots treated with paraquat in 1984. Corn yields were much lower; the average yield in suppressed sods was only 28% of the yield in chemically killed sod.

We attributed the low corn yields to one or more of the following:

M. A. Schultz is a graduate assistant, A. E. Erickson is a professor, and J. A. Bronson is a field research technician, Crop and Soil Sciences Department, Michigan State University, East Lansing, 48824.

► Corn seedlings could not compete with the residual legume canopy when inappropriate herbicide rates or adverse weather conditions resulted in inadequate initial legume suppression.

► Pest damage associated with rodents and insects resident in the forage sod.

► Legume regrowth following adequate temporary suppression and the resulting competition for water and nutrients.

We attributed the low legume survival to two factors: the reapplication of systemic herbicides, made necessary by inadequate intial suppression, and unforeseen additional stress in the form of insect infestation and water shortage.

We did not achieve predictable legume suppression with glyphosate and 2,4-D. Paraquat and mowing gave reliable early suppression and were frequently associated with the higher corn and legume yields. Banding was associated with increases in both corn population and yield. The amenity of bands was reduced by failure to ensure adequate contact between spray and leaves and failure to control perennial weeds in the band. Weeds became a problem when ground cover was removed temporarily. Alachlor plus simazine, used for weed control in alfalfa, did not visually damage crownvetch and birdsfoot trefoil stands.

We found no evidence supporting the hypothesis that corn obtained N from an associated, actively growing legume. Nitrogen content of corn tissue suggested that actively regrowing legumes actually can deplete the soil N pool. Corn response to fertilizer N suggested that N was transferred from associated legumes only when the legume was severely repressed.

Results of acetylene reduction analysis of N fixation in the surface 3 inches of soil varied considerably and were not representative of the profiles' N-fixing potential. Under the alfalfa sod, a peak of N-fixing activity occurred between 6 and 10 inches deep.

Based on our results, the system that would give the best combination of corn yield and late-season legume regrowth includes the following steps:

1. Removal of perennial weeds prior to the year of intercropping is essential for success; any measures taken later will also eliminate the legume.

2. Farmers should mow the legume prior to planting corn if they anticipate the legume will be more than 6 inches tall on the proposed planting date. Paraquat may be used instead of mowing. This serves to eliminate bulky material and remove pest habitat. Mowing should be done about two weeks prior to planting to allow for removal of the clippings if desired, application of residual herbicides, and regrowth of the legume. Regrowth is desirable at banding to absorb foliar herbicides.

3. Residual, soil-active herbicides should be applied as soon as possible after mowing for annual weed control. Recommended rates for control of annual broadleaves and grasses in corn should not permanently damage the perennial legumes. Simazine plus alachlor is suitable.

4. At planting time, establish killed, legume bands (minimum of 12 inches wide) using directed spray nozzles attached to the planter. Appropriate herbicides should be selected on the basis of whether leaf material is present. Positioning spray nozzles in front of the planting shoes will en-

sure that foliar-absorbed herbicides are applied to leaves and not to a bare strip temporarily created by the shoe.

5. Apply a starter fertilizer to the corn, including N to enhance the corn seedlings' chances of outgrowing the legume.

6. Directed or broadcast sprays of 2,4-D amine can be applied after corn emergence if legume regrowth is excessive. The rate depends upon the legume species and the corn growth stage. A single application of 0.28 kg/ha should temporarily suppress the legume when applied under suitable weather conditions.

REFERENCE

1. Schulz, M. A. 1986. *Intercropping corn and forage legumes: Development of a cropping system.* M.S. thesis, Mich. State Univ., East Lansing.

Conservation tillage systems for green pea production in the Pacific Northwest

R. E. Ramig

Peas (*Pisum sativum* L.) are grown annually on about 114,000 acres in the Pacific Northwest (*3*). About 80,000 of those acres are cropped in a pea-winter wheat rotation in the Palouse and Nez Perce Prairies major land resource areas of northeastern Oregon and southeastern Washington. Soils in this region are mainly loessial, and slopes vary from almost level to 30% or more. The silt loam-textured soils are permeable, well-drained, and deep enough to adequately store precipitation. Soil organic matter varies from 1.5% to 4%.

Although highly productive, these soils also are highly erodible by winter rainfall when there is little soil cover and soils may be frozen. Climatic conditions in this area often limit pea production due to limited precipitation and temperatures over 95°F during pod filling and harvest (*2*). Power and associates (*1*) concluded that before farmers could make significant progress in using legumes in conservation tillage systems research was needed to catalog water requirements of legume cultivars in various climates, soils, and cropping systems.

Herein, I summarize a 13-year study (1974-1986) of the effects of four tillage systems in a pea-winter wheat cropping sequence on water conservation and use, yields, water use efficiency, and the changes in weed populations associated with the tillage systems.

Study methods

The study compared four tillage systems in a pea-wheat cropping sequence in an experimental design with four replications. The design allowed for growing each crop annually in each tillage system. The primary tillage method used for the pea-wheat crops were (1) fall rototill after wheat, sweep after peas; (2) fall plow-plow; (3) spring plow-plow; and (4) no-till-sweep. The fall plow-plow system is the conventional tillage system in this region. Moldboard plowing was 6 to 7 inches deep, rototilling was 5 to 6 inches deep, and sweeping was subsurface tillage with a blade at a depth of 3 inches. MCPB [4-(4-chloro-2-methylphenoxy)butanoic acid], applied post-emergence when the peas were 4 to 6 inches tall and weeds were small, controlled broadleaf weeds in the pea crops. Weeds were controlled in the wheat crops with diuron [3-(3,4-dichlorophenyl)-1, 1-dimethylurea] applied post-emergence prior to wheat tillering.

I measured soil water by neutron moderation to a depth of 8 feet at the start of wheat growth following winter dormancy in early March, at wheat harvest in July, at pea planting in early April, and at pea harvest in late June. The Walla

R. E. Ramig is a soil scientist with the Agricultural Research Service, U.S. Department of Agriculture, P.O. Box 370, Pendleton, Oregon 97801.

Walla silt loam soil is a Typic Haploxeroll with a slope of 1% to 2%.

Results

Water storage on land on which wheat stubble was left standing overwinter averaged 10% more than on fall-tilled stubbleland (Table 1). However, I observed considerable variation in water stored between years. There were no differences in water storage among tillage systems in warm,

Table 1. Long-term effects of four tillage systems for peas after wheat in a pea-wheat sequence on water conservation, green pea yield, water use, and water use efficiency, Pendleton, Oregon, 1974-1986.

Item (units)	Fall Rototill	Fall Plow	Spring Plow	Spring No-till	Standard Error
	Primary Tillage for Peas after Wheat				
Water stored (inches)*	8.34	8.11	9.07	9.02	0.91
Water stored (percent of precipitation)	56	54	61	61	4
Pea yields (pounds/acre)	3,187	3,189	2,995	2,844	393
Soil water use (inches)	5.29	5.36	5.82	5.40	0.51
Total water use (inches)†	8.49	8.56	9.02	8.60	0.67
Water use efficiency (pounds/acre/inch)	375	373	332	331	33

*Average precipitation during the 258-day storage period was 14.89 inches
†Sum of soil water plus 3.20 inches of rain during the pea-growing season.

Table 2. Long-term effects of four tillage systems for wheat after peas in a pea-wheat sequence on water conservation, wheat yield, water use, and water use efficiency, Pendleton, Oregon, 1974-1986.

Item (units)	Sweep	Plow	Plow	Sweep	Standard Error
	Primary Tillage for Wheat after Peas				
Water stored (inches)*	6.70	7.24	6.38	5.91	0.90
Water stored (percent of precipitation)	48	52	46	42	4
Wheat yield (bushels/acre/inch)	64.8	62.1	64.7	59.4	6.3
Soil water use (inches)	8.78	8.94	8.77	8.56	0.45
Total water use (inches)†	13.59	13.75	13.58	13.27	0.84
Water use efficiency (bushels/acre/inch)	4.8	4.5	4.8	4.5	0.3

*Average precipitation during the 264-day storage period was 13.91 inches
†Sum of soil water plus 4.81 inches of rain during the wheat-growing season.

Table 3. Changes in weed populations in the pea crop after 12 years of differential tillage for peas and wheat in a pea-wheat sequence, Pendleton, Oregon, 1986.

Weed Species	Fall Rototill	Fall Plow	Spring Plow	Spring No-till
	Primary Tillage for Peas after Wheat			
	weeds/square yard			
Downy bromegrass	2	3	3	18
Lambsquarters	3b*	7b	227a	5b
Red root pigweed	2	0	0	2

*Figures in a row followed by the same or no letter are not signficantly different at the 0.05 level of probability.

wetter-than-average winters, but land with standing stubble stored up to 15% more water than clean, fall-tilled land in cold, drier-than-average winters.

The shallow-rooted pea crop used about 60% of the stored soil water, leaving about 3 inches of available water in the lower 4 feet of the 8-foot soil profile (Table 1). Water use efficiency was slightly greater with fall tillage. Farmers in the area conventionally fall plow stubbleland because of labor distribution, earlier spring seeding, no crop residue problems during seeding, and reduced root disease problems.

Water storage on land conventionally plowed after pea harvest was slightly greater where the wheat stubble had been plowed the previous fall rather than in the spring (Table 2). Water storage was lowest on land where no-till-seeded pea land was swept. There were no significant differences in wheat yields among tillage systems. The deep-rooted wheat crops used all of the water stored over winter after pea harvest and wheat planting, plus an additional 2.20 inches that had not been used by the prior pea crop. Soil water use and water use efficiency by the wheat crop did not differ signficantly among tillage treatments.

Weed infestations in peas have shifted after 12 years of differential tillage (Table 3). Early, post-harvest tillage of wheat stubble (fall rototill or fall plow) buried green seed production and reduced lambsquarters populations to 3 to 7/square yard. Spring plowing allowed seed maturation, planted the seeds, and resulted in 227 lambsquarters/square yard. No-till allowed weed seed production but did not plant them, with a resulting population of 5 lambsquarters/square yard. Populations of other weed species, such as downy bromegrass, red root pigweed, coast fiddleneck, tansy mustard, and Russian thistle, did not change due to different tillage.

Conservation tillage systems for a pea-wheat rotation in the Pacific Northwest enhance water conservation, especially in dry years, which can result in a 20% increase in pea yields and a 5% increase in wheat yields. However, long-term effects are not consistent because of problems of crop residue handling, successful stand establishment, and root diseases associated with concentrated residues at the soil surface or within the root zone (1).

REFERENCES

1. Power, J. F., R. F. Follett, and G. E. Carlson. 1983. *Legumes in conservation tillage systems: A research perspective.* J. Soil and Water Cons. 38(3): 217-218.
2. Pumphrey, F. V., R. E. Ramig, and R. R. Allmaras. 1979. *Field response of peas (*Pisum sativum *L.) to precipitation and excess heat.* J. Am. Soc. Hort. Sci. 104(4): 548-550.
3. United States Department of Agriculture. 1985. *Agricultural Statistics 1985.* Washington, D.C.
4. Wilkins, D. E., R. R. Allmaras, J. M. Kraft, and R. E. Ramig. 1985. *Machinery systems for pea and winter wheat production in the Pacific Northwest.* In *International Conference on Soil Dynamics, Proceedings. Volume 3. Tillage machinery systems as related to cropping systems.* Auburn Univ., Auburn, Ala. pp. 582-591.

Influence of green-manured, hayed, or grain legumes on grain yield and quality of the following barley crop in the northern Great Plains

D. W. Meyer

Several researchers have reported the value of including legumes in crop rotations (1, 2, 3, 4, 5). But most studies have emphasized legume effects on corn (*Zea mays* L.) and wheat (*Triticum aestivum* L.) productivity. I sought to evaluate legume effects on spring barley (*Hordeum vulgare* L.) productivity and quality because (a) the majority of the six-rowed malting barley is produced in the northern Great Plains and (b) little information is available on cropping sequences that use legumes to replace summerfallow on set-aside acreage required by federal farm programs.

Several legumes were clear-seeded in early spring at Prosper, North Dakota, in 1984 and 1985 and at Fargo, North Dakota, in 1985. Legume stands generally were excellent. The field design was a randomized, complete block with three replicates. Forage legumes were either green-manured (all growth incorporated) or harvested for maximum hay production in the seeding year. Grain legumes were harvested for seed. All aboveground grain legume residues were removed in 1984 and incorporated in 1985. I also used two check treatments: wheat fertilized with 100 pounds/acre N and fallow. All treatments were fall-rototilled to incorporate remaining residues. Hazen barley was seeded at 70 pounds/acre during 1985 and 1986 across all cropping system treatments. Urea was broadcast at 0, 67, and 133 pounds/acre N on all cropping treatments in a split-plot arrangement. Grain was harvested with a Hege plot combine, cleaned, and expressed on 12% moisture basis. I determined grain N on all 1985 samples.

Green-manured legume treatments produced unfertilized barley grain yields generally equivalent to yields following fallow but significantly higher than yields following wheat in all environments (Table 1). Yields following hairy vetch (*Vicia villosa* Roth.) were the highest of the green-manured treatments, averaging 7% to 10% higher than yields following fallow across all environments. Barley yields on the forage and grain legume treatments generally were less than those on fallow and green-manured treatments, but generally were higher than yields on the wheat check. Barley yields following sweetclover [*Melilotus officinalis* (L.) Lam.] and hairy vetch harvested for hay averaged 92% to 107% of yields following fallow across all environments.

Fertilized (67 pounds/acre N) barley grain yields tended to be higher following legume treatments than following the wheat check (Table 1). Fertilized grain yields averaged from 12% to 15% higher after forages and grain legumes than after wheat. Grain yields after fertilized green-manured and fallow

D. W. Meyer is a professor, Agronomy Department, North Dakota State University, Fargo, 58105.

Table 1. Barley grain yield in three environments as affected by the previous cropping system and nitrogen level.

Location and Previous Crop	Number of Treatments	Grain Yield (bushels/ acre) by Nitrogen Rate (pounds/acre)		
		0	67	133
Prosper, 1985				
Wheat (check 1)	1	59.3	78.7	98.2
Fallow (check 2)	1	92.7	102.6	103.9
Green manure legumes	7	85.5	96.0	105.7
Forage legumes	4	72.3	88.4	90.6
Grain legumes	7	74.8	89.3	105.0
LSD (0.05)		16.8	21.8	NS
Prosper, 1986				
Wheat (check 1)	1	62.8	69.2	79.7
Fallow (check 2)	1	77.0	76.6	83.3
Green manure legumes	7	77.1	80.8	84.6
Forage legumes	6	67.4	78.1	80.8
Grain legumes	6	69.2	79.7	83.4
LSD (0.05)		12.4	NS	NS
Fargo, 1986				
Wheat (check 1)	1	41.9	70.4	83.6
Fallow (check 2)	1	74.3	84.7	80.2
Green manure legumes	7	77.4	83.0	80.0
Forage legumes	6	70.3	81.2	80.9
Grain legumes	6	70.1	79.7	77.8
LSD (0.05)		12.4	5.9	NS

treatments were nearly equal.

Barley grain yields of unfertilized legume treatments generally were equivalent to the wheat check fertilized with 67 pounds/acre N. Yields of unfertilized green-manured treatments generally approached yields following the wheat check fertilized with 133 pounds/acre N. These data indicate that including 1-year forage or grain legumes in the rotation had the equivalent effect on barley production of adding up to 70 pounds/acre N.

Grain N of unfertilized legume treatments was equivalent to or higher than N of the wheat check fertilized with 67 pounds/acre N. Grain N levels of unfertilized green-manured sweetclover and hairy vetch treatments were equivalent to or higher than grain N of the wheat or fallow checks fertilized with 133 pounds/acre N. Grain N of grain and forage legume treatments generally was equal to the wheat control; however, grain N level depended upon the individual grain or forage legume included.

These data suggest that including legumes in crop rotations increased the yield and protein concentration of the subsequent barley crop with reduced N fertilizer inputs. Green-manured treatments generally were equal to fallow in grain yield, indicating that producers should consider growing legumes on set-aside acreage to reduce erosion and possibly take advantage of haying provisions of the farm program.

REFERENCES

1. Badaruddin, M., and D. W. Meyer. 1986. *Influence of including several legumes in wheat cropping systems in eastern North Dakota.* In *1987 Crop Production Guide.* N. Dak. Agr. Assoc., Fargo. pp. 170-172.
2. Bailey, L. D. 1982. *Nitrogen fixation and legumes in crop rotations.* Agdex 537/121. Manitoba Agr., Winnipeg.
3. Hesterman, O. B., C. C. Shaeffer, D. K. Barnes, W. E. Lueschen, and J. H. Ford. 1986. *Alfalfa dry matter and nitrogen production, and fertilizer nitrogen response in legume-corn rotations.* Agron. J. 78: 19-23.
4. Meyer, D. W. 1987. *Sweetclover: An alternative to fallow for set-aside acreage in eastern North Dakota.* N. Dak. Farm Res. 44(5): in press.
5. Peterson, A. E., and D. A. Rohweder. 1983. *Value of cropping sequences in crop production for improving yields and controlling erosion.* In Proc., Am. Forage Grassland Conf., Am. Forage Grassland Cong., Lexington, Ky. pp. 102-105.

Legumes as a green manure in conservation tillage

Greg D. Hoyt

Farmers have used legume cover crops for many years as a green manure source for cultivated cropping systems. Farm managers have often placed legume cover crops in a crop rotation to produce a substantial quantity of biologically fixed N and to recycle other plant-essential nutrients—P, K, Ca, Mg—in the soil. The introduction of conservation tillage brought the ability to continue cover crop growth past the normal spring plowdown. This provides additional biomass and increases the quantity of N and other plant nutrients accrued into the cover crop.

Growth of legume cover crops depends upon geographic location and climatic conditions (1). Cropping sequence also plays a major role in biomass and nutrient accumulation. Farmers who wish to plant summer crops (corn) early in the spring are limited to a few legume species that will provide substantial biomass and nutrients. Growers producing grain sorghum, tobacco, or vegetables, such as tomatoes, squash, snapbeans, or sweetcorn, require warmer soil and air temperatures before planting or transplanting. This delayed planting enables them to choose from a wider selection of winter legumes that produce sufficient biomass and increase nutrient accruement by late spring.

Various legumes and grass cover crops have been planted in western North Carolina at elevations of 2,000-3,000 feet from 1982 to 1985. I measured biomass and nutrient accumulation for these cover crops before use in a tobacco, corn, or vegetable conservation tillage system. Plant foliage was collected from May 1 to May 20, depending upon the desirable time of planting for the subsequent crop. For example, potatoes, cabbage, broccoli, and corn crops normally were planted by May 10, with cover crops desiccated 10-15 days previous. Tobacco and tomatoes were set the last week in May, allowing the cover crops to grow until May 20 before desiccation. Biomass means (Table 1) reflect various harvest dates of the cover crops and provide a comparison of the legumes listed with a standard grass species (rye). Other legume and grass species were planted in these various tests, but those listed have been selected due to their productivity and use in this area of North Carolina.

Aboveground plant biomass measurements (Table 1) showed rye cover crops had the highest accumulation of organic matter (5,608 pounds/acre). Crimson clover and Austrian winter peas generally had excellent biomass for residue and averaged 4,243 and 4,114 pounds/acre, respectively. Hairy vetch had the lowest biomass measurements of these selected covers, but still provided ample residue for conservation tillage. These measurements represent normal growing cycles of each species, with rye producing sufficient

Greg D. Hoyt is an assistant professor, Department of Soil Science, North Carolina State University, Mountain Horticultural Crops Research and Extension Center, Fletcher, North Carolina 28732-9628.

ground cover in the fall, providing excellent soil coverage through the winter, and continued earlier growth in the spring. Crimson clover also provided some soil protection during the winter months. Peas and vetch generally provided less soil coverage.

Two important plant nutrients that are accrued and recycled efficiently by cover crops are N and K. Under many cropping systems high fertilizer inputs and low summer crop use results in N and K remaining in the soil and susceptible to fall leaching. Both legume and grass cover crops remove high quantities of N and K from the soil. Legumes accrue more K than grasses. Legumes also exceed grass cover crops in accumulation of N, with a large proportion of that N generally supplied by symbiotic N fixation. Hairy vetch and Austrian winter peas provided the highest quantity of N in the aboveground portion of the plant. Crimson clover tended to have lower quantities of N in the plant, but still higher than the grass species. Calcium seemed to be readily taken up by legumes, with well over twice as much Ca in legumes than rye. Less P and Mg accrued in the plant for both the legume and grass species, with little differences among species.

Predicting nutrients in cover crops

An ultimate goal of a grower using cover crops as a green manure is to predict the amount of nutrients in the cover, the percentage that will decompose that summer, and then reduce accordingly the soil test recommendation of fertilizer for the following summer crop. This prediction normally requires three measurements: the amount of biomass (dry weight), the elemental composition of the cover crop, and the decomposition rate of the cover crop during the summer for release of the nutrients. Biomass measurements are relatively easy and require little time or expense. Drying and weighing the cover crop requires only a microwave or normal oven and a small balance.

I measured plant moisture content of the four cover crops from one location (May 15 harvest). The moisture content

Table 1. Biomass yield and nutrient accruement by selected cover crops.

Cover Crop	Rep	Biomass*	N	K	Ca	P	Mg
			\multicolumn{5}{c}{pounds/acre}				
Hairy vetch	34	3,260	141	133	52	18	11
Crimson clover	33	4,243	115	143	62	16	11
Austrian winter peas	16	4,114	144	159	45	19	13
Rye	50	5,608	89	108	22	17	8

*Dry weight of aboveground plant material.

Table 2. Nutrient ratios of selected cover crops.

Cover Crop	N	K	P	Ca	Mg
	\multicolumn{5}{c}{Nutrient Ratio*}				
Hairy vetch	1	.96	.12	.39	.08
Crimson clover	1	1.26	.14	.53	.10
Austrian winter peas	1	1.24	.14	.53	.09
Rye	1	1.30	.20	.25	.10

*Ratio of nutrient to N

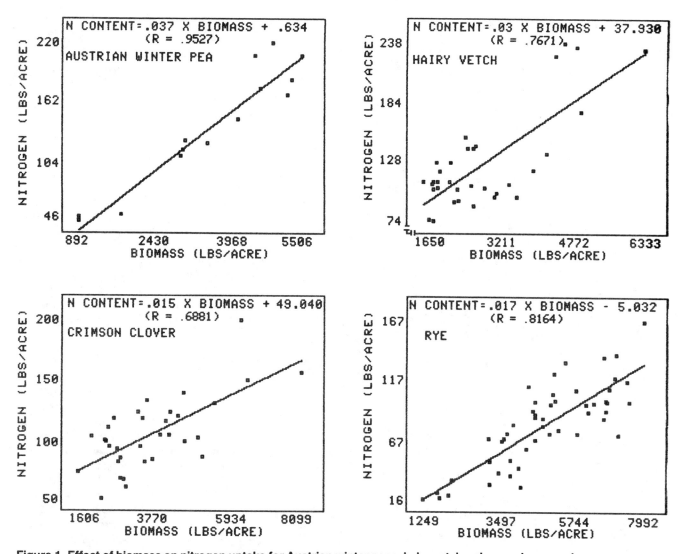

Figure 1. Effect of biomass on nitrogen uptake for Austrian winter peas, hairy vetch, crimson clover, and rye cover crops.

of rye was 53%; crimson clover, 73%; Austrian winter peas, 84%; and hairy vetch, 85%. Plant water content should be higher in the spring and decrease with the age of the cover crop. Thus, measurements taken in early or late spring should not be generalized for calculations. Plant elemental composition requires more time and cost for analysis. Determining decomposition rate involves great expense and lots of measurements, but has been measured by a few researchers.

Calculating plant nutrients available from a cover crop can be done as follows:

(a) Remove a square foot, yard, or meter of cover crop from the field, dry it and weigh it or (b) weigh the fresh weight, take a subsample of the cover crop, dry the subsample, and use the following equations:

$$\text{Percent plant moisture content} = \frac{(\text{wet weight + bag weight}) - (\text{dry weight + bag weight})}{(\text{wet weight - bag weight})} \times 100$$

Remove the moisture content from the cover crop by:

Dry plant weight = (wet weight - bag weight) × [1 - (moisture content ÷ 100)]

Calculate the biomass/acre (or hectare) by one of the following equations:

Square yard of cover crop
Dry weight in pounds × 4,840 = pounds dry biomass/acre
Square foot of cover crop
Dry weight in pounds × 43,560 = pounds dry biomass/acre
Square meter of cover crop
Dry weight in kg × 10 = kg dry biomass/ha

Once the weight of dry biomass/acre has been calcuated, N and other nutrients can be calculated by using this value in one of the four equations in figure 1 and the nutrient ratios in table 2.

For example, for hairy vetch (from figure 1):

N content = .03 × plant biomass (pounds/acre) + 37.9

Using a realistic value of 3,000 pounds of vetch cover/acre, the results would be as follows:

N content = .03 × 3,000 + 37.9 = 127.9 pounds/acre N.

If metric units are used (kg/ha), the line intercept will be different and the following equations should be used:

Hairy vetch : N content = .03 × plant biomass + 42.52

Austrian winter peas : N content = .037 × plant biomass + .71

Crimson clover: N content = .015 × plant biomass + 54.98

Rye : N content = .017 × plant biomass - 5.64

Continuing with table 2 then supplies the quantity of K, P, Ca, or Mg in a vetch cover crop:

127.9 pounds N × .96 = K content = 122.8 pounds/acre K
127.9 pounds N × .12 = P content = 15.3 pounds/acre P
127.9 pounds N × .39 = Ca content = 49.9 pounds/acre Ca
127.9 pounds N × .08 = Mg content = 10.2 pounds/acre Mg

REFERENCE

1. Hoyt, G. D., and W. L. Hargrove. 1986. *Legume cover crops for improving crop and soil management in the southern United States.* HortScience 21: 397-402.

Forage contributions of winter legume cover crops in no-till corn production

J. F. Holderbaum, A. M. Decker,
F. R. Mulford, J. J. Meisinger, and
L. R. Vough

We conducted three field experiments on a Matapeake silt loam in eastern Maryland between 1983 and 1986 to determine the effects of harvest management of winter cover crops on both spring forage and summer corn production. Harvest management schedules of cover crops included simulated grazing (clippings removed or left in situ), spring silage, and no harvest. A no-cover check was also included. We applied fertilizer N at rates of 0 and 80 pounds/acre. Crimson clover, crimson-ryegrass, arrowleaf clover, and subclover were tested. Total forage yields were greatest when both the cover crop and corn crop were harvested as silage. A signficant amount of N was still available to the subsequent corn crop after the cover was removed as silage.

Study methods

We seeded winter cover crops in mid-September into plots that were previously in corn production. We made harvest management cuttings of the cover crops to simulate grazing early the following spring. We made cuttings again for the grazing simulations and spring silage just before knockdown herbicide application. About 10 days prior to corn planting, we applied paraquat plus residual herbicides into the cover crop residues. We applied 80 pounds/acre N to half of each plot when corn was at the 5- to 7-leaf stage. We measured legume dry matter yields and total N content, corn silage and grain yields, corn N uptake, and total silage production.

Results

Although not presented here, results with arrowleaf clover in 1983 were similar to crimson clover. Sub clover did not lend itself well to pasture treatments due to its low growth stature and inadequate regrowth, and that data is not presented here. Data for the third year is being compiled and analyzed at this writing.

Application of 80 pounds/acre N increased corn grain yields regardless of harvest management treatment (Table 1). Corn grain yields were greatest when the cover crop was

J. F. Holderbaum is a graduate assistant, Agronomy Department, University of Florida, Gainesville, 32611; A. M. Decker is a professor, Agronomy Department, University of Maryland, College Park, 20742; F. R. Mulford is manager of the Poplar Hill Research Farm, Quantico, Maryland, 21856; J. J. Meisinger is a soil scientist, Agricultural Research Service, U.S. Department of Agriculture, Beltsville, Maryland, 20705; and L. R. Vough is an associate professor, Agronomy Department, University of Maryland, College Park, 20742. This paper is a contribution from .the Maryland Agricultural Experiment Station; work was supported by USDA/SAE/ARS Grant No. 58-32U4-2-424.

Table 1. Effects of harvest management of a crimson clover cover on subsequent corn grain, corn silage, and total silage yields.

Year and Harvest Management	Yields by N Fertilizer Rate (pounds/acre)					
	Corn Grain		Corn Silage		Total Silage	
	0	80	0	80	0	80
	bushels/acre		— tons dry matter/acre —			
1983-1984						
No harvest	112	151	7.1*		7.1*	
Spring silage	88	137	6.2		8.9	
Clippings removed	91	145	6.3		7.9	
No cover	64	123	5.9		5.9	
LSD (0.5)	11		0.8		0.9	
1984-1985						
No harvest	104	124	7.7	8.9	7.7	8.9
Spring silage	90	122	6.1	8.1	7.9	9.8
Clippings left	108	125	7.5	8.7	7.5	8.7
Clippings removed	82	113	5.9	7.6	7.4	9.0
No cover	27	84	3.0	6.5	3.0	6.5
LSD (.05)	15		0.5		0.5	

*Mean of N rate.

Table 2. Effects of harvest management of a crimson clover cover on corn N uptake.

Year and Harvest Management	Cover Crop N Content	Corn N Uptake by N Fertilizer Rate		
		0	80	Mean
		pounds/acre		
1983-1984				
No harvest	161	95	139	117
Spring silage	161	79	127	103
Clippings removed	120	75	132	104
No cover	—	73	116	95
LSD (.05)		9.8		15
1984-1985				
No harvest	109	121	188	154
Spring silage	109	88	161	124
Clippings left	115	117	194	155
Clippings removed	115	83	149	115
No cover	—	47	125	87
LSD (.05)		9.8		13

not removed—the no-harvest and clippings-left-in-situ treatments. Nitrogen contribution by a legume cover when left in place has been documented by other researchers (3, 4, 5). Grain yields also were signficantly greater when the cover crop was removed than when no cover crop was present.

Data on corn N uptake, based on the no-cover treatment data, revealed that available soil N was greater in the first year than in the second (Table 2). This influenced the degree of yield response obtained with corn following harvest management treatments. Clearly, however, even when the cover crop was harvested, the crop made significant N contribution to the subsequent corn crop. We attributed this to the N that was present in the belowground portion of the cover crop that was released upon decay of the legume roots and nodules. Earlier work by Bowen (1) and Butler and associates (2) support this theory.

Our most striking observation was the significant increase in total silage production when the cover crop was used as spring silage (Table 1). Previous research on the effects of removal of legume cover crops on subsequent nonlegume

crop yields has made little mention of the forage value of the removed cover crop herbage. Taylor and associates (6) did note the potential for vetch cultivars to serve as both cover crops and forage crops. In terms of dry matter yields alone, our results show that the value of the cover crop herbage removed is greater than the resulting reduction in corn silage yields. As well as increasing the total silage produced, legume cover herbage is also higher in quality, in terms of protein, than corn silage, as is evidenced by its relatively high N concentration.

We concluded that harvesting legume cover crops can be a beneficial management option in a forage production system when the cover crop itself is used as a silage. For corn grain production, on the other hand, legume covers are most valuable as a source of N and make the greatest contribution when not removed.

REFERENCES
1. Bowen, G. D. 1959. *Field studies on nodulation and growth of* Centrosema pubescens *Benth.* Queensland. J. Agr. Sci. 16(4): 253-265.
2. Butler, G. W., R. M. Greenwood, and K. Soper. 1959. *Effects of shading and defoliation on the turnover of root and nodule tissue of plants of* Trifolium repens, Trifolium pratense, *and* Lotus uliginosus. N.Z. J. Agr. Res. 2: 415-426.
3. Ebelhar, S. A., W. W. Frye, and R. L. Blevins. 1984. *Nitrogen from legume cover crops for no-tillage corn.* Agron. J. 7,651-7,655.
4. Hargrove, W. L. 1986. *Winter legumes as a nitrogen source for no-till grain sorghum.* Agron. J. 78: 70-74.
5. Mitchell, W. H., and M. R. Teel. 1977. *Winter annual cover crops for no-tillage corn production.* Agron. J. 69: 569-573.
6. Taylor, R. W., J. L. Griffin, and G. A. Meche. 1982. *Yield and quality characteristics of vetch species for forage and cover crops.* Prog. Rpt., Clover and Spec. Purpose Legumes Res. Dept. Agron., Univ. Wisc., Madison. V.15: 53-56.

Crownvetch in no-till silage corn

Francis L. Zaik

In southern New England silage corn is the predominant row crop grown for the dairy industry, with acreages of 58,000 in Connecticut, 37,000 in Massachusetts, and 4,000 in Rhode Island. The glaciated uplands soils of the Northeast are thin and of varying productivity. While erosion control is important for land users to protect their soil resource, average annual erosion rates exceed twice the tolerable limit on more than 30% of the corn land in Connecticut.

Because agricultural land has been and continues to be in strong demand for uses other than crop production, the majority of the dairy operators rent the additional crop acres needed for production of feed for expanded dairy herds. These rented acres often are highly erodible and in need of erosion control measures to support intensive crop production. Yet, the land users are reluctant to implement expensive, needed measures because most lands are rented for short duration or without extensive leases.

The scarcity of good agricultural land also prohibits removal of this land from crop production for installation of conservation measures or for growth of an erosion-controlling sod crop.

Cereal rye, fall-planted after both conventional and no-till corn harvest, is used for erosion control and to improve soil tilth. Late planting dates, poor fall weather, and early freeze-up often result in inadequate winter cover. Land users, university researchers, and Soil Conservation Service conservationists have explored uses of other plants and alternative seeding dates for better cover establishment.

Crownvetch (*Coronilla varia* L.), as a living mulch, shows promise to control erosion, add N for the corn crop, and provide high quality forage for milk production. Crownvetch is a perennial legume adapted to the temperate Northeast. Two cultivars, Penngift from Pennsylvania and Chemung from New York, are hardy, deep-rooted, and have been used successfully for years for stabilizing roadsides, steep banks, and as a protective cover on open areas where the objective is to prevent growth of woody vegetation.

In 1981 Connecticut SCS conservationists worked with landowners in each of the eastern and western uplands to establish separate 3-acre plantings of crownvetch as a living mulch in silage corn. The eastern planting was broadcast at 5 pounds/acre in May and planted to silage corn at the same time. Growth was average, but the corn harvesting equipment damaged the plants through compaction and uprooting. By the end of 1982 cover had improved prior to harvest, but again experienced damage during harvest. Cover remained at 10% as it did for the previous year. By the end of 1983 cover increased to 35%. The landowner decided to till under the crownvetch during the spring of 1984.

The western planting was drilled in June at 5 pounds/acre and planted to silage corn at the same time. Ground cover was 25% after corn harvest. Wet weather delayed corn planting in 1982 until early July. There was lush growth of crownvetch, which the landowner cut off in mid-June. This reduced the plant vigor, and there was only a 20% plant cover after fall corn harvest. Stand density was sparse during 1983 and fell to 10% after fall corn harvest that year. The landowner tilled under the crownvetch in the spring of 1984.

At the University of Connecticut Agronomy Research Farm, Robert A. Peters, professor of crop ecology, established three replications in 1984. He drilled crownvetch into a prepared seedbed at 5 pounds/acre at the end of July. The plots were planted to silage corn in early June 1985. Each plot was divided into six strips for five preemergent herbicide treatments. The sixth strip was left untreated as a control. In 1986 corn was drilled into the plots at the end of May. After treatment with glyphosate on the same date, the plots were divided into three strips and treated with herbicides 1 week later.

At the University of Rhode Island Agronomy Farm, W. Michael Sullivan, professor of agronomy, established a 1.1-acre plot of crownvetch in late April 1983 by planting with a brillion seeder into a prepared seedbed. In addition to fertilizer at planting time, the crownvetch received a topdressing in late August. At the end of the growing season ground cover was 80%. In 1984 five corn strips were delineated and received four different herbicide treatments; one strip was left as the control. At corn harvest yields ranged from 8 tons/acre for the control to a high of 26 tons/acre for one herbicide treatment. Crownvetch vigor and coverage was best on the control and worst on the corn plot that yielded 26 tons. In 1985 the crownvetch was harvested for hay. The first two cuts yielded 3 tons/acre each. In 1986 corn again was planted and yielded an average of 22 tons/acre for all varieties tested. However, the crownvetch density fell to 20% cover.

Several conclusions can be drawn from these trials:

▶ Good land use management is a primary requirement for successful stand establishment and maintenance. There is little room for error in timing weed control and knockdown herbicides.

▶ For the best growth and vigor, growers should plant crownvetch no later than August 1 in the Northeast into killed sod (alfalfa) with a no-till drill. Corn should be planted the following year.

▶ One plant/square yard appears to be adequate for erosion control after any required treatments. That will provide adequate recovery and vigor.

▶ Early grass and weed control is essential for establishment of a good stand.

▶ High P and K levels are important for good establishment.

▶ For maintaining stand integrity, farmers should not harvest the crownvetch after August 1, and they should make only one cut of hay each year.

▶ Crownvetch can work well as a living mulch for erosion control in silage corn, may add some N to the soil during the early years, and fits well into a rotation system of corn and hay.

Francis L. Zaik is state resource conservationist for the Soil Conservation Service, U.S. Department of Agriculture, Storrs, Connecticut 06268-1299.

Use of red clover in corn polyculture systems

Thomas W. Scott and Robert F. Burt

Farmers grow about 1.4 million acres of field corn (*Zea mays* L.) annually in New York State. This represents an 86% increase in corn acreage since 1954. Of these acres, 51% are harvested for silage, and little residue is returned to the soil surface. Much of this corn is produced on sloping land, greater than 3%. During the same period, New York hay production has dropped 27%. An additional 244,000 acres of cropland are being used for vegetable and other row-crop production, such as potatoes and dry beans. Although soil erosion losses from New York cropland average only 4.4 tons/acre/year (*4*), the thin solums of most of these soils cause concern about erosion.

The best means to control soil erosion by water is by maintaining a year-round vegetative cover or using crop rotations. Before the common use of herbicides, many researchers were interested in seeding forages in corn (*2, 5*). With the widespread acceptance and use of herbicides, continuous corn culture has become an accepted practice. The persistance of some of these herbicides in the soil restricts the use of cover crops and perennial forage crops in a cropping sequence. A legume intercrop in corn may provide an N benefit (*1*). Thus, we sought to evaluate the feasibility of establishing short-term, perennial forage hay crops through intercropping in corn with less persistent herbicides. We also sought to determine what ground cover and N benefits a legume crop might provide to subsequent corn crops.

Study methods

In 1981 we established Arlington medium red clover (*Trifolium pratense* L.) as an intercrop in corn and without a companion crop (clear seeding) in the spring at research farms at Aurora and Mt. Pleasant, New York. The two methods of establishment were repeated in 1982. The soil at the Aurora farm is a Lima loam (a fine-loamy, mixed, mesic Glossoboria Hapludalf). Elevation is 800 feet. The soil at Mt. Pleasant is a Mardin silt loam (coarse loamy, mixed, mesic, Typic Fragiochrept). Elevation is 1,600 feet.

We seeded intercrops when corn was 6 to 12 inches high. Corn row spacing was 30 inches. At Cornell a seeder was developed by combining a three-point-hitch, row-crop cultivator; seed box; and cultipacker sections (Figure 1). Seeding by this technique resulted in excellent stands provided there was adequate soil moisture at seeding. We controlled weeds by preplant incorporating Eradicane (EPTC + safener). In addition, an 8-inch band of atrazine and alachlor was applied over the corn rows. These herbicides,

Thomas W. Scott is a professor of soil science and Robert F. Burt is a research support specialist, Department of Agronomy, Cornell University, Ithaca, New York 14853. This paper is a contribution from the Department of Agronomy, Cornell University, Ithaca, 14853, as Agronomy Series Paper 1609.

plus cultivaiton as part of the interseeding process, provided acceptable weed control. We established clear seedings with recommended practices and a conventional grain drill.

We sidedressed N at four different rates to corn grown without intercrops to determine the N response.

Clear seedings were harvested during the establishment year. We used a three-cutting management system to harvest red clover forage in 1982 and 1983. In 1982 some of the intercropped and clear-seeded red clover treatments were plowed and planted with corn. We planted additional treatments from each establishment method with corn in 1983 and 1984. We measured ground cover using a beaded wire method (*3*). We also measured red clover hay, corn grain and silage yields, and ear leaf N.

Results

In both 1982 and 1983 red clover established as intercrops in corn yielded 87% as much forage as red clover established

Figure 1. Seeder developed for establishing intercrops during the 6- to 18-inch growth stage of corn.

Table 1. Red clover hay yields established as clear seedings and intercrops at Aurora and Mt. Pleasant, New York, 1981-1983.

| Establishment Year and Method | Red Clover Hay Yields (tons/acre)* | | | | | |
| | Aurora | | | Mt. Pleasant | | |
	1981	1982	1983	1981	1982	1983
1981						
Clear seeding, hay 1981	1.83			.80		
Clear seeding, hay, 1981-1982	2.11	4.88		.86	2.68	
Clear seeding, hay, 1981-1983	1.85	4.92	2.64	.58	3.70	1.00
Intercrop in corn, hay 1982		4.26			3.36	
Intercrop in corn, hay 1982-1983		4.21	2.29		3.40	1.17
LSD 5%	NS	NS	NS	.22	NS	NS
1982						
Clear seeding, hay 1982		2.11			1.53	

*Yield at 15% moisture; legume component for hay yields; grassy and broadleaf weeds not included.

Table 2. Corn grain yields as influenced by plowdown of legumes and sidedressed fertilizer nitrogen.

| Establishment Year and Method | Corn Grain Yields (bushels/acre)* | | | | | | | |
| | Aurora | | | | Mt. Pleasant | | | |
	1981	1982	1983	1984	1981	1982	1983	1984
1981								
Clear seeding, hay 1981		109	78	87		80	86	33
Clear seeding, hay, 1981-1982			95	97			106	38
Clear seeding, hay, 1981-1983				111				70
Intercrop in corn†	126	70	67	79	86	59	72	36
Intercrop in corn, hay 1982†	131		89	84	95		109	35
Intercrop in corn, hay 1982-1983†	137			104	102			67
1982								
Clear seeding hay 1982†	134		92	99	92		104	40
Intercropped in corn†‡	136	101	70	75	90	87	77	45
No red clover pounds/acre sideressed N§								
0	110	78	68	78	91	67	67	39
35	122	93	81	113	85	65	77	52
85	145	114	83	136	97	94	95	77
135	151	104	82	143	94	96	90	80
LSD 5%	19	17	18	22	NS	27	13	14

*Yield of shelled grain at 15% moisture.
†Sidedressed with 68 pounds/acre N in 1981.
‡Sidedressed with 85 pounds/acre in 1982.
§Sidedress rates for Mt. Pleasant in 1982 were 0, 23, 57, and 90 pounds/acre N instead of 0, 35, 85, and 135 pounds/acre N due to an error in calculation.

by clear seeding at the Aurora Farm. At the Mt. Pleasant farm red clover established as an intercrop yielded as much or slightly more forage than clear seeding (Table 1). During the year of intercrop establishment, corn yields were similar for all treatments.

Ground cover measurements in November of the establishment year indicated that both seeding methods were effective in establishing a cover that would minimize erosion. Red clover established by intercropping resulted in a total cover of greater than 50% while clear seeding provided 70% cover. Corn residue and weeds contributed about 15% of the total cover with intercrop treatments.

Red clover established as an intercrop of clear seeding was plowed under the year following establishment, then corn was planted. The corn experiments were fertilized with 15 pounds/acre N in a band at planting. Some treatments (Tables 2 and 3) were sidedressed with N. There was not a corn grain response on plots where red clover established as an intercrop had been plowed under (Table 2). This is consistent with data from several other experiments on the same farms. Where clear seedings had been plowed under the year following establishment, there was a corn grain yield response comparable to 35 and 85 pounds/acre of sidedressed N. We observed the greatest corn grain yields where corn followed a year of red clover harvested for forage. After a full hay year, there were no differences in corn yields, regardless of whether the red clover had been established by clear seeding or by intercropping.

The highest corn yields in 1983 occurred following red clover established in 1981 and harvested as hay in 1982 (Table 2). In 1984 corn grain yields were highest following 2 years of hay at both farms. At Aurora grain yields were about the same for the second year of corn following the plowdown of red clover hay. Yields at Mt. Pleasant were generally poor in 1984 because of late planting and unusually wet weather.

Table 3 shows percent N in the corn ear leaf. Nitrogen

levels were higher following 1 year of hay from either intercropped or clear-seeded red clover. These N levels were comparable to those from the corn that received the two highest rates of sidedressed N.

Summary

Establishing red clover or other forage legumes by intercropping in corn may prove to be a beneficial management

Table 3. Percent corn ear leaf nitrogen as influenced by legumes and sidedressed fertilizer nitrogen.

| Establishment Year and Method | Percent Ear Leaf N (%) | | | | | |
| | Aurora | | | Mt. Pleasant | | |
	1982	1983	1984	1982	1983	1984
1981						
Clear seeding, hay 1981	2.19	1.72	1.73	2.15	2.04	1.66
Clear seeding, hay, 1981-1982		2.40	1.77		2.43	1.86
Clear seeding, hay, 1981-1983			2.01			2.28
Intercropped in corn	1.76	1.57	1.71	1.79	1.78	1.85
Intercropped in corn, hay 1982		2.44	1.72		2.51	1.92
Intercropped in corn, hay 1982-1983			1.96			2.11
1982						
Clear seeding, hay 1982		2.46	1.91		2.50	1.86
Intercropped in corn*	2.21	1.73	1.60	2.62	1.87	2.00
No red clover pounds/acre sidedressed N						
0	1.67	1.47	1.61	1.82	1.76	1.77
35	2.10	1.92	2.22	2.02	1.82	2.41
85	2.51	2.19	2.49	2.66	2.15	2.72
135	2.24	2.23	2.57	2.77	2.32	2.71
LSD 5%	.25	.20	.22	.49	.33	.29

*Sidedressed with 85 pounds/acre N in 1982.

practice in terms of additions of N and organic matter and the reduction of soil erosion. Good stands (greater than 30% soil cover) of red clover can be obtained consistently by cultivating corn grown in 30-inch rows and broadcasting seed with or without soil firming. Nonpersistent herbicides plus cultivation provided acceptable weed control and caused no or minimal injury to the intercrops. Highest corn yields occurred following the plowdown of 1 year of red clover hay. Nitrogen benefits from the red clover were evident in the second year of corn following the plowdown of the hay.

REFERENCES

1. Olson, R. A., W. R. Raun, Yang Show Chun, and J. Skopp. 1986. *Nitrogen management and interseeding effects on irrigated corn and sorghum and on soil strength.* Agron. J. 78: 856-862.
2. Schaller, F. W., and W. E. Larson. 1955. *Effects of wide spaced corn rows on corn yields and forage establishment.* Agron. J. 47: 271-276.
3. Sloneker, L. C., and W. C. Moldenhauer. 1977. *Measuring the amounts of crop residue remaining after tillage.* J. Soil and Water Cons. 32: 231-236.
4. Soil Conservation Service, U.S. Department of Agriculture. 1982. *Basic statistics for New York State in National Resources Inventory.* Syracuse, N.Y.
5. Tesar, M. B. 1957. *Establishment of alfalfa in wide-row corn.* Agron. J. 49: 63-68.

Crimson clover and corn: A conservation tillage system that works in Georgia's Coastal Plain

M. B. Leidner

Farmers in the Coastal Plain area of Georgia have not widely accepted conservation tillage cropping systems. Although conservation tillage is recognized as a soil-conserving practice, farmers often cite problems with weed and disease control and the cost of specialized equipment as reasons for not adopting conservation tillage. A frost-free period 230 to 260 days increases the probability of weed, insect, and disease problems in row crops. Complex cropping rotations with peanuts, corn, cotton, soybeans, and winter grains are difficult to manage under conservation tillage.

One promising conservation tillage system is a winter legume cover crop followed by corn. Farmers used winter legumes widely in the Coastal Plain as green manure crops until the late 1940s when N fertilizers became available. Interest in winter legumes is increasing as the costs of those fertilizers rise. Winter legumes are challenging to grow in the Coastal Plain. Fall plantings are subject to long periods of hot, dry weather. Soil conditions are usually dry. The sandy nature of the soils and naturally low levels of organic matter inhibit the growth of the legumes and *Rhizobium* bacteria. Corn is planted in early March in the southern half of the Coastal Plain. Thus, a winter legume must have good fall and winter growth to provide adequate cover and N benefits.

I initiated a field demonstration in October 1981 to look at the feasibility of legume-corn conservation tillage systems. Previous conservation tillage corn systems in the Coastal Plain depended upon rye as the cover crop. The demonstration site is a 2.4-ha field in Cook County, in southcentral Georgia about 80 km north of the Georgia-Flordia state line. Soil type is a Tifton loamy sand with a 3% slope. Erosion is a concern in the field. The soil and field conditions are representative of better cropland in the Coastal Plain. There is a 250-day frost-free period. Average rainfall is 122 cm. Rainfall amounts are lowest from September 15 through December 15 in most years. Frequent, short-duration dry periods are common during the corn growing season. Erosion is most probable from March through August, when high-intensity thunderstorms are common.

The demonstration addressed three concerns of local farmers: (a) Are legume varieties available that are reliable for fall plantings and provide good fall and winter growth (November-February)? (b) How much N will be available for the corn crop? (c) What is the cost of the legume-corn system compared to the traditional rye-corn system?

For the first year rye, Tibbee crimson clover and Vantage vetch were planted as cover crops. Cover crops were

M. B. Leidner is a conservation agronomist, Soil Conservation Service, U.S. Department of Agriculture, Tifton, Georgia 31793.

chemically killed in early March as corn was planted. The rye-corn plot received 168 kg/ha N. The legume-corn areas received either 34 kg/ha or 101 kg/ha N.

Table 1 shows the costs of each crop system for the first year. The first year results showed both legumes would provide at least 67 kg/ha N for corn. Yields on all plots were excellent in 1982.

The demonstration continued on the same field through the 1986 crop year. Yearly results have been similar to those shown in table 1, with little change in the cost of fertilizer or legume cover crops. Corn yields with the legume cover crops have averaged 6,649 kg/ha for the 1982-1985 crop years. The 1986 corn yield was 3,660 kg/ha because of an extended drought throughout the state.

Throughout the 5-year period, I observed cover crops for ease of establishment, fall and winter growth, winter hardiness, disease problems, and ease of kill with herbicides when corn was planted (Table 2).

Both crimson clover-corn and the rye-corn cropping systems are reliable and suited for Coastal Plain farmers. The crimson clover-corn system is slightly more cost-effective than the rye-corn combination. Should the cost of N fertilizer rise in the future, the crimson clover-corn combination may have a real cost advantage.

The vetch-corn combination was cost-effective, but the vetch was disappointing as a cover crop. It is much slower to establish a ground cover and is prone to fungal diseases in late winter. The vetch varieties are not as cold-hardy as the clover. Winter kill was a problem in 2 of the 5 years. Seed availability was a problem with the vetch.

After 5 years, the demonstration showed:

► Crimson clover is a good cover crop for the Coastal Plain. Although the Tibbee variety was the only one used in the demonstration, other early maturing varieties may work as well.

► Both vetch and crimson clover provided 67-101/kg/ha N for the corn crop. The amount of N available varied with weather conditions.

► The cost information slightly favors the legume-corn combination at the present time. Winter legumes are more expensive to establish than rye. The value of N provided by the legumes offset the establishment cost.

Table 1. Cost comparison for cover crops and nitrogen requirements, 1982.

| | Costs by Cover Crop and N Rate (kg/ha) | | | | |
| | Rye | Crimson Clover | | Vetch | |
	168	101	34	101	34
Cover crop costs ($/ha)					
Seed	18.50	46.43	46.43	33.84	33.84
Inoculant	—	9.83	9.83	9.83	9.83
Corn fertilizer costs ($/ha) 600 pounds					
5-10-15	103.74	103.74	103.74	103.74	103.74
N solution	81.07	40.63	—	40.63	—
Total costs	203.31	200.63	160.00	188.04	147.41
Corn yield (kg/ha)	9,150.0	8,448.5	8,241.1	8,887.1	8,680.3

Table 2. Relative rating of cover crops

| | Rating by Cover Crop* | | |
Characteristic	Crimson Clover	Vetch	Rye
Easy to establish	1	3	1
Fall-winter growth	2	3	1
Winter hardiness	2	3	1
Ease of chemical kill	2	1	2
Disease prone	1	2	1
Availability of seed	2	3	1

*Rating: 1 = most desirable, 3 = least desirable

Effect of soil surface color on soybean seedling growth and nodulation

P. G. Hunt, M. J. Kasperbauer, and
T. A. Matheny

Spectral composition of light, especially in the blue, red, and far-red regions of the spectrum, acts through photoreceptors within the growing plant and functions in natural bioregulation of physiological processes to control growth and development (2, 3, 4, 9, 10). For example, the far-red relative to red ratio, acting through the phytochrome system in the developing plant, can regulate chloroplast ultrastructure (5); photosynthetic efficiency (3); and partitioning of photosynthate among leaves, stems, and roots (7). Also, blue light regulates some physiological responses, such as stem length suppression (10). Subtle differences in quantity and quality of light reflected from different colored soils can alter seedling environment sufficiently to affect establishment and early growth of soybeans. This might contribute to differences in seedling establishment between bare- and plant residue-covered seed beds, as has been noted among various geographic areas (1, 8) with different soil colors.

Table 1 shows the variation in the spectral composition of light reflected from soil surfaces (6). These measurements were made with a LiCor spectroradiometer at a height of 10 cm above the soil surface. The photosynthetic photon flux density values of reflected light decreased as follows: white > red > black. Crop residue cover modified the quantity of reflected light. For example, residues over dark soils more than doubled the photosynthetic photon flux density, whereas residues over gray-white soils decreased the quantity of reflected light. Far-red/red light ratio differences over the various colored soils and residues were small. However, the reflected blue light was much greater over the light colored soils than over the dark colored ones.

We investigated the effects of surface color on soybean seedling growth in a greenhouse study. White, red, and black soils were placed to a depth of 2 cm on a 1-m² plywood box. The plants grew through tubes that were inserted through the soil box into growth containers attached to the bottom of the boxes. We altered the air flow beneath the black soil boxes enough to make root zone temperature of the black-soil-surface treatment similar to those of the other treatments even at peak temperatures.

Leaf area, tap root weight, and nodule number were not signficantly affected by soil surface color. Total shoot weight and leaf weight were greater for plants grown over white soil compared with the other surfaces (Table 2). Stem weight and leaf weight/area were greater for plants grown on the straw-residue-covered soil and white soil than on the red or

P. G. Hunt is a soil scientist, M. J. Kasperbauer is a research plant physiologist, and T. A. Matheny is a soil scientist, Coastal Plains Soil and Water Conservation Research Center, Agricultural Research Service, U.S. Department of Agriculture, Florence, South Carolina 29502-3039.

black soils. However, residue-covered soil surfaces affected stem length differently than white soil. Plants in the residue-covered soil were the tallest, and plants on the white soil were the shortest. Plants grown over white soils had much greater weights of lateral roots than plants grown over the black, red, or residue-covered soils. The weight of lateral nodules on plants grown over white soil also was significantly greater than those of plants grown over the other surfaces. Plants grown over the white soil surface had a lower shoot:root ratio than plants grown over the residue-covered soil.

Plant growth was significantly altered by the soil surface color even though root temperatures as well as moisture and nutrition were not significantly different for the different surface color treatments. We could not consistently relate the grouping of plant growth responses to the total reflected light, blue light enrichment, or far-red/red ratio. These observations suggest that the measured growth parameters were sensitive to different aspects of reflected light. The quantity and spectral distribution of reflected light that would be best for plant growth and yield in a particular environment would vary

Table 1. Light reflection upward to a point 10 cm above the soil.

Light Parameter and Soil Color	Soil Surface		
	Dry	Wet	Under Residue Corn
Photosynthetic photon flux density (μmol/m²/s)*			
Black	82	65	161
Brick-red	164	114	189
Gray-white	355	261	217
Far-red/red ratio relative to ratio in direct sunlight			
Black	1.15	1.10	1.19
Brick-red	1.18	1.23	1.21
Gray-white	1.09	1.14	1.19
Blue (400 to 500 nm) as percent of blue in direct sunlight			
Black	5.2	4.4	8.0
Brick-red	6.6	5.6	8.9
Gray-white	20.8	15.0	11.4

*Flux density of direct sunlight was 1,670 μmol/m²/s.

Table 2. Shoot and root growth and nodule characteristics of soybean seedlings grown over different colored soils in a greenhouse.

Plant Parameter	Soil Surface Color			
			Black	
	White	Red	Bare	Residue Covered
Shoot				
Leaf weight (mg)	289a	234b	224b	238b
Leaf denisty (mg/cm²)	2.1a	2.0b	2.0b	2.1a
Stem length (cm)	12.6c	13.9bc	14.1b	15.9a
Stem weight (mg)	137a	120b	120b	136a
Root				
Tap length (cm)	16b	19a	18a	18a
Lateral weight (mg)	220a	151b	161b	156b
Tap:lateral weight ratio	.18b	.27a	.26a	.25a
Total weight (mg)†	293a	213b	222b	217b
Lateral nodule weight (mg)	29a	17b	17b	19ab
Shoot:root weight ratio	1.49b	1.69ab	1.57ab	1.75a

*Means on the same line followed by the same letter are not significantly different by the LSD test at the P level of 0.05.
†Total root weight includes nodules.

with management. Thus, a better understanding of the effects of soil color on seedling establishment is needed to capitalize on potential benefits for conservation tillage, crop production, and resource management.

REFERENCES

1. Campbell, R. B., D. L. Karlen, and R. E. Sojka. 1984. *Conservation tillage for maize production in the U.S. Southeastern Coastal Plain.* Soil Tillage Res. 4: 511-529.
2. Downs, R. J., S. B. Hendricks, and H. A. Borthwick. 1957. *Photoreversible control of elongation of pinto beans and other plants.* Bot. Gaz. 188: 119-208.
3. Kasperbauer, M. J., and D. E. Peaslee. 1973. *Morphology and photosynthetic efficiency of tobacco leaves that received end-of-day red or far-red light during development.* Plant Physiol. 52: 440-442.
4. Kasperbauer, M. J., H. A. Borthwick, and S. B. Hendricks. 1964. *Reversion of phytochrome 730 (Pfr) to P660 (Pr) in* Chenopodium rubrum *L.* Bot. Gaz. 125: 75-80.
5. Kasperbauer, M. J., and J. L. Hamilton. 1984. *Chloroplast structure and starch grain accumulation in leaves that received different red and far-red levels during development.* Plant Physiol. 74: 967-970.
6. Kasperbauer, M. J., and P. G. Hunt. 1987. *Soil color and surface residue effects on seedling light environment.* Plant and Soil 97: 295-298.
7. Kasperbauer, M. J., P. G. Hunt, and R. E. Sojka. 1984. *Photosynthate partitioning and nodule formation in soybean plants that received red or far-red light at the end of the photosynthetic period.* Physiol. Plant 61: 549-554.
8. Phillips, R. E., R. L. Blevins, G. W. Thomas, W. W. Frye, and S. H. Phillips. 1980. *No-tillage agriculture.* Science 208: 1,108-1,113.
9. Tanada, T. 1984. *Interactions of green or red light with blue light on the dark closure of* Albizzia *pinnules.* Physiol. Plant. 61: 35-37.
10. Thomas, B. 1981. *Specific effects of blue light on plant growth and development.* In H. Smith [ed.] *Plant and the Daylight Spectrum.* Academic Press, New York, N.Y. pp. 455-456.

White lupines as a rotation alternative with winter wheat in conservation tillage systems

Bohn D. Dunbar and David C. Nielsen

Lupines are an extremely diverse group of species, several of which have been adapted for agronomic use. Farmers have grown *Lupinus alba* in the Great Lakes region of the United States with some success. *Lupinus angustifolia* is grown extensively in Australia, with more than 1 million acres planted in some years. Lupine seed is high in protein with a content equivalent to soybeans at 30% to 40%. White lupine can be ground and fed directly to livestock without treatment because the lines of lupines available have been bred to keep toxic alkaloids out of the seeds.

A legume crop that could be added to the minimum tillage rotation with wheat and corn would be extremely desirable. Soybeans are marginal in this rotation except under irrigation. Lupines may be adapted to fit this role.

Scientists at the Central Great Plains Research Station have studied lupines since 1984. The results reported herein are some of the general findings of this 3-year study.

In April 1984 we planted 200 pounds of *Lupinus alba* (Ultra) on 1 acre under a solid-set-gradient irrigation system with a disk drill to observe its potential as a crop for the area. Treflan and Lorox were used singly and in combination on the area. The spring was wet and cold with normal precipitation during the rest of the season.

Yield on the nonirrigatied area was about 30 bushels/acre, with an increase up to 50 bushels/acre with full irrigation. Treflan stunted lupine root growth and noticeably reduced yield in the nonirrigated areas. Lorox did not appear to effect yield, but several species of weeds were not controlled.

With encouraging results in this preliminary trial, during the next winter we conducted herbicide screening trials in the greenhouse to determine the type and amount of herbicide that could be used on lupines. All of the triazines, in quantities that would control weeds, damaged or totally killed the lupine. We observed the same root stunting effect with Treflan. Lorox and Lasso did not appear to effect the lupines up to a 4-pound-per-acre rate. An experiment was designed with three varieties of lupines (Kiev, Ultra, and Multilupa), three planting dates determined by the first day that the ground temperature was 40°F and 7 and 14 days thereafter, and Lorox and Lasso alone or in combination at 1 quart/acre and 0.5 quart/acre, respectively. The spring of 1985 was extremely wet and the lupines germinated well, but the herbicide rates were not enough to control weeds. The entire experiment was abandoned after attempts to use postemergent herbicides failed. To date, nothing we have tried will kill weeds and not the lupines.

Bohn D. Dunbar and David C. Nielsen are research agronomists, Central Great Plains Research Station, Agricultural Research Service, U.S. Department of Agriculture, Akron, Colorado 80720.

In 1986 we conducted a similar but scaled-down experiment on two sites. The varieties 471, Kiev mutant, and 410 were added and Muliplupa was deleted. We planted three small plots to *Lupinus angustifolia* because only a few ounces of seed were available for the varieties available, Yandee and Illyarri. This was a dry spring and the lupines emerged poorly, resulting in a poor stand. The site with the best stand was completely hailed out in June. The yield on the one remaining site was reduced by a late-season infestation of sandbur. The early herbicide treatment, even with 2 pounds Lorox and 2 pounds of Lasso, did not control this grassy weed. The highest yield was 14 bushels/acre. The angustifolia did not set seeds and is apparently not adapted to grow in the 7.5 to 8 pH range soil in this area. It also is a winter annual and the vernal requirement may not have been met even with the early planting. This is not likely because this plant was grown in the greenhouse in 6 pH soil with no cold period and pod set was achieved.

David Nielsen has carried out other experiments in the same years at the Akron Station. He planted lupines in the spring of 1985 (Ultra) and 1986 (Gutwein 471). Plant populations were 219,700 plants/ha in 1985 and 387,350 plants/ha in 1986. Row spacing was 17.8 cm in 1985 and 38.1 cm in 1986. The plot area was not fertilized. The seed was inoculated before planting. Evapotranspiration was calculated by the water balance method from weekly measurements of soil water made with a neutron probe. Irrigations were applied via a line-source, gradient irrigation system. Irrigations in 1985 ranged from 0.7 to 9.6 cm. Irrigations in 1986 ranged from 0.0 to 12.6 cm. Precipitation was 23.4 cm in 1985 and 15.3 cm in 1986.

The lupines in both years responded well to supplemental, limited irrigation (Figure 1). The significantly lower slope of the yield-evapotranspiration relationship in 1985 probably was due to the lower plant population attributable to poor germination and emergence. The lower amount of vegetative cover resulting from the lower plant population probably caused lower interception of solar radiation, lower photosynthesis/unit ground area, higher solar radiation levels at the soil surface, and higher evaporative losses from the wet soil surface after an irrigation or precipitation event. This resulted in decreased water use efficiency (lower slope of the regression line) in 1985 than in 1986. Another explanation for the differences in slope may be due to variety differences in water use efficiency.

Although the lupines responded well to irrigation, yield levels were not high. Water application throughout the growing season may have been insufficient; pod-bearing branches were virtually nonexistent. The strong response of yield to irrigation indicates that these varieties of lupine are neither drought tolerant nor drought avoiding and, hence, may not be suitable as dryland crops for this region.

Analysis of water extraction patterns showed that root activity was limited primarily to the upper 90 cm of the soil profile. Wilting point appeared to be fairly high at about 12%. Hence, lupines may not be a good crop to follow another irrigated crop in a rotation to take advantage of stored soil water at lower soil depths that may result from excessive irrigations. This water would be unavailable to the lupine crop due to shallow rooting and high wilt point.

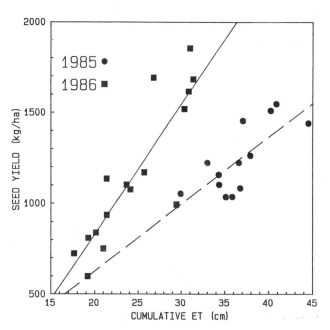

Figure 1. Lupine seed yield-cumulative evapotranspiration relationship.

Visual examination of roots for nodulation showed prolific, healthy nodules in both years.

The varieties of lupines grown in these studies were not well adapted to this, area and available herbicides are not adequate or economically feasible. Before abandoning this research, several other species of lupines will be evaluated and new herbicides screened.

Benefits from pulses in the cropping systems of northern Canada

A. T. Wright and E. Coxworth

Producers in northern Canada frequently report high yields from cereals grown on pulse residues. Researchers have assumed this yield enhancement is due to additional N from pulse residues. However, the additional N in these residues is usually less than 25 pounds/acre. Others have suggested that the yield benefit is due to pulse residues decomposing more rapidly, thereby releasing the N sooner than cereal residues. Therefore, in the second year yield on cereal residues should be greater. Accordingly, we undertook a study to investigate the yield and N response to barley and wheat 1 and 2 years after pulse crops, respectively. We analyzed the cropping systems data for energy use and economic return as well as agronomic performance.

Each cycle of this study involved 3 years: faba beans, field peas, lentils or barley in year one; barley fertilized with various levels of N and minimum or no-till seeded into residues of preceeding crops in year two; and no-till-seeded wheat in year three. Minimum tillage refers to fall incorporation of crop residues. In year two we broadcast ammonium nitrate fertilizer prior to seeding barley at rates of 0, 45, 90, 135, and 180 pounds/acre N. In 1986 four cycles of this experiment were completed; the other two cycles have completed year two.

Overall, barley yields on pulse residue increased and yields on barley residue decreased on the no-till treatments compared to the minimum tillage system. However, yields varied considerably among sites. Under conditions of moisture stress, no-till seeding into barley residue resulted in the highest yield. When barley yield declined with no-till, the decline probably was related to lowered soil temperature, retarded plant development, and increased levels of foliar disease.

Over the five cycles of this experiment, barley grain yield on pulse residues averaged 21% greater than on barley residue. Differences among pulse residues were frequently small, but overall yields were higher on faba beans and field pea residues than on lentil residue. Based on soil tests we could not attribute yield differences among crop residues to differences in soil N levels at time of seeding.

Nitrogen fertilizer equivalence values showed that to bring yields on barley residue up to the check yield on pulse residues would require 105, 85, and 50 pounds/acre N for faba beans, field peas, and lentils, respectively. These values were too high to provide a reasonable estimate of the N contribution of pulse residue. This indicated that factors in addition to N contributed to the high yields on pulse residues.

In both 1985 and 1986 the response of barley to N was greater on the gray-wooded, low-organic matter soil. Overall,

the response to N was similar on pulse and barley residues, though there was considerable fluctuation in relative response from site to site. Under conditions of moisture stress, the response to N was usually greater on barley than pulse residues, but N fertilizer did not bring the yield on barley residue up to the maximum yield attained on pulse residues.

Pulse residues also affected barley quality, increasing kernel weight and percent plump kernels by 5% and protein content by 0.6 percentage point. Increases in kernel size increase both nutritive and malting quality. But the increase in protein content, while nutritionally beneficial, may reduce the chance of producing high-quality malting barley.

In the second year following pulses the dry matter yield, grain yield, and N uptake of wheat was 15% higher than in the continuous cereal sequence. The carryover effect from N fertilizer applied in year two of each cycle resulted in higher wheat yields in year three. The response to this N is described by the equation, Wheat Yield $= 1,900 + 1.25$ N ($R^2 = 0.96**$), where N $=$ pounds/acre of N fertilizer applied to barley the previous year. Absence of a significant crop by N interaction indicated a similar use of N fertilizer by barley grown on pulse and barley residues.

Analysis of three completed cycles showed that cropping sequences that included pulses were considerably more productive than the continuous cereal sequence in terms of net energy production and economic gross margin to cash costs. Field peas were the most effective first-year crop in terms of net energy production, followed respectively by faba beans, lentils, and barley. Maximum net energy production was achieved with 90 pounds/acre N applied to barley sown on pulse residues. Barley succeeding barley required 135 pounds/acre N to achieve maximum net energy production. At the optimum N fertilizer in year two of the sequence, field peas-barley-wheat and barley-barley-wheat produced the same amount of gross food energy, 3.93×10^6 Kcal/acre/year. However, the sequence including field peas used only 59% of the production energy for fuel, fertilizer, and herbicides as the barley-barley-wheat sequence. Tillage methods made little difference to net energy production in any of the sequences evaluated.

A. T. Wright is a field crop scientist, Agriculture Canada Research Station, Melfort, Saskatchewan S0E 1A0, and E. Coxworth is principal biomass resource research scientist, Saskatchewan Research Council, Saskatoon, S7N 2X8.

Cropping practices using crownvetch in conservation tillage

Nathan L. Hartwig

Crownvetch (*Coronilla varia* L.) is a perennial legume adapted to the temperate climates in the northern two-thirds of the United States. It is a native of central and southern Europe, North Africa, and the Near East. Chance introduction of crownvetch in the late 1800s and early 1900s as a contaminant in other seed led to three naturalized varieties: Penngift from Pennsylvania, Chemung from New York, and Emerald from Iowa (5). These hardy, deep-rooted varieties are valuable in soil stabilization and erosion control on roadsides, steep banks, and other noncropland areas. Newly discovered as a perennial living mulch for conservation tillage crop production, crownvetch provides permanent, living ground cover. Crops are seeded into it, as if it were not there. Such a living mulch solves some of farmers' age-old soil erosion problems (7, 8, 9, 12, 13, 15).

Benefits of a crownvetch living mulch

Erosion control. A perennial crownvetch living mulch reduces water runoff, soil loss, and pesticide loss from sloping land (6). It provides a permanent cover for cushioning raindrop impact and holding the soil in place. It slows the rate of water runoff, thus improving moisture infiltration.

Improved fertility. A crownvetch living mulch increases soil fertility by limiting nutrient loss in surface runoff. In addition, the legume fixes N from the air. Our research indicates that 100-bushel-per-acre corn yields are possible in 3- to 5-year-old stands of crownvetch without added N (2). Each additional bushel of potential yield requires about 1 pound/acre N. Some growers of crownvetch seed have planted corn to clean out weeds and use the excess N.

Crownvetch produces more N than it needs for optimum seed production. Minimum tillage or no-till corn seems the ideal crop to use this excess N. Farmers have had good yields with no additional N, but 10 to 20 pounds applied in the row at planting time may optimize yields. This application enables the corn to better compete with the crownvetch.

Crownvetch will not contribute more than 25 to 50 pounds/acre N to second-year corn. But this amount saves $5 to $10/acre when N is 20 cents/pound. For third-year corn, crownvetch apparently contributes no N, and fertilizer N must be applied. Additional P or K fertilizer is not needed unless the crownvetch is pastured or removed for haylage.

Establishing living mulch

Crownvetch seedling growth is very slow. It normally requires 2 to 3 years to produce a well-established, vigorous

Nathan L. Hartwig is an associate professor of weed science, Department of Agronomy, Pennsylvania State University, University Park, 16802.

stand. Because few growers can afford to take a field out of production during establishment, the best time to seed crownvetch is in corn or after a small-grain harvest (4, 16). If following small grains, farmers should make no-till seedings at least 6 weeks before the first expected light frost, about August 1 in central Pennsylvania. This will allow plants to become sufficiently established for winter survival (16). Winter small grains should be harvested for grain early enough to allow for no-till crownvetch seeding. Spring-sown small grains should be harvested for silage in time to allow crownvetch establishment in late summer.

Winter survival in conventional crownvetch seedings is generally slightly less than in no-till seedings. To compensate, farmers should plant conventional seedings 1 month earlier than no-till seedings. For strong winter survival from conventional crownvetch seedings, the crownvetch should be seeded with corn or after harvesting small grains for silage (except early varieties of winter barley).

One year before seeding crownvetch, growers should apply lime to raise pH to 6.2 to 7.0 and apply P and K for both the small grain crop and crownvetch. At the time of crownvetch seeding the grower should apply 200 pounds/acre of 10-30-10 starter fertilizer in the row. No-till seedings should be made with any drill that will seed about ¼-inch-deep. Conventional crownvetch seedings may be made with any legume drill or seeder. A seeding rate of 5 pounds of inoculated seed/acre in 7- or 8-inch rows gives a solid stand in 1 or 2 years. A rate of 1 pound/acre in rows as wide as 36 inches provides good stands in 2 or 3 years. To obtain this rate growers should use a corn planter and the insecticide box as a seed metering device. A crownvetch stand of 1 plant/square yard after the first winter should be a success. Once well established, the crownvetch should last indefinitely given the right treatment. If it is necessary to destroy the crownvetch stand, growers can apply dicamba (Banvel) or 2,4-D for 2 or 3 consecutive years.

First-year rotations

No-till corn. Crownvetch seedings are sensitive to herbicides and crop competition during the first year. No-till corn is the best crop to plant following crownvetch seeding. Growers should avoid 2,4-D, dicamba (Banvel), or simazine (Princep) the first year. A mixture of Bladex plus Bicep at rates labeled for the soil type may be used safely without severely injuring crownvetch (4).

Forages. Seeding a legume, such as alfalfa, at the same time as crownvetch seldom works because alfalfa does not allow crownvetch establishment. If crownvetch is seeded in late summer or early fall, another legume or legume-grass mixture should not be seeded into crownvetch until the following spring. Growers should not use 2,4-DB on seedling crownvetch. It will cause as much injury as 2,4-D.

Rotations in established crownvetch

Forages after corn or small grain. Established crownvetch will compete excessively with a new legume or legume-grass seeding unless it is severely suppressed. Corn usually provides sufficient suppression. But following small grains, additional suppression of crownvetch may be necessary by harvesting, grazing, or applying paraquat.

When following small grains, legume or grass seeding should be made with any no-till drill that will seed ¼-inch deep no later than August 15 in central Pennsylvania. Growers should seed the legume in the spring when following corn. In the mid-South where corn is harvested early, fall seedings may be possible.

Once the newly interseeded legume or grass reaches the second or third leaf stage, growers should apply 2,4-DB to suppress crownvetch further and to control seedling broadleaf weeds. The forage mixture should be managed according to the dominant species, ignoring the presence of crownvetch. Growers may use any herbicide labeled for the forage mixture without losing the crownvetch. Crownvetch will persist throughout the life of the forage stand or rotated crop.

Corn after forage. When rotating from hay back to corn, growers should be careful not to eliminate the crownvetch, which will have nearly disappeared. In research trials at Penn State, crownvetch persisted through 3 years of alfalfa and alfalfa-grass mixtures and was still present when the field was rotated back to corn (*15*). Whether crownvetch can survive 4 to 5 years in a hay stand is unknown.

Because quackgrass usually invades hay stands in Pennsylvania, a split triazine treatment of atrazine at 2 pounds/acre in late March followed by atrazine plus simazine at 1 + 1 pound/acre at planting time is recommended for corn the first year (*9*). If perennial grasses are not present in the hay stand, growers can forgo the early spring atrazine treatment. Alfalfa, other legumes, and biennial and perennial broadleaf weeds with foliage can be controlled by adding dicamba (Banvel) to atrazine plus simazine at planting time (*1, 12, 13*). If applied when dandelion begins to bloom, Banvel will not severely injure crownvetch because the crownvetch is still dormant or just breaking dormancy and has little foliage. If dandelion is a major problem, growers can add a small amount of 2,4-D ester (0.25 pints/acre) along with the Banvel (*12, 13*). Growers should not use 2,4-D indiscriminately; crownvetch is very sensitive to it. Crownvetch will be severely suppressed the first year. In fact, it may appear to be killed completely, but in our research regrowth always occurred.

Corn after corn. The herbicide program for weed control in the second year of corn must be less harmful to crownvetch so it has a chance to recover. A mixture of cyanazine (Bladex) plus atrazine plus metolachlor (Dual) is ideal. If crownvetch appears to be growing too much (more than 6 inches when corn is 6 inches), it should be suppressed with dicamba (Banvel) at 0.25 pounds/acre (0.5 pint).

Crownvetch also provides some competitive weed control. Yellow nutsedge has been controlled very nicely by ground cover competition (*3, 11*). Competition from crownvetch will reduce corn yields 5% to 10%. But this loss has to be compared with the value of additional weed control, soil erosion control, and the feeding value of fall pasture or stalkage after corn grain harvest (*2, 3, 10, 17*).

REFERENCES

1. Cardina, J., and N. L. Hartwig. 1980. *Suppression of crownvetch for no-tillage corn.* Proc. Northeast. Weed Sci. Soc. 34: 53-58.
2. Cardina, J., and N. L. Hartwig. 1981. *Influence of nitrogen and corn population on no-till corn yield with and without crownvetch.* Proc. Northeast. Weed Sci. Soc. 35: 27-31.
3. Cardina, J., and N. L. Hartwig. 1982. *The influence of nitrogen, corn population and crownvetch cover crop on weed-corn competition.* Abst. Weed Sci. Soc. Am. No. 115.
4. Gover, A. E., and N. L. Hartwig. 1986. *Crownvetch establishment two years after seeding in a small grain-corn rotation.* Proc. Northeast. Weed Sci. Soc. 40: 18, 19.
5. Grau, F. V. 1968. *In the beginning crownvetch was a void.* In G. W. McKee, and M. L. Risius [eds.] *Second Crownvetch Symposium.* Agron. Mimeo No. 6. Pa. State Univ., Univ. Park. pp. 2-5.
6. Hall, J. K., N. L. Hartwig, and L. D. Hoffman. 1984. *Cyanazine losses in runoff from no-tillage corn in "living mulch" and dead mulches vs. unmulched conventional tillage.* J. Environ. Qual. 13(1): 105-110.
7. Hartwig, N. L. 1974. *No-tillage and crownvetch—The latest in soil conservation practices.* In Proc., Third Tech. Conf. Keystone Chapter, Soil Cons. Soc. Am., Ankeny, Iowa. pp. 44-45.
8. Hartwig, N. L. 1974. *Crownvetch makes a good sod for no-till corn.* Crops and Soils 27(3): 16-17.
9. Hartwig, N. L. 1975. *Crownvetch - A perennial legume cover crop for no-tillage corn.* Abst. Weed Sci. Soc. Am. No. 8.
10. Hartwig, N. L. 1976. *Legume suppression for double cropped no-tillage corn in crownvetch and birdsfoot trefoil removed for haylage.* Proc. Northeast. Weed Sci. Soc. 30: 82-85.
11. Hartwig, N. L. 1977. *Nutsedge control in no-tillage corn with and without a crownvetch cover crop.* Proc. Northeast. Weed Sci. Soc. 31: 20-23.
12. Hartwig, N. L. 1983. *Crownvetch—A perennial legume "living mulch" for no-tillage crop production.* Suppl. to Proc. Northeast. Weed Sci. Soc. 37: 28-38.
13. Hartwig, N. L. 1985. *Crownvetch and no-tillage crop production for soil erosion control.* Proc. Northeast. Weed Sci. Soc. 39: 75.
14. Hartwig, N. L., and L. D. Hoffman. 1975. *Suppression of perennial legume and grass cover crops for no-tillage corn.* Proc. Northeast. Weed Sci. Soc. 29: 82-88.
15. Hoffman, L. D., and N. L. Hartwig. 1975. *Perennial soil conserving cover crops for no-till corn.* Proc. Northeast. Weed Sci. Soc. 29: 89.
16. Hynes, E. M., and N. L. Hartwig. 1985. *Crownvetch establishment following small grains and preceding winter wheat.* Proc. Northeast. Weed Sci. Soc. 39: 34-38.
17. Mayer, J. B., and N. L. Hartwig. 1986. *Corn yields in crownvetch relative to dead mulches.* Proc. Northeast. Weed Sci. Soc. 40: 34, 35.

Potential use of annual legumes in a winter-rainfall, California-type climate for a cereal-ley farming system

W. L. Graves, B. L. Kay, M. B. Jones, W. M. Jarrell, and J. C. Burton

Several million California acres are dry farmed to cereals each year during the winter and spring rainy season. This demands a large amount of fossil fuel energy for N fertilizer and machinery. Introducing a ley-farming system with specific lines of annual legumes would considerably reduce these nonrenewable energy inputs.

The 1973 oil embargo reminded farmers that oil and associated products are limited (4). These limitations renewed our interest in legumes and their role in cropping systems. Although researchers have shown that natural biological-N fixation is highly effective, only a small percentage of the legume family's 20,000 species has been studied for that capability (2). Studies show annual legume introductions and varietal improvements since the 1930s have lead to doubled cereal and animal production in southern Australia (5). This cereal/annual legume rotation system, ley-farming, occurs in minimum tillage farming. We are evaluating annual legume germplasm on this farming system in a similar climate of California and Baja Mexico.

The Australian experience has guided us on types of annual legume germplasm applicable to a ley-farming system. Australian research also helped us evaluate these plant materials and their adaptation to this system (1, 3).

Sub clovers (*Trifolium subterraneum*) and medics (*Medicago* sp.) from the Mediterranean region are the annual legumes most easily adapted to our region. The plant exploration program collected these legumes on a trip to Tunisia in 1981 and to Morocco in 1983 (6). This collection continues to increase through exchanges with Australian researchers, Mediterranean researchers, and the U.S. Department of Agriculture Plant Introduction Program.

The ability to carry over viable seed through the cereal cropping phase is essential in a ley-farming system. Medics possess a considerable degree of hardseededness, thus adapting well to basic soils. Because the majority of California cereal farming soils are much closer to neutral or acid types, we chose the sub clover germplasm to screen for hardseededness. The program began in the fall of 1983 at the Hopland Field Station in California. Table 1 shows first year yield results and third year hardseededness for 16 experimental strains and two commercial varieties of sub clover.

First-year yields showed adequate establishment with no

W. L. Graves is a farm advisor, University of California Cooperative Extension, San Diego, 92123; B. L. Kay is a specialist and M. B. Jones is an agronomist, Agricultural Experiment Station, University of California, Davis, 95616; W. M. Jarrell is a professor of soils and environmental scientist, University of California, Riverside, 92521; and J. C. Burton is a microbiologist (retired), NIFTAL, University of of Hawaii, Honolulu, 96844.

Table 1. Evaluation of sub clover strains for first year establishment and beginning third year hardseededness at Hopland, California.

Sub Clover Strain	Dry Matter Yield (g/square foot)	Percent Hardseed (10/85)
SA15077	85	82
Tooday B	75	82
WA 65234-J	91	80
Baulkamaugh	80	74
Grayland	92	73
SA 65324-L	94	67
WA 65321-A	85	66
WA 65331-D	77	63
Trus-400	106	62
WA 47275	82	53
WA 65322-E	81	53
WA 65320	94	53
Trsu-405	76	52
Bellevue	66	51
Collie B	74	49
Mulwala	84	33
Woogenellup	101	20
Enfield	96	8

significant differences between strains. Hardseeded differences began to show by the beginning of the third season between the experimental strains and the standard commercial varieties, Woogenellup and Enfield. All 16 experimental strains demonstrated higher hardseededness than the 2 commercial strains, and 7 of these 16 strains demonstrated hardseeded of 66% or more. Other similar screening trials using sub clover introductions in varying rainfall zones of California and Baja California show encouraging results.

Our next step is to test the most promising hardseededness annual legume lines in a ley-farming situation. The lines will be evaluated on their regeneration ability under a periodic cereal cropping system. The integration of livestock (sheep and/or cattle) will be necessary to optimize land use, grain and meat/fiber outputs, and N cycling. Grazing during the growing season enhances the establishment and persistence of legume components. Also, proper grazing management can partially control undesirable weeds, optimize seed production, and increase N recycling.

REFERENCES

1. Crawford, E. J. 1970. *Variability in a large Mediterranean collection of introduced lines of* Medicago truncatula Gaertn. In Proc., Eleventh Int. Grassland Cong. Int. Grassland Cong., Supfer's Paradise, Australia. pp. 188-192.
2. Food and Agriculture Organization. 1984. *Legume inoculants and their use.* Rome, Italy.
3. Gladstone, J. S. 1967. *Naturalized subterranean clover: Strains in Western Australia: A preliminary agronomic examination.* Aust. J. Agr. Res. 18: 713-731.
4. Green, Maurice B. 1978. *Eating oil.* Westview Press, Boulder, Colo.
5. Puckridge, D. W., and R. J. French. 1983. *The annual legume pasture in cereal-ley farming systems of southern Australia. A review.* Agr., Ecosystems and Environ. 9: 229-267.
6. Rumbaugh, M. D., and W. L. Graves. 1984. *Foreign travel report: Collecting germplasm in Morocco.* In Proc., 29th Alfalfa Imp., Conf. Agr. Res. Serv., U.S. Dept. Agr., St. Paul, Minn., pp. 92-97.

WEED CONTROL

Legume effects on weed control in conservation tillage

A. Douglas Worsham and Randall H. White

Little is known about the effects legumes have on weed control in conservation tillage. Most research on legumes in conservation tillage systems reported in the literature deals with aspects other than control of the legume cover crops and subsequent weed control. In fact, there are few published papers dealing with these subjects. Work on these aspects has been started, however, in many states.

There is considerable research and experience throughout the United States on weed management in conservation tillage and no-till cropping systems in a wide variety of situations, including small-grain cover crops, previous crop residues, stale seedbeds, and others. Once a legume cover crop is killed by a herbicide(s) or is allowed to mature naturally, we will assume for purposes of this paper that subsequent weed management is similar to these other situations where much has been published (22, 27) and will not be considered in detail here.

Legume cover crop kill

The key to successful no-till or conservation tillage crop productivity is management, especially weed management (22, 27). Cover crop management is one aspect that is of importance if growers are to realize the many benefits of a legume mulch. All legume management schemes employed in no-till or conservation tillage systems involve chemcial control of the cover crop with one exception, which will be discussed later (9). Triplett (22) concluded that the cover

A. Douglas Worsham is a professor of crop science, North Carolina State University, Raleigh, 27695-7620, and Randall H. White is a graduate assistant, Agronomy Department, University of Illinois, Urbana, 61801. The authors acknowledge the help of many colleagues in the United States for sharing research information and personal experiences that made possible the writing of this paper.

crop must be killed completely for satisfactory no-till corn production. Stand reductions have been reported when grain sorghum was planted into live legumes (16). Thus, inadequate control of the cover crop usually reduces the effectiveness of this no-till system.

Herbicide treatments applied to control cover crops and existing vegetation before no-till or conservation tillage planting usually include nonselective herbicides, such as paraquat or glyphosate, with an additional herbicide(s) that provides residual control. Some of these herbicides, such as the triazines or ureas, may provide additional control activity (21, 26, 27). Many researchers have found it difficult to control legumes adequately in no-till systems. They have suggested that this inadequate control of legume cover crops is a causative factor for reduced crop stands and early season crop growth (3, 9, 26).

White (26) evaluated eight herbicide treatments from 1984 to 1986 for legume cover crop control in no-till corn and cotton in North Carolina. Herbicide treatments of paraquat; 2,4-D; dicamba; and cyanazine were applied alone or in various paraquat or glyphosate combinations. A separate herbicide tank-mix for each crop was included with every burndown treatment for residual control. Crimson clover and hairy vetch cover crops were sprayed the same day as corn planting and 2 to 4 weeks prior to cotton planting. Herbicide treatments used to control the legume cover crops before planting cotton were tank-mixed with fluometuron and herbicides applied to corn with metolachlor and atrazine. Because the postemergent activity of the tank mixes differ, desiccation ratings for both no-till crops were not combined.

Results showed that paraquat alone or in combination with 2,4-D or dicamba provided rapid control of both legumes in corn (Table 1). Single or combination treatments of glyphosate, 2,4-D, and dicamba provided significantly slower control than the paraquat treatments; glyphosate alone provided the slowest mulch desiccation. All paraquat applications continued to provide better control than the other herbicide treatments, and kill from glyphosate alone remained significantly lower than most other treatments. In both years all treatments adquately controlled vetch, except for glyphosate. However, vegetational regrowth occurred in vetch plots treated with paraquat alone in 1986 (Table 1). Hairy vetch was in an active vegetative growth stage when treated, which could have reduced the effectiveness of contact herbicides, such as paraquat, that do not translocate to the growing points of the plant and thereby remove the potential for regrowth. The addition of 2,4-D or dicamba apparently alleviates this problem. Glyphosphate at the rate of 1.5 pound/acre active ingredient was not enough for adequate legume control. Higher rates probably would have provided greater efficacy but may be uneconomical considering the alternatives available. Combinations of paraquat or glyphosate with 2,4-D or dicamba provided better control than single treatments of each herbicide. No difference was detected in results between 2,4-D or dicamba.

Initial cover crop control ratings in cotton were similar to those made on the same legume in the corn plots (Table 2). No differences were found between 2,4-D and cyanazine alone or in combination with paraquat or glyphosate. Legume control ratings 4 weeks after application were not as similar to those obtained in corn. Clover was controlled adequately by all treatments in 1986, but percent kill ranged from poor to fair to good in 1985. Vetch regrowth occurred in all paraquat treatments in both years. The efficacy of the residual herbicide tank mixes may have influenced the regrowth of vetch in cotton plots compared to corn. Cyanazine or 2,4-D alone or in combination with glyphosate provided the best hairy vetch control. Glyphosate alone did not adquately kill the vetch plots in either year.

In 1983 J. L. Griffin and R. W. Taylor of the Louisiana State University Rice Research Station evaluated several herbicides at various rates for control of subterranean clover, crimson clover, and common vetch (unpublished progress report). Paraquat was most effective on crimson clover and vetch. With a spray volume of 40 gallons/acre, 0.5 pound active ingredient/acre provided excellent control (Table 3). They found a rate of 2.67 pounds/acre of glyphosate or SC 0224 (an experimental herbicide that is another salt of glyphosate) was needed to control subterranean and crimson clovers. Both herbicides were weak on vetch even at high rates. Glufosinate gave excellent control of all legume species. A rate of 1.3 pounds active ingredient/acre controlled subterranean clover, but 0.7 pound/acre was adequate

Table 1. Percent control of crimson clover and hairy vetch from herbicides applied at time of planting corn (26).

Herbicide Treatments*	Herbicide Rate (pounds/acre)	Percent Legume Control							
		10 Days After Application				30 Days After Application			
		1985		1986		1985		1986	
		Clover	Vetch	Clover	Vetch	Clover	Vetch	Clover	Vetch
Paraquat	0.5	83a†	88a	76a	70a	88a	96a	93a	85c
Paraquat + 2,4-D	0.5 + 0.5	86a	88a	78a	78a	92a	100a	94a	94ab
Paraquat + dicamba	0.5 + 0.25	85a	88a	88a	77a	89a	97a	95a	90b
Glyphosate	1.5	22b	31c	23c	39b	65c	80b	75c	62d
Glyphosate + 2,4-D	1.5 + 0.5	29b	56b	48b	56ab	73b	100a	81bc	99a
Glyphosate + dicamba	1.5 + 0.25	40b	56b	41bc	59ab	76b	99a	83b	94ab
2,4-D	1.0	33b	53b	45bc	66a	70bc	98a	83b	100a
Dicamba	0.5	35b	53b	38bc	60ab	72bc	97a	84b	94ab
LSD (.05)		24				7			

*Each treatment was tank mixed with 1.0 pounds/acre atrazine plus 2.0 metolachlor and 0.5% (v/v) X-77 surfactant. Paraquat treatments were applied in 40 gallons of water/acre; other treatments were applied in 18.3 gallons of water/acre.
†Means within columns followed by the same letter are not signficantly different (Fisher's LSD Test).

Table 2. Percent control of crimson clover and hairy vetch from herbicides applied about 2 weeks before planting cotton (26).

Herbicide Treatments*	Herbicide Rate (pounds/acre)	Percent Legume Control							
		10 Days After Application				30 Days After Application			
		1985		1986		1985		1986	
		Clover	Vetch	Clover	Vetch	Clover	Vetch	Clover	Vetch
Paraquat	0.5	86a†	73a	78a	63a	89a	51c	98a	79abc
Paraquat + 2,4-D	0.5 + 0.5	88a	88a	73a	63a	89a	78ab	96a	71bc
Paraquat + cyanazine	0.5 + 0.75	84a	81a	69abc	61a	84ab	94abc	94a	81abc
Glyphosate	1.5	21c	36b	24d	24b	50c	69bc	89a	59c
Glyphosate + 2,4-D	1.5 + 0.5	52b	72a	44bcd	48ab	73abc	98a	98a	99a
Glyphosate + cyanazine	1.5 + 0.75	53b	86a	39cd	54ab	74abc	88ab	93a	86ab
2,4-D	1.0	37bc	71a	39cd	44ab	63bc	98a	88a	99a
Cyanazine	1.5	36bc	73a	33d	59a	86ab	87ab	94a	89ab
LSD (.05)		30				25			

*Each treatment was tank mixed with 2.0 pounds/acre active ingredient fluometuron and 0.5% (v/v) X-77 surfactant. Paraquat treatments were applied in 40 gallons of water/acre; other treatments were applied in 18.3 gallons of water/acre.
†Means within columns followed by the same letter are not signficantly different (Fisher's LSD Test).

for control of crimson clover and vetch. Dicamba gave unacceptable control of subterranean and crimson clover. But at the rate of 0.5 pound/acre, it did provide excellent control of vetch (Table 3).

Griffin and Taylor (9) conducted further studies in 1984 and 1985 with the same herbicides and legumes with two herbicide application dates. Sub clover was the most difficult to control and vetch was the least difficult (Table 4). Paraquat applied at 1.0 pound/acre in either 20 or 40 gallons of water/acre provided at least 97% sub clover control for the late application and 80% control at the early application (Table 4). Although more sub clover biomass was present at the late application, Griffin and Taylor attributed the increased control to warmer temperatures that resulted in greater paraquat activity. They obtained poor sub clover

control with both glyphosate and SC 0224 at 1.0 and 2.5 pounds/acre active ingredient. Glufosinate at 1.68 pounds/acre active ingredient controlled sub clover at both application dates. At the early application date paraquat at 0.5 and 1.0 pound/acre in either 20 or 40 gallons of water/acre and glufosinate at 0.75 and 1.5 pounds/acre provided at least 95% crimson clover control. Glyphosate and SC 0224 provided poor crimson clover control. Clover had matured already at the time of the late herbicide application so control ratings were not made.

Vetch control was at least 85% for paraquat at 0.5 and 1.0 pound/acre in both 20 and 40 gallons of water/acre and for glyphosate at 0.75 and 1.5 pounds/acre for both the early and late applications. Glyphosate and SC 0224 at 2.0 pounds/acre provided at least 93% vetch control at the late application, but only about 50% control at the early application date.

In Florida, Gallaher (8) found that a minimum of 0.37 pound/acre paraquat was necessary to completely desiccate a crimson clover cover crop. It appeared, from the 1-year study, that mixing 2,4-D and paraquat together was ineffective compared with using paraquat alone. Use of 2,4-D alone up to 1.0 pound/acre gave slow and incomplete kill. Diquat, at a rate of up to 0.5 pound/acre was not adequate for clover kill, and regrowth occurred quickly. There appeared to be no benefit in adding 2,4-D to paraquat (8).

Glyphosate or glyphosate plus 2,4-D was extremely effective in complete kill of crimson clover. The addition of 2,4-D to glyphosate had a positive effect on clover kill (8).

In Florida studies of crimson clover, arrowleaf clover, subterranean clover, white clover, alfalfa, vetch, and Austrian winter peas, crimson and arrowleaf clovers have been the most difficult to control, while vetch and Austrian winter peas have been the easiest to kill. Paraquat and glyphosate often have provided less than adequate control. A mixture of atrazine plus paraquat or atrazine plus 2,4-D has been more effective than any of the herbicides applied

Table 3. The effect of herbicide treatments on control of Woogenellup subterranean clover, Tibbee crimson clover, and Cahaba White vetch grown as cover crops, Rice Research Station, Crowley, Louisiana, 1983-1984*

Herbicide†	Herbicide Rate (pounds/acre)	Spray Volume (gallons/acre)	Sub Clover	Crimson Clover	Vetch
Paraquat	0.05	20	33	82	88
Paraquat	1.00	20	53	95	100
Paraquat	0.50	40	43	93	92
Paraquat	1.00	40	67	100	100
Glyphosate	1.40	6	58	47	53
Glyphosate	2.67	6	97	90	77
SC 0224	1.40	6	83	50	25
SC 0224	2.67	6	97	88	78
Glufosinate	0.70	6	68	100	98
Glufosinate	1.30	6	95	100	98
Dicamba	0.25	6	42	18	82
Dicamba	0.50	6	43	27	93
Untreated Check	—	—	0	0	0
LSD (0.05)			24.5	14.2	13.1

*Adapted from J. L. Griffith and R. W. Taylor, annual report.
†Herbicides applied April 18, 1984.
‡Control ratings made May 9, 1984.

Table 4. The effect of herbicide treatments on control of Mt. Barker subterranean clover, Tibbee crimson clover, and hairy vetch grown as cover crops, Rice Research Station, Crowley, Louisiana, 1985*.

Herbicide†	Rate (pounds/acre)	Spray Volume (gallons/acre)	Early Application Sub Clover	Early Application Crimson Clover	Early Application Vetch	Late Application Sub Clover	Late Application Crimson§ Clover	Late Application Vetch
Paraquat	0.50	20	30	95	93	82	NR	100
	1.00	20	78	98	98	100	NR	100
	0.50	40	40	97	83	44	NR	100
	1.00	40	80	98	95	97	NR	100
Glyphosate	1.50	6	22	32	57	13	NR	78
	2.50	6	25	30	50	23	NR	93
SC 0224	1.50	6	17	25	53	17	NR	83
	2.50	6	25	33	62	23	NR	95
Glufosinate	0.75	6	100	100	100	79	NR	100
	1.50	6	100	100	100	90	NR	100
LSD (0.05)			15	10	17	15		6

*Adapted from J. L. Griffith, R. J. Habetz, and R. P. Regan, annual report.
†Herbicide treatments applied early on April 2 and late on May 7.
‡Cover crop control ratings made 16 and 20 days after treatment for early and late applications, respectively.
§Crimson clover not rated (NR) since it had matured prior to herbicide application.

alone (David Wright, North Florida Research and Education Center, Quincy, Florida, personal communication).

In Alabama no-till cotton studies, paraquat plus cyanazine at 0.5 plus 0.5 pound/acre were applied 14 days prior to planting and immediately after planting in covers of rye and vetch. Other treatments included a preemergence application of prometryn plus pendimethalin plus paraquat at 2.5 + 0.5 + 0.25 pounds/acre. Cotton stand and/or yield declined when no early treatment was applied to kill the cover crops. It was more important to kill the vetch early as compared to the rye (R. H. Walker and Brian Gamble, Agronomy Department, Auburn University, Auburn, Alabama, personal communication).

In other legume cover crop no-till studies in Alabama, paraquat generally has given better vetch control than glyphosate and glyphosate has given better crimson clover control. The Alabama researchers suggested use of paraquat plus cyanazine or glyphosate plus cyanazine to kill vetch for no-till cotton planting (R. H. Walker and B. Norris, Agronomy Department, Auburn University, Auburn, Alabama, personal communication).

Paraquat plus a number of residual herbicides has provided good legume cover crop kill and residual weed control in legume studies in Maryland (A. Morris Decker, Agronomy Department, University of Maryland, College Park, personal communication).

Although most work with legumes in conservation tillage systems has been with annuals, Moomaw and Martin (*18*) noted that 2,4-D with dicamba provided the most consistent control of alfalfa and that spring applications were better than fall applications. Triplett (*22*) reported that the addition of paraquat or dinoseb increased the amount of 2,4-D required to control alfalfa, indicating decreased effectiveness of 2,4-D in the presence of the contact herbicides. He suggested applying 2,4-D and contact herbicides in separate applications. Tank mixes of paraquat and dicamba, however, performed satisfactorily in the alfalfa kill studies. White (*26*) did not find an antagonistic effect with paraquat plus 2,4-D or dicamba mixtures in control of crimson clover and hairy vetch, which are annual legumes. In fact, the addition of the systemic herbicide to paraquat reduced legume regrowth over paraquat alone.

Time of kill

It is difficult to determine the optimum time at which to kill a cover crop before planting a summer crop. Killing time is influenced by the legume, the summer crop, and spring weather conditions. Killing the cover crop too far in advance of planting the summer crop will curtail N production by the legume. Also, the mulch biomass may be insufficient to provide adequate soil moisture conservation later in the growing season. On the other hand, high-producing legume cover crops decrease soil moisture. Thus, killing the cover crop at time of planting the summer crop in a dry spring often leads to stand and yield losses in the summer crop (*26*). That same drying effect, however, could also be advantageous during a wet spring.

It is evident that scientists need to do much more work in evaluating different herbicide treatments to provide effi-

cient, fast, and thorough kill of annual legume cover crops and in determining timing of herbicide applications. Such information is pertinent to determining the influence that legume control, and subsequent weed control in the planted crop, has on crop stand establishment, growth, and yield.

Summer crops in a living mulch

Most researchers have indicated that a complete kill of winter annual or perennial legumes is necessary for maximum yield of the summer crop. But Hartwig has developed a system of planting crops, such as no-till corn, into a living but suppressed sod of crownvetch (*12, 13*). The living mulch provides all of the desirable characteristics of a dead mulch, such as reduced soil erosion and rainwater runoff, increased infiltration, and reduced surface soil evaporation losses, without the need for reestablishing a cover crop each year.

Competition from living crownvetch significantly reduced yellow nutsedge as long as herbicide suppression of the legume was not too great. Where corn was planted the year after seeding crownvetch and where corn follows corn in crownvetch, Hartwig recommended a mixture of cyanazine plus atrazine plus metolachlor. This mixture suppresses the crownvetch and controls annual broadleaf and grass weeds and yellow nutsedge. If crownvetch growth becomes excessive—more than 6 inches of growth when corn has 6 leaves—it can be suppressed with dicamba at 0.25 pound/acre active ingredient.

From Hartwig's work it appears that competition from the living crownvetch reduces corn yields 5% to 16%. This loss, however, should be compared with the value of additional weed and soil erosion control and the feeding value of fall pasture or stalkage after corn grain harvest (*12, 13*).

Rotations including use of other legumes, legume-grass mixtures, and small grains in the living crownvetch sod have also been used (*12, 13*).

Other researchers have reported poor weed control where cover crops have not been completely killed (*4, 9, 16, 22, 26*). Some have suggested that living mulches could only be maintained where high management levels, especially irrigation, are available.

Another system developed in the Southeast does not involve herbicide kill of the legume cover crop or annual reseeding. In this system a crop, such as grain sorghum, is planted into crimson clover after the clover has matured and shed seed (*16*). This system would be suitable only for summer crops that could be planted later than corn, such as grain sorghum, soybeans, or sunflowers.

Effects of legumes on weed management

One of the first references to weed suppression by legume cover crops was by Hardy, who in 1939 stated that leaving a mulch or disking the winter legume resulted in little weed growth until June (*11*).

White and associates (*25*) studied the effects of N fertilizer applied in February in a pecan orchard on legume forage, N production, and weed growth. Early (October) and late-planted (December or January) cahaba white vetch and Amclo arrowleaf clover were used as the cover crops.

Nitrogen application caused severe yield decline in the late-planted vetch and clover. With N applied in February, most of the forage produced was weeds. Although significant weed growth resulted on all plots, the late-planted legume treatments with little cover produced the most weed growth.

In the studies on legume cover crop kill in North Carolina previously discussed, White (26) found that in the blocks planted to corn differences in early season annual weed control between vetch and clover cover crop plots or among herbicide treatments were either nonsignficant or more pronounced in the mid-season ratings. Weed control ratings in the corn and cotton plots were not combined due to the differing tank mixes. Annual broadleaf weeds included redroot pigweed, common lambsquarters, and common ragweed. Annual grasses rated were large crabgrass and fall panicum. Morningglory control was rated when these species were present.

Weed control was generally the same in both legume mulches for all weed species present. Annual grass and morningglory control was better in the conventional tillage treatments than the no-till plots. The presence of a legume mulch provided more effective broadleaf weed control than the conventionally tilled plots.

Control of all weeds in the 1985 clover plots was significantly lower in plots treated with glyphosate alone. The increased weed interference may be explained by the lack of desiccated mulch cover on the soil surface (mulch effect), resulting in warmer soil temperatures that created conditions more conducive to weed seed germination. Lower amounts of decomposing debris on the soil surface would also reduce the potential of allelopathic interactions. These same weed control responses occurred in vetch. The mulch effect was clearly demonstrated in desiccated vetch, which forms a completely flat, thick mat of decomposing vegetation. Due to the lack of soil moisture, weed populations were very low in the 1986 growing season and no morningglory species were present. There were few differences between herbicide treatments and all ratings were above 85%.

In cotton weed control was similar for all herbicide treatments. However, weed control ratings varied between tillage systems. In 1985 there were no differences in morningglory control. Grass control in no-till vetch plots was significantly lower than in clover or conventional plots. Both annual grass and broadleaf weed populations were lower in the no-till mulch. In no-till plots weeds only emerged from the narrow planting furrow compared to emergence over the entire plots in the conventional treatments. Grass control was slightly better in clover than vetch no-till plots, and White found no differences in weed control between tillage systems at a second location (26).

Johnson and Webb (14) conducted field experiments in 1985 and 1986 to evaluate the ability of cover crops to suppress problem weeds in the esablishment of full-season, no-till soybeans in Delaware. Both winter rye and Austrian winter peas provided excellent suppression of horseweed, common ragweed, and fall panicum.

Combinations of paraquat (0.5 pound/acre) plus linuron (0.75 pound/acre), paraquat (0.5 pound/acre) plus metribuzin plus Chlorimuron (0.25 pound/acre), paraquat (0.5 pound/acre) plus linuron (0.75 pound/acre) plus chlorimuron (0.5 pound/acre), and paraquat (0.5 pound/acre) plus alachlor (2.0 pounds/acre) plus linuron (0.75 pound/acre) provided complete burndown of the covers and excellent control of horseweed in 1985. Horseweed control was poor to fair in 1986 because of drought.

Webb (23, 24) has also stated that, from many years of experience, legume covers especially benefit corn because of the N production, but only benefit no-till soybeans because of improved weed control.

In a Missouri study Aldrich (1) evaluated 11 fall-planted cover crop species for their effectiveness in providing ground cover and weed suppression the following spring. Four species—red clover, perennial ryegrass, spring oats, and annual ryegrass—provided more than 90% ground cover the next spring. Of these, the grasses, but not red clover, appeared to greatly reduce annual weeds. Other legume species that also did not suppress annual weeds were field peas, soybeans, and white lupine. Annual alfalfa and rape appeared to reduce the number of weeds in the spring.

Brown and Whitwell (4) conducted studies in Alabama on minimum-till cotton using crimson clover, hairy vetch, and rye as cover crops. The results indicated that successful weed control programs in minimum-till cotton should include paraquat or glyphosate plus fluometuron and possibly other residual herbicides. Preemergence combinations of foliar-active materials and fluometuron may suffice in locations or years of light annual weed pressure. Increased weed pressure and/or use of cyanazine preemergence would dictate the addition of postemergence herbicides applied as directed sprays.

Brown and Whitwell observed several other problems associated with the legume cover crops: they were difficult to desiccate, adversely affect cotton stands, reduced weed control (especially annual grasses), and delayed crop maturity. The researchers concluded that successful dessication of legumes prior to planting might overcome or reduce these problems (4).

On the other hand, Ebelhar (7) reported results of studies on winter annual legume crops with no-till corn in Kentucky in which all the cover crops seemed to help control weed populations to some extent. There were no significant differences, however, among cover crops as to their effect on weed control in corn.

Potential allelopathic effects

In the past most researchers have attributed reduced growth of a crop planted into a legume cover crop to competition between the legume and the subsequent crop for resources, especially soil moisture, because of inadequate legume control. A few researchers also have suggested that allelopathy may play a role (2, 6).

Many legume species contain secondary plant products capable of producing allelopathic effects (20). Even so, limited research has been conducted specifically on legumes and their potential beneficial or adverse effects on other plant species for agronomic purposes, excluding N production. Water extracts of sweetclover, a legume that synthesizes the phenolic compound coumarin, have been found to inhibit germination and growth of corn, wheat, and sorghum seedlings (10, 17). Lettuce bioassay results indicated the presence of phytotoxic substances in decomposition products of vetch extract (*19*). Clover soil sickness has been suggested in berseem clover and Persian clover fields (*19*). Isoflavonoids released by red clover leaves and stems have been isolated and shown to decompose to phenolic compounds, accumulate in the soil, and further inhibit red clover growth (*5, 20*). However, aqueous extracts of red clover and ladino clover did reduce germination and growth of other forage crops (*20*).

The potential for allelochemical production is present in legumes that are genetically similar to crimson clover and hairy vetch. Field observations support the possibility of an interaction between these legumes and other plants.

Drake (*6*) suggested that allelopathy may aid or negate some potential combinations of legumes and corn, sorghum, or other crops. He found that oats were a good nurse crop for legumes, such as alfalfa and sweet clover. Drake suggested that these combinations were successful because of allelopathy toward weeds. He also suggested that where nurse crops were especially successful there was the possibility of synergistic allelopathy where two species coordinate chemically to exclude other species. This might also be a factor when some crops are planted into cover crops.

Allelopathic interactions resulting from the release of phytotoxic chemicals from decomposing plant debris can result in reduced crop stand, growth, and yield (*10, 17, 19*). In work with winter annual legume crops in no-till corn and cotton systems in North Carolina, we have made similar observations and also found reduced broadleaf weed populations in summer crops planted no-till into killed legume cover crops. We considered that allelopathy may have influenced both crop and weed growth.

We, therefore, conducted several field and greenhouse tests to define the potential allelopathic interactions for clover and vetch in no-till cropping systems with crimson clover and hairy vetch (*26*). We conducted three different experiments: (a) a field soil-root core bioassay, (b) effects of aqueous legume extracts on seed germination and seedling growth, and (c) effects of legume residue—incorporated or on the soil surface—on seedling emergence and growth.

Results from the soil-root core bioassay showed that soil samples collected from beneath field-grown vetch and clover plants and containing root biomatter did not affect emergence or growth of corn or cotton compared with soil cores removed from adjacent areas absent of legumes.

Germination and radicle length of corn and Italian ryegrass and radicle plus hypocotyl length of cotton, morningglory, and wild mustard progessively declined when exposed to increasing concentrations (1.1 to 4.5 ounces of debris/gallon) of aqueous clover and vetch extract. Undiluted extract inhibited almost completely the germination and growth of mustard and ryegrass. Bioassay responses to polyethylene glycol solutions suggested that extract osmotic potentials did not cause the germination and growth reductions. In both studies bioassay species exhibited greater phytotoxic responses to hairy vetch debris and aqueous extract than crimson clover.

Field-grown crimson clover and hairy vetch debris reduced corn, cotton, and pitted morningglory seedling emergence and dry weight accumulation when incorporated into the soil in pots in the greenhouse. Legume debris located on the surface of the soil demonstrated both inhibitory and stimulatory

responses, depending upon the bioassay species and growing conditions. Increasing amounts of clover or vetch (0.013 to 0.1 ounces of debris/pound of soil) generally augmented these effects by various degrees. Cotton and morningglory exhibited greater sensitivity to both legume species and debris location. Soil-incorporated legume debris inhibited cotton and morningglory emergence and growth, but cotton was not affected significantly when the debris was placed on the soil surface. Vetch debris on the soil surface produced a quadratic growth response in morningglory. Some stimulatory growth effects on corn were noted from legume debris.

These studies demonstrate that aboveground hairy vetch and crimson clover debris may contain allelopathic chemicals (*26*). However, the impact of potential allelopathic interactions in no-till cropping systems appeared to be limited if legume debris was left on the soil surface. The release of these phytotoxic substances appeared to be facilitated by soil incorporation of the legume debris, an alternative production practice not associated with no-till systems and probably not with conservation tillage. But incorporation does augment organic N availability from decomposing legume debris.

Phytotoxic interactions occur more frequently in debris decomposition under conditions of low aeration and excessive soil moisture. The aqueous extract study demonstrated that water-soluble constituents released from the legume debris could cause phytotoxicity to certain plants. These results might be applicable to field situations if saturated soil moisture conditions are present during crop and weed germination and seedling emergence. However, this potential interaction depends upon the concentration of the phytotoxic substances released and the duration of interaction. Incorporating legume no-till cover crops into weed control strategies in conservation tillage systems seems possible because small-seeded weeds appear to be sensitive to lower concentrations of water-soluble debris constituents; cotton also appeared to be more sensitive than corn. Under field conditions, it seems that allelopathic suppression of some weed species would be more likely than suppression of the crops planted into a killed-legume mulch. Root debris had no obvious effect on weeds or crops under the experimental conditions reported here.

REFERENCES

1. Aldrich, R. J. 1986. *The relative effectiveness as ground cover of species seeded alone and in combination.* N. Cent. Weed Control. Conf. Res. Report 43: (in press).
2. Berger, D. A., and S. M. Dabney. 1985. *Retardation of germination and early growth of corn planted in sub clover cover crop.* In Proc., S. Region No-Till Conf. Agr. Exp. Sta., Univ. Ga., Athens. pp. 54-58.
3. Breman, J. W., and D. L. Wright. 1984. *Using winter legume mulches as a nitrogen source for no-tillage corn and grain sorghum production.* In Proc., S. No-Till Systems Conf. Agr. Exp. Sta., Auburn Univ., Auburn, Ala. pp. 6-17.
4. Brown, S. M., and Ted Whitwell. 1985. *Weed control programs for minimum tillage cotton* (Gossypium hirsutum). Weed Sci. 33: 843-847.
5. Chang, C. F., A. Suzuki, S. Kumai, and S. Tamura. 1969. *Chemical studies on "clover sickness." Part II. Biological functions of isoflavonoids and their related compounds.* Agr. Biol. and Chem. 33: 398-408.
6. Drake, L. D. 1976. *Prairie models for agricultural systems.*

In Proc., 5th Midwest Prairie Conf. Iowa State Univ., Ames. pp. 226-230.

7. Ebelhar, S. A. 1981. *Nitrogen from winter-annual legume cover crops for no-tillage corn.* M.S. Thesis. Univ. Ky., Lexington.

8. Gallaher, R. N. 1986. *Studies of chemical combinations and rates used to convert a living crimson clover cover crop to a mulch for no-tillage planting of summer crops.* Agron. Res. Rpt. AX-86-07. Agron. Dept. Univ. Fla., Gainesville. 11 pp.

9. Griffin, J. L., and R. W. Taylor. 1986. *Evaluation of burndown herbicides in no-till systems with legume cover crops.* Abstr., Tech. Papers, S. Assoc. Agr. Sci. Agron. Div., S. Branch, Am. Soc. Agron., Madison, Wisc. No. 13, page 16.

10. Guenzi, W. D., and T. M. McCalla. 1962. *Inhibition of germination and seedling development by crop residues.* Soil Sci. Soc. Am. Proc. 26: 456-458.

11. Hardy, M. B. 1939. *Cultural practices for pecan orchards.* Proc. S. E. Pecan Growers Assoc. 33: 58-64.

12. Hartwig, N. L. 1976. *Nutsedge control in no-tillage corn with and without a crownvetch cover crop.* Proc. Northeast Weed Sci. Soc. 31: 20-23.

13. Hartwig, N. L. 1984. *Crownvetch and no-tillage crop production for soil erosion control.* Coop. Ext. Serv., Penn. State Univ., Univ. Park. 8 pp.

14. Johnson, Q. R., and F. J. Webb. 1986. *Weed suppression from cover crops in full season no-till soybeans.* Proc. Northeast Weed Sci. Soc. 41: (In Press).

15. Katznelson, J. 1972. *Studies in clover soil sickness. I. The phenomenon of soil sickness in berseem and Persian clover.* Plant Soil. 36: 379-393.

16. Martin, G. W., and J. T. Touchton. 1982. *Nitrogen requirements for till and no-till grain sorghum doublecropped with various winter legumes.* Grain Sorghum Newsletter 25: 67-68.

17. McCalla, T. M., and F. L. Duley. 1948. *Stubble mulch studies: Effect of sweet clover extract on corn germination.* Science 108: 163.

18. Moomaw, R. S., and A. R. Martin. 1976. *Herbicides for no-tillage corn in alfalfa sod.* Weed Sci. 24: 449-453.

19. Patrick, Z. A., T. A. Toussoum, and W. C. Snyder. 1963. *Phytotoxic substances in arable soils associated with decomposition of plant residues.* Phytopathol. 53: 152-161.

20. Rice, E. L. 1984. *Allelopathy.* Academic Press, New York, N.Y. 422 pp.

21. Robinson, L. R., and H. D. Wittmuss. 1973. *Evaluation of herbicides for use in zero and minimized till corn and sorghum.* Agron. J. 65: 283-288.

22. Triplett, G. B. Jr. 1985. *Principles of weed control for reduced-tillage corn production.* In A. F. Wiese [ed.] *Weed Control in Limited Tillage Systems.* Mono. No. 2. Weed Sci. Soc. Am., Champaign, Il.

23. Webb, F. J. 1981. *Full-season no-tillage soybean production.* EB 121. Univ. Del. Coop. Ext. Serv., Newark. 4 pp.

24. Webb, F. J. 1982. *The recipe for no-tillage corn production.* EB 114. Univ. Del. Coop. Ext. Serv., Newark. 10 pp.

25. White, Jr., A. W., E. R. Beaty, and W. L. Tedders. 1981. *Legumes as a source of nitrogen and effects of management practices on legumes in pecan orchards.* Proc. S. E. Pecan Growers Assoc. 74: 97-106.

26. White, R. H. 1987. *Control of legume cover crops in no-till corn* (Zea mays *L.) and cotton* (Gossypium hirsutum *L.) and allelopathic potential of legume debris and aqueous extracts.* M.S. Thesis. Crop Sci. Dept., N. C. State Univ., Raleigh. 110 pp.

27. Worsham, A. D., and W. M. Lewis. 1985. *Weed management: Key to no-tillage crop production.* Proc. S. Region No-Till Conf. Agr. Exp. Sta., Univ. Ga., Athens. pp. 177-188.

Legume cover crops for no-till corn

Philip L. Koch

For several years soil conservationists at the Soil Conservation Service's Rose Lake Plant Materials Center near East Lansing, Michigan, have been working with legume cover crops in no-till corn.

The project, now in its third year, is designed to deal with erosion associated with inadequate residue or soil cover when corn has been harvested for silage. The objective of the project is to select and evaluate various legumes used with selected herbicide mixtures to determine their potential for maintaining permanent cover crops to control erosion in no-till corn while producing reasonable corn yields.

We selected nine perennial legumes for use with nine herbicide mixtures. The legumes include peak alfalfa (*Medicago sativa*), arlington red clover (*Trifolium pratense*), white clover (*T. repens*), mammoth clover [*T. pratense* (H.V.)], ladino clover [*T. repens* (H.V.)], Mackinaw birdsfoot trefoil (*Lotus corniculatus*), narrowleaf trefoil (*Lotus tenius*), emerald crownvetch (*Cornilla varia*), and lutana cicer milkvetch (*Astragalous cicer*).

Cover crops were established in 1983 and 1984 on a Boyer sandy loam in strips that were 216 feet long and 21 feet wide. Most of the herbicides were applied in late April in long, 21-foot-wide strips perpendicular to the cover crop strips. Mixtures with Tandem were applied in mid-May. The entire block resembles a giant checkerboard, consisting of 81 plots, each measuring about 21 feet by 21 feet. The blocks were replicated three times and randomized by varying the arrangement of the cover crops and herbicide strips from replication to replication.

Corn was planted in early May with a modified Buffalo, two-row, no-till planter that planted the corn in 30-inch rows and applied the insecticide and liquid N. Planting and application rates were 20,000 kernels of corn, 9 pounds of granular insecticide, and 65 pounds of N/acre. In late September we harvested and weighed corn silage from a 1/500-acre sample from each plot to estimate yields.

We divided the cover crops into three groups based on performance. The poorest performing group was the clovers. The clovers had 85% to 99% ground cover, with excellent vigor before the 1985 herbicide applications. After the 1985 harvest, the clovers plots had either good corn yields with very little ground cover or good ground cover with poor yields. By the end of the 1986 season the clover stands were nonexistent.

The most promising group of cover crops were the two trefoils and the two vetches, which had excellent vigor at the start of the study. Trefoil ground cover was between 40% and 75% and vetch ground cover was generally under 40% prior to the 1985 herbicide applications. By the end of the

Philip L. Koch is a soil conservationist with the Rose Lake Plant Materials Center, Soil Conservation Service, U.S. Department of Agriculture, East Lansing, Michigan 48823.

Table 1. Average no-till corn silage yields and percent ground cover of the four best performing perennial legumes, 1986.

Herbicide Combination	Average Corn Silage Yields (tons/silage/acre, field weight)				Percent Ground Cover at Corn Planting*				Percent Ground Cover at Corn Harvest†			
	Birdsfoot Trefoil	Crownvetch	Narrowleaf Trefoil	Cicer Milkvetch	Birdsfoot Trefoil	Crownvetch	Narrowleaf Trefoil	Cicer Milkvetch	Birdsfoot Trefoil	Crownvetch	Narrowleaf Trefoil	Cicer Milkvetch
Atrazine, 1 pound/acre; metolachor, 2.5 pounds/acre, clean-tilled	8.7	6.4	6.0	7.7	0	0	0	0	0	2	0	0
Atrazine 1.5 pounds/acre, metolachor 2 pounds/acre, paraquat 1 quart/acre	10.0	5.0	9.0	8.9	53	45	37	43	63	73	12	48
Cyanazine 1 pound/acre, cyanazine 2.5 pounds/acre, paraquat 1 quart/acre	3.4	1.3	5.9	7.1	90	82	33	48	87	88	55	70
Atrazine 1.5 pounds/acre, cyanazine 2 pounds/acre, paraquat 1 quart/acre	8.7	9.8	8.6	8.0	47	37	38	27	55	52	8	25
Atrazine 1.5 pounds/acre, DOWCO 356 0.5 pound/acre, Crop oil 1 quart/acre, parquat 1 quart/acre	11.5	11.1	8.1	9.4	80	60	57	60	55	52	10	57
Cyanazine 1.5 pounds/acre, DOWCO 356 0.5 pound/acre, paraquat 1 quart/acre	6.9	8.3	11.9	9.8	93	87	93	72	75	88	47	70
Atrazine 0.75 pound/acre, cyanazine 0.75 pound/acre, DOWCO 356 0.5 pound/acre, paraquat 1 quart/acre	11.6	11.0	10.6	10.6	88	82	83	62	80	82	27	65
Simazine 1.5 pounds/acre metolachlor 2.5 pounds/acre, paraquat 1 quart/acre	10.2	3.4	11.3	7.2	75	63	67	47	77	92	40	73
Atrazine 2 pounds/acre paraquat 1 quart/acre	11.1	6.4	10.5	6.6	55	40	37	37	43	53	8	25

*Includes cover crop, weeds, residue, etc.; evaluated April 4, 1986, average of three replications.
†Ground cover of the cover crop; evaluated November 5, 1986, average of three replications.

1986 season most plots containing these trefoils or vetches had increased ground cover to between 50% and 80%.

Success or failure of each plot is based on crop yield and percent ground cover, both of which are functions of the interaction of the cover crops with the herbicide treatments (Table 1). Birdsfoot trefoil has performed well with most of the treatments and seems to perform better with the lower atrazine rates. Crownvetch is very competitive and needs either the higher atrazine rates or atrazine plus an additional herbicide, such as Tandem or Bladex, that will help suppress the crownvetch. Narrowleaf trefoil did not tolerate atrazine, but performed very well with two of the three herbicide treatments that did not contain atrazine. Cicer milkvetch performed very well with the herbicide treatments containing Tandem.

Performance of alfalfa fell between the clover group and the trefoil/vetch group. Alfalfa started the study with excellent stands, which have been slowly deteriorating. Corn yields have been fair.

Based on corn silage yields and legume cover, we rated the cover crops in the following order: birdsfoot trefoil, crownvetch, narrowleaf trefoil, cicer milkvetch, and alfalfa. The four clovers have not shown potential for use as perennial cover crops.

The project is being carried out in cooperation with the Crop and Soil Sciences Department of Michigan State University and will be evaluated for another year to determine if the trends noted continue.

Evaluating cover crops for no-till corn and soybeans with banded herbicides

Steven B. Bruckerhoff and Jimmy Henry

Farmers in the Midwest are planting more and more corn (*Zea mays* L.) and soybeans (*Glycine max.*) with no-till techniques. They are using different techniques, herbicides, and types of equipment with varying degrees of success. As farmers gain experience, their demands also increase for more refined techniques and interest in the use of cover crops. Several legumes can add N for corn production. Studies have also shown that crownvetch can be used as a cover crop in no-till crop production for erosion control. The purpose of our study is to find a cover crop that can be maintained as a living mulch in a crop rotation. We are testing silage corn and soybeans, both leaving minimal residue, in conjunction with three banded herbicide treatments. Our goal is to suppress the cover crop to a point that it is not competitive with the grain crop but not to the point that it would not recover and provide ground cover for the following spring.

Study methods

The study, conducted at the Soil Conservation Service Plant Materials Center in Elsberry, Missouri, is in the second year of a planned five-year study. We initially planted six cover crops: Spreader 2 alfalfa (*Medicago sativa* L.), cicer milkvetch (*Astragalus cicer* L.), emerald crownvetch (*Coronilla varia* L.), hairy winter vetch (*Vicia villosa* L.), bigflower vetch (*Vicia grandiflora* Scop.), and downy brome (*Bromus tectorum*). Corn and soybeans were no-till planted into the cover crops. We are testing three different herbicide treatments. The plot design is a randomized complete block with three replications. There are a total of 54 plots in each block. Plot size is 20 feet by 40 feet.

Among the six cover crops, three are winter annuals (downy brome, bigflower vetch, and hairy winter vetch) and three are perennials (crownvetch, milkvetch, and alfalfa). Cover crop plots were planted with a Brillion[1] seeder on a conventional seed bed. Planting date was September 4, 1984.

The grain crops were planted in 30-inch rows with a Buffalo no-till slot planter. We planted full-season corn on May 10, 12, and 23, 1985, and May 5, 1986, at a population of 22,700 seeds/acre. We planted full-season soybeans June 9, and 21, 1985, and June 2, 1986, at a rate of 62 pounds/acre.

For the herbicide treatments we used a banded technique, with the nozzle directly over the row spraying a different chemical or rate than the nozzles between the row. A sprayer designed for banded herbicides was attached directly to the Buffalo no-till planter. We applied all herbicides, except the paraquat, at planting time. Paraquat was applied 4 to 5 days after planting. Table 1 shows the herbicide applications.

Steven B. Bruckerhoff is a conservation agronomist and Jimmy Henry is a soil conservationist, Soil Conservation Service, U.S. Department of Agriculture, Elsberry, Missouri 63343.

Yield was measured by harvesting two inner rows of corn 29 feet long (.0033 acre) and six rows of soybeans 29 feet long (.01 acre). We also made ratings for weed competition, percent ground cover, response to herbicides, stand counts, and percent of cover crops during the year.

We fertilized corn with 100 pounds/acre N 7 days after planting. The soybeans received no additional fertilizer.

Results and discussion

Corn yield averages for 1985 and 1986 varied between cover crops (Table 2). Yields in 1985 would have been somewhat higher in the downy brome and alfalfa cover crops if replication three could be omitted. Yields were lower in these replications because wet field conditions forced a later planting date.

Stands of crownvetch and milkvetch were weak and sparse at corn planting time in 1985 so another year for establishment was needed. Most plots had very dense stands in the spring of 1986. Where no herbicide was applied between the rows, the cover crops were too competitive for the corn. Milkvetch plots that received a half rate of herbicide between

[1]Mention of brand names does not imply endorsement by the Soil Conservation Service.

Table 1. Herbicide application treatments.

| Crop | Herbicide Treatments | |
	Over the Rows	Between the Rows
Corn	Bronco, 5 quarts/acre plus Attrex, 1.5 pounds active ingredient/acre	No herbicide
	Bronco, 5 quarts/acre plus Attrex, 1.5 pounds active ingredient/acre	½ of over-the-row rate
	Bronco, 5 quarts/acre plus Attrex, 1.5 pounds active ingredient/acre	2 pints paraquat/acre
Soybeans	Bronco, 5 quarts/acre plus Sencor, 1.5 pounds active ingredient/acre	No herbicide
	Bronco, 5 quarts/acre plus Sencor, 1.5 pounds active ingredient/acre	½ of over-the-row rate
	Bronco, 5 quarts/acre plus Sencor, 1.5 pounds active ingredient/acre	2 pints paraquat/acre

Table 2. Corn yield average by cover crop and herbicide treatment.

| | Corn Yields (bushels/acre) | | | | | | | | |
| | No Treatment Between Rows | | | One-half the Over-the-Row Rate | | | Two Pints Paraquat/Acre | | |
Cover Crop	1985	1986	Avg.	1985	1986	Avg.	1985	1986	Avg.
Downy brome	33.7	79.0	56.4	57.7	69.3	63.5	59.7	93.7	79.7
Crownvetch		18.3	18.3		61.0	61.0		50.3	50.3
Milkvetch		4.3	4.3		56.3	56.3		118.0	118.0
Alfalfa	54.7	23.7	39.2	69.0	76.0	72.5	31.7	44.7	38.2
Bigflower vetch	84.0		84.0	101.0		101.0	89.7		89.7
Hairy vetch	96.0		96.0	88.0		88.0	94.0		94.0

Table 3. Soybean yield averages by cover crop and herbicide treatment.

| | Soybean Yields (bushels/acre) | | | | | | | | |
| | No Treatment Between Rows | | | One-half the Over-the Row Rate | | | Two Pints Paraquat/Acre | | |
Cover Crop	1985	1986	Avg.	1985	1986	Avg.	1985	1986	Avg.
Downy brome	17.3	20.7	19.0	28.7	27.7	28.2	22.5	30.3	26.4
Crownvetch		1.0	1.0		3.0	3.0		6.0	6.0
Milkvetch		2.7	2.7		4.3	4.3		3.3	3.3
Alfalfa	12.3	7.3	9.8	11.0	6.3	8.7	17.3	9.3	13.3
Bigflower vetch	16.7		16.7	22.7		22.7	24.3		24.3
Hairy vetch	29.0		29.0	27.7		27.7	35.7		35.7

the rows had excellent yields. Yields on all other plots were variable.

The best corn yields occurred on the bigflower vetch and hairy vetch plots. Both species provided good ground cover, but failed to reseed themselves.

Corn yields in the downy brome varied over all the plots. Average yields were well below average crop yields on these soils. Downy brome did provide excellent ground cover and reestablished itself. Herbicides did not control alfalfa adequately, resulting in too much competition for the corn. Corn ears had poor fill, resulting in poor yield. On plots that received paraquat and on which corn was planted early, some success was achieved. The alfalfa provided excellent ground cover but recovered too well from the herbicide.

Soybean yields also varied greatly (Table 3). Average soybean yields with conventional tillage ranged from 30 to 40 bushels/acre. Soybean yields in hairy vetch were average to slightly below average. Yields in the downy brome plots with chemical control between the rows were slightly below average. All other yields ranged from below average to very poor.

Soybeans usually are not considered a good choice for a legume cover crop because they fix their own N. The reason here is if a living mulch cover crop can persist for several years, can crops be rotated between corn and soybeans?

This is the second year of a 5-year study and we have reached few conclusions. Bigflower vetch and hairy vetch perform well as cover crops but not as living mulches. It is possible that they are too viney to escape any of the chemicals applied over the row, resulting in a total kill. Cold winter nights with insufficient snow cover also may have contributed to the low cover. Winter wheat froze the same year, which is not normal. Crownvetch can provide both a cover crop and a living mulch, but not with this herbicide treatment. Alfalfa becomes too competitive if not controlled adequately, especially if wetness delays planting. Downy brome, although it contributed no N, has possibilities if better weed control and yields can be attained.

This study will be moved to an upland site in the future. Also, new legumes and grasses will be added into the project.

Efficacy of burndown herbicides on winter legume cover crops

S. M. Dabney and J. L. Griffin

The cost of herbicides for cover crop burndown limits their use in conservation tillage production systems. We conducted studies from 1984 to 1986 at the Rice Research Station, Crowley, Louisiana, and during 1985 and 1986 at the Idlewide Research Station, Clinton, Louisiana, to evaluate herbicide efficacy on several legume cover crops.

Study methods

At Crowley we applied glyphosate at 1.5 and 2.5 pounds of active ingredient/acre, paraquat at 0.5 and 1.0 pounds active ingredient/acre, and glufosinate at 0.75 and 1.5 pounds active ingredient/acre on Tibbie crimson clover (*Trifolium incarnatum*), Woogenellup or Mt. Barker sub clover (*T. subterraneum*), and Cahaba white common vetch (*Vicia sativa*) or hairy vetch (*Vicia villosa*) during April and May of each year.

At Clinton, we tested six rates of glyphosate (0.5, 0.75, 1.0, 1.5, 2.0, and 2.5 pounds active ingredient/acre) with the same two rates of paraquat and glufosinate used at Crowley. At both locations we applied paraquat in a spray volume of 20 gallons of water/acre; glyphosate and glufosinate were applied in 6 to 10 gallons of water/acre. We added nonionic surfactant to all spray mixtures at a rate of 0.5% by volume.

The herbicide treatments along with a mowing treatment (2 inches above the soil surface) were imposed at Clinton in mid-April each year on the same legumes considered at Crowley. At Clinton we also used Austrian winter peas (*Pisum arvense*), rough pea (*Latharys hisutus*), Woodford bigflower vetch (*V. grandiflora*), Bigbee berseem clover (*T. alexandrinum*), Chesapeake red clover (*T. pratense*), Segrest ball clover (*T. nigrescens*), Amclo arrowleaf clover (*T. vesiculosum*), Gulf annual ryegrass (*Lolium multiflorum*), and Coker 916 wheat (*Triticum aestivum*). In 1986 we also included Chief crimson clover, and Meteroa sub clover.

At Clinton we also evaluated volunteer vegetation on fallow areas, dominated by cutleaf evening primrose (*Oenothera laciniata* Hill). We noted growth stage and took samples for biomass estimation prior to herbicide application at both locations. At Crowley we arranged 5- by 30-foot herbicide plots in a split-plot randomized complete block design with three replications. At Clinton we used 7- by 7-foot herbicide and mowed plots in a strip-block design with four replications.

S. M. Dabney is an assistant professor, Department of Agronomy, Louisiana State University, Baton Rouge, 70803 and J. L. Griffin is an associate professor, Rice Research Station, Crowley, Louisiana 70527. Both are employees of the Louisiana Agricultural Experiment Station, Louisiana State University Agricultural Center. Approved for publication by the Director of the Louisiana Agricultural Experiment Station as manuscript number 86-09-0374.

We rated percent cover crop kill 15 to 20 days after treatment.

Results and discussion

Response of the cover crops to mowing and herbicide treatments was variable but consistent between locations. A single mowing in April essentially killed some cover crops, notably the vetches, crimson clover, and winter wheat, at Clinton (Table 1). Most of the other clovers and ryegrass were more difficult to control. Among the herbicides, glufosinate was more effective than glyphosate or paraquat.

At Clinton 0.75 pound/acre of glufosinate was less than 90% effective only on arrowleaf clover and two sub clover cultivars. Glufosinate at 1.5 pounds/acre provided at least 93% control on all cover crops except Mt. Barker sub clover. At Crowley 0.75 pound/acre of glufosinate controlled all the cover crops except Woogenellup sub clover (Table 2).

Cover crop response to glyphosate and paraquat treatments varied significantly. At rates of 1 pound/acre at Clinton, paraquat was more effective than glyphosate on most of the vetches, as well as on ryegrss and crimson, arrowleaf, and ball clovers (Table 1). Glyphosate was more effective than paraquat on sub clovers. But neither herbicide acceptably

Table 1. Growth stage of winter cover crops, biomass at time of treatment, number of years observed, and percent control 2 weeks after mowing or herbicide application, Clinton, Louisiana, 1985-1986.

Cover Crop	Growth Stage*	Biomass (pounds/acre)	Number of Years Observed	Percent Control† by Herbicide Treatment‡										
				0	1	2	3	4	5	6	7	8	9	10
Austrian winter peas	1	2,510	2	86	47	53	81	83	81	91	100	100	99	100
Caley peas	1	3,470	2	82	25	37	51	54	41	62	99	99	89	91
Hairy vetch	1	3,710	2	95	30	45	48	63	56	73	99	99	86	94
Bigflower vetch	4	3,280	2	61	35	49	61	81	81	86	100	100	95	98
Common vetch	2	3,600	2	85	46	72	85	88	91	86	100	100	97	97
Tibbie crimson clover	4	5,629	2	95	60	82	83	87	95	90	100	100	99	99
Chief crimson clover	3	5,063	1	77	56	75	87	97	94	99	100	100	97	97
Berseem clover	0	3,500	2	27	62	70	73	95	94	96	99	98	46	74
Red clover	2	3,470	2	11	17	33	39	51	52	63	94	97	34	50
Ball clover	2	3,470	2	44	36	57	61	76	87	89	98	98	69	77
Arrowleaf clover	1	4,930	2	59	23	35	47	67	64	78	86	93	60	75
Mt. Barker sub	3	4,250	2	27	21	25	36	45	59	68	70	77	17	29
Woogenellup sub	4	3,940	2	18	59	64	83	89	95	96	84	94	44	63
Meteroa sub	3	3,940	1	14	38	53	67	81	91	97	97	100	39	44
Fallow§	2	1,680	2	18	28	36	55	83	75	88	100	100	80	95
Wheat	3	3,740	2	90	99	99	89	100	100	100	97	100	99	100
Ryegrass	2	4,030	2	15	36	53	74	91	96	98	94	96	97	100

*Growth stage ratings: 0 = vegetative, 1 = less than 10% bloom, 2 = less than 50% bloom, 3 = full bloom, 4 = late bloom seed development.
†Percent control ratings based on a scale of 0 to 100; 0 = no kill, 100 = complete kill.
‡Herbicide treatments:

   0 = mowing only, no herbicide   4 = glyphosate, 1.5 pounds ai/acre   8 = glufosinate, 1.5 pounds ai/acre
   1 = glyphosate 0.50 pound ai/acre   5 = glyphosate 2.00 pounds ai/acre   9 = paraquat 0.5 pound ai/acre
   2 = glyphosate 0.75 pound ai/acre   6 = glyphosate 2.50 pounds ai/acre   10 = paraquat 1.0 pound ai/acre
   3 = glyphosate 1.00 pound ai/acre   7 = glufosinate 0.75 pound ai/acre
§Growth stage and percent control ratings refer to cutleaf evening promrose.

Table 2. Winter legume cover crop growth stage, biomass, number of years observed, and percent control 2 week after herbicide treatment, Crowley, Louisiana, 1984-1986.

Cover Crop and Month	Growth Stage*	Biomass (pounds/acre)	Number of Years Observed	Percent Control† by Herbicide Treatment‡					
				4	6	7	8	9	10
Hairy vetch									
April	0	1,592	2	47	49	99	100	87	98
May	3	2,852	2	83	91	100	100	100	100
Cahaba white common vetch									
April	2	2,305	1	53	77	98	98	88	100
May	—	—	0	—	—	—	—	—	—
Tibbie crimson clover									
April	2	3,178	3	34	51	97	100	91	98
May§	3,618	2		—	—	—	—	—	—
Mt. Barker sub clover									
April	2	1,500	2	33	39	96	100	34	78
May	3	4,166	2	51	57	90	95	91	100
Woogenellup sub clover									
April	2	2,290	1	58	97	68	95	33	53
May	—	—	0	—	—	—	—	—	—

*Growth stage ratings: 0 = vegetative, 1 = less than 10% bloom, 2 = less than 50% bloom, 3 = full bloom, 4 = late bloom seed development.
†Percent control ratings: 0 = no kill, 100 = complete kill.
‡Herbicide treatments:

   4 = glyphosate, 1.5 pounds ai/acre   7 = glufosinate 0.75 pound ai/acre   9 = paraquat 0.5 pound ai/acre
   6 = glyphosate 2.50 pounds ai/acre   8 = glufosinate 1.5 pounds ai/acre   10 = paraquat 1.0 pound ai/acre
§Crimson clover not rated since it had matured prior to herbicide application.

controlled the cover crops at the 1 pound/acre rate. We observed similar trends at Crowley, although paraquat appeared more effective on sub clover when applied in May than in April (Table 2).

We found large differences in response to glyphosate between the sub clover cultivars at both locations. Mt. Barker was much more difficult to kill than Woogenellup. We also observed this difference for the other herbicide treatments tested at Clinton.

Our results indicate that rather high rates of currently labeled herbicides are needed to effectively control a number of legume cover crops in no-till systems. The costs of herbicide treatments must be considered in any economic analysis of such cropping systems. Differential responses to herbicides between cultivars of a given species could be an important selection criteria. Mowing may be an economical alternative to herbicide use for certain cover crops. If glufosinate is labeled, it could play an important role in future conservation tillage plantings involving a wide range of legume cover crops. More research is warranted on the efficacy of burndown compounds in combination with preemergence herbicides. Tank mixtures of herbicides may enhance control, possibly allowing for both reduced rates and costs.

Interim legume crops and herbicides for replacing endophytic tall fescue with new cultivars

H. D. Kerr, A. M. Bagegni, and D. A. Sleper

Forage quality of tall fescue infested with the endophytic fungus *Acremonium coenophialum* is such that grazing animals perform poorly in daily weight gains, milk production, and general health. Pastures started with seeds free of this fungus are of higher quality. Animal performance on fungus-free pastures measurably improves compared to infected pastures of the same cultivar.

Animal weight losses alone in Missouri are valued at $50 million annually for beef producers. About 80%, or 5 million acres, of tall fescue pastures in Missouri are infected with the fungus. Pastures in other states are in a similar status because tall fescue pastures were started in most of the growing areas before the endophytic fungus was known to be a problem in tall fescue. Over 14 million acres are probably infested in the United States.

Seed handling and long-term storage methods will alleviate the problem of viable seed infection. Forage grass breeders have developed new, higher quality cultivars that are free of the fungus. The main problems in converting pastures to the better cultivars are costs, the labor and time needed, pasturage down time, and potential soil erosion from tilling sloping lands now under tall fescue. Replacing present pastures of tall fescue with new cultivars must be a reliable procedure because conversion costs are the major limiting factor. Missouri provides monetary assistance for helping producers defray costs, about $35/acre for conservation tillage emphasis.

Herbicides are now available to effectively control grass and dicot species of unwanted vegetation even in growing pasture and legume crops. Again, the problem is cost of the operation. Use of an interim cash or herbage crop can defray the cost of converting fungal-infected pastures to clean, higher quality tall fescue pastures. Protecting the remaining topsoil from erosion and providing a smooth, timely transition from low- to high-quality forage is desirable in such operations. Our research is directed at developing a conversion system that will protect the environment and soil.

Fall sprays of glyphosate or paraquat controlled old tall fescue stands quite well. Planned liming and fertilization should be done during the season prior to spraying the endophytic tall fescue. Maximum use of the pasture in summer and early fall helps weaken the old tall fescue, making it easier to eliminate with herbicides. Fall rains renew the grass growth; growers should apply paraquat or glyphosate when the regrowth is about 6 to 8 inches tall and growing well. A first treatment should be made in October with 1

H. D. Kerr is an associate professor, A. M. Bagegni is a research associate, and D. A. Sleper is a professor, Department of Agronomy, University of Missouri, Columbia, 65211.

pound/acre active ingredient glyphosate or 0.5 pound/acre paraquat. A second treatment about 3 weeks after the first with paraquat at 0.25 pound/acre usually is needed to adequately control tall fescue. One spray of glyphosate is usually adequate.

Red clover or soybean legumes are reliable crops for seeding into sprayed tall fescue sod. Herbicides are labeled for both fall and spring soil treatments for soybean plantings. But clover drilled into dead sod in February or March did not require spray for weeds as soybeans did. Endophyte infected seedlings from tall fescue stands were controlled in red clover with 0.125 pounds/acre fluazifop and in soybeans at planting with 0.125 pounds/acre imazaquin.

Soybeans planted into fall-treated tall fescue sod yielded 47.3 bushels/acre as endophytic tall fescue was eliminated in preparation for planting fescue-free cultivars in the fall after soybean harvest. The soybeans fixed extra N to support the fall-sown grass, and the stubble was an ideal seedbed for drilling without erosive tillage. Red clover produced 1.5 tons of high quality hay while allowing complete herbicidal control of the infected tall fescue with both fall and summer herbicide use. Like soybeans, the red clover stubble was a proper seedbed for the tall fescue drilling in the fall, and the clover fixed adequate N for establishment of high-quality tall fescue cultivars free of the endophyte fungus.

Legume effects on soil erosion and productivity

R. R. Bruce, S. R. Wilkinson, and G. W. Langdale

To address the effects that legumes may have on soil erosion and productivity, it first is necessary to describe the soil erosion-productivity problem. Defining the influence of erosion on soil characteristics and the consequent impact on soil productivity will provide a basis for analyzing legume effects.

We define soil productivity as the potential rate that a soil system can accumulate energy in the form of vegetation. Productivity, therefore, reflects future production. We use crop yield for the most part as a useful proxy for soil productivity. Although crop yield reflects historical production, it is measured readily, and when accompanied by adequate assessments of soil, climate, vegetation, slope, and management variables, soil productivity may be evaluated.

The soil erosion-productivity problem

L. D. Baver addressed the question "How serious is soil erosion?" in a general session of the Soil Science Society of America in 1950 (10). Some of his remarks were as follows: "The erosion problem lends itself either to a calm, considerate, and factual analysis or to a blatant expression of half-truths for the purpose of exciting people to action... For example, it does take nature several thousands of years to build topsoil to plow depth, when nature starts from scratch on bedrock. This statement of truth becomes a half-truth even an untruth when applied to all exposed subsurface material.... When we ask the question, 'How serious is soil erosion?', we probably have phrased our query wrongly. A better question would be, 'How serious is poor soil management?' Poor soil management affects more people than soil erosion.... If there

R. R. Bruce, S. R. Wilkinson, and G. W. Langdale are soil scientists, Southern Piedmont Conservation Research Center, Agricultural Research Service, U.S. Department of Agriculture, Watkinsville, Georgia 30677.

is continued erosion under the best management practices in agriculture then erosion becomes so serious as to enter the alarm or disaster stage." Somehow these words seem appropriate.

Our discussion is restricted to the influence of erosion by water on field soils and the behavior of soils as media for crop growth. Soil erosion researchers have been preoccupied with the volume or mass of sediments removed from fields. Although the quantity of sediment has important off-field consequences, the consequences to productivity are less certain (63, 64). The effects of soil erosion on the productivity of a field, using crop yield as an index, is difficult to assess, not only because erosion occurs nonuniformly and is a selective transport process, but also because of inherent soil variability and few, if any, historical references or benchmark conditions. It therefore is necessary to adhere to unifying principles of soil and crop science in the context of climate, hydrology, and geology to avoid conflicting interpretations. Although erosion influences are global, research results can be extrapolated only when soil and climate information accompany crop culture and species information.

Crop yield and productivity. Within the past 5 years literature on the subject of erosion-induced soil productivity losses has been reviewed thoroughly (53, 63, 72). Included among the research reports are results of artificial simulation of erosion by desurfacing a specified depth of soil as well as natural erosion. Most investigators have expressed the soil loss-yield relationship either as an absolute yield decline per unit of soil loss or as a proportional decline in relation to some measure of progressive erosion. The Soil Conservation Service bases its classification of erosion on the cumulative fraction of topsoil removed (Table 1) (66).

Using data from experiments relating yield to specified depths of soil removed by artificial means to describe natural erosion's relation to crop yield can be misleading. Soil erosion selectively transports soil material during each runoff event and may seldom involve an entire field. Between events, a variety of operations may occur in crop culture that affect future runoff events and soil transport. Among the most important operations are tillage and crop residue handling. Removing soil to a specified depth in a specified manner does not allow for studying the consequences of progressive soil loss under a given set of crop cultural operations. To avoid misinformation and dissemination of half-truths researchers must exercise care in framing research questions and in interpreting experimental results. Farmers will adopt alternative cropping procedures only when scientists are able to make clear statements about the influence of erosion on crop production economics.

Some reviewers have chosen to compare the erosion-induced losses in crop yield among soil orders, particularly Alfisols, Mollisols, and Ultisols, on which most of the research has been done. We believe the soil order is too inclusive for such a comparison and can lead to contradictory data interpretations. For example, Ultisols include a range in soil water and temperature regimes, resulting in humus-poor and humus-rich suborders in addition to other characteristics that evolve (Table 2). Therefore, we suggest that a classification level should identify soil features useful in describing the potential effect of soil erosion on productivity; that is, soil family.

One of the early investigations of soil erosion and productivity almost uniquely provides yield and soil loss data over 10 years with accompanying soil organic matter content (15). These data show a cumulative soil loss of 5.5 inches or 894 tons/acre, a 97% decline in corn yield, and a steady decline in soil organic matter from 2.05 to 0.5% (Table 3). This investigation was conducted on a Muskingum silt loam (fine-loamy, mixed, mesic Typic Dystrochrept) without fertilizer applications. Although it is risky to interpret the rate of yield decline without weather data, the rapid decline in yield and soil organic matter content during the first 4 years is noteworthy. Over the same period the researchers noted 15% less soil was lost and average yield was 27% higher with applications of 150 pounds/acre of 4-10-6 fertilizer.

The Palouse Basin is one of the most erodible areas in the United States, and investigations there have yielded insights into the complexity of the erosion-soil productivity relationship. With slopes of up to 50% and slope lengths of 165 to 330 feet, these loessial hills are devoted primarily to winter wheat production, with secondary crops of spring barley, dry peas, and lentils. In a representative area in eastern Whitman County, Washington, where Haploxerolls and Argixerolls in the fine-silty, mixed, mesic families predominate, average annual erosion in the county was estimated at 12 tons/acre/year, but varies considerably with land capability class. Winter wheat yield varies with topsoil depth (Figure 1). Despite the yield-depressing effect of erosion, yields have increased with time but more so on deep topsoils than on shallow topsoils. To separate the effect of erosion from the yield-increasing impact of technology, the researchers combined these two response functions to permit yield projections over time for each land capability subclass. For land classes II and III yield reductions were small and easily masked by technology gains. On the other hand, erosion took a heavy yield toll on class IV and VI lands (Figure 2) (57).

A study in the Southern Piedmont near Watkinsville, Georgia, showed that soybean yields on moderately and severely eroded sites on 40 farm fields were 25% to 50% lower than on slightly eroded sites, respectively (70). Furthermore, 42% of the area of an average field was either moderately or severely eroded. Yields on those areas, therefore, could be 25% to 50% lower (Table 4). In another study Langdale and associates (51) observed a 42% corn yield reduction due to historic erosion. These observations were on Cecil or Pacolet soil series (clayey, kaolinitic, thermic Typic Hapludults). Studies in the North Carolina Piedmont (21, 65) on Cecil and Georgeville soil series (Typic

Table 1. Definitions of soil erosion classes used by the Soil Conservation Service (66).

Class	Definition
Slightly eroded or class 1	Soils that have lost less than 25% of the original A horizon or of the uppermost 8 inches (0.2 m) if the original A horizon was less than 8 inches (0.2 m) thick.
Moderately eroded or class 2	Soils that have lost between 25% and 75% of the original A horizon. Ap in most cultivated areas consists of a mixture of A horizon and material below.
Severely eroded or class 3	Soils that have lost 75% or more of the original A horizon. Material below the A horizon is exposed at the surface in cultivated area.

Table 2. A few suborder specifics of Ultisols (66).

Aquults	Udults	Humults	Ustults	Xerults
Aquic	Udic		Ustic	Xeric
Thermic or warmer	Mesic or warmer Humus poor	Mesic or warmer Humus rich	Thermic or warmer Low organic matter	Mesic or thermic Low to moderate organic matter
Atlantic and Gulf Coast	Southeast United States	Mountain areas of United States with high rainfall	Few in the United States or Puerto Rico	Not extensive in the United States

Table 3. Corn yields, cumulative soil loss, and soil organic matter on an eroding Muskingum silt loam on a 12% slope and with no fertilizer, 1933-1942 (15).

	Corn yield	Cumulative Soil Loss	Soil Organic Matter
Year	(pounds/acre)	(tons/acre)	(%)
0	3,420	0	2.05
1	1,482	36	—
2	1,786	119	1.75
3	955	220	1.26
4	1,089	308	
5	839	426	
6	607	559	1.00
7	143	662	
8	304	793	
9	89	894	0.50

Hapludults) demonstrated the potential confounding effect of landscape position in assessing the soil erosion-productivity relation. These studies, compared to the Watkinsville, Georgia, studies, also introduce the modifying influence of crop culture upon the yield-soil erosion class relationship. In many areas of the North Carolina Piedmont it seems that greater tillage depths have persisted in concert with more sod crops than in Georgia. This may explain the greater Ap horizon depths and less difference in clay content among the erosion classes in North Carolina.

McDaniel and Hajek (54) reported the effects of past erosion on crop yields and several properties of soils in the Lower Coastal Plain, Black Belt Prairies, and Tennessee Valley of Alabama. Their work was on farm fields (cotton, corn, or soybeans) where slight and moderate erosion sites of a given soil series could be identified. The 3-year study included 12 soil series and 131 fields representing 4 soil orders and 6 suborders, most of which were Paleudults. Yields declined on moderately eroded sites in 65% of the fields; average yield reduction was 22%. Surface soil thickness and organic matter declined, clay and Fe_2O_3 increased, and surface soil P was generally lower on areas where yields were likely to be lower.

Lal (50) reported some dramatic effects of soil loss on maize and cowpea yields over a 10-year period in southwest Nigeria on an Oxic Paleustalf with a medium to light textured surface and a sandy clay B21 horizon. Yields declined exponentially with soil loss on 1% to 15% slopes. In that study 107 pounds/acre N, 23 pounds/acre P, and 27 pounds/acre K were applied, and no mechanical land preparation was done after deforestation.

Battiston and associates (9) studied the effects of erosion

Table 4. Average land area of 40 farm fields occupied by each erosion class of Cecil-Pacolet soils near Watkinsville, Georgia.

Erosion Class	Land Area (%)	Range Among Fields (%)
Slight	56.7	32.7-80.4
Moderate	32.4	14.1-58.3
Severe	9.4	1.4-23.1
Alluvial	1.5	0-14.8

on soil productivity in Waterloo County, Ontario. Their work involved a range in erosion influence on several soil series with a range in subsoil and surface soil texture. These soils are mostly sandy to fine clayey, mixed, or illitic calcareous, mesic Typic Hapludults. Corn yields on severely eroded sites were 34% lower than on slightly eroded sites in 1982 and 43% less in 1983. Yields of spring oats and barley on severely eroded plots averaged 50% lower than on slightly eroded plots. All areas of each field were tilled the same, planted in one day, and received the same applications of fertilizer and pesticide.

Yield reductions resulting from soil loss that occurs on sloping lands under cultivation vary, but frequently are large enough to have economic impact at some time. The relationship between yield decline and soil loss probably is exponential (15, 50, 63, 64). However, this relationship is conditioned by the specific soil, climate, initial conditions, crop and culture, or soil and crop management system employed. Analyzing erosion's influence will only be useful when conditioned by these factors.

Soil volume characteristics. Hall and associates (27)

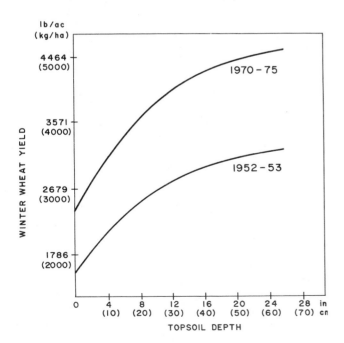

Figure 1. Wheat yield-topsoil depth response functions fit to sample observations made in the early 1950s and again about 20 years later (57). Both curves are Mitscherlich-Spillman functions.

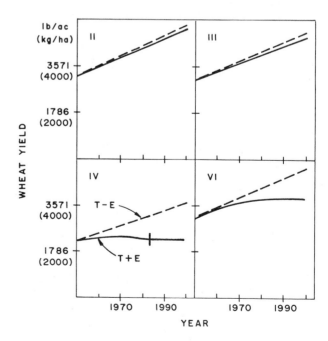

Figure 2. Maximum potential wheat yields assuming historic technology and no erosion (T-E) compared with yields assuming historic technology (T+E) for land capability classes of the Palouse, projected for 1950-2000 (57).

discussed soil-forming processes relative to erosion. They emphasized the complexity of soil formation and the hazards in generalizing about the impacts of erosion. As dramatic as erosion damage can be, the recovery of crop production on much eroded land also can be dramatic. Incorporation of organic matter in the surface of a parent material is usually the first indication of soil formation (27). One study showed that A horizons can form under forest and grass in less than 50 years (27). In fact, in many strip mine spoils high productivity is realized in 5 or 10 years (62).

Stocking and Peake (64) concluded that productivity decline by soil erosion is influenced by (a) changes in soil and profile characteristics that affect crop water and nutrient supply, (b) climatic characteristics of the zone in question, and (c) the type of crop considered. Although they elaborate further, we believe it is possible to commit the influence of soil erosion upon productivity to greater definition by focusing on the soil volume dimensions and status that determine crop performance in a given set of weather conditions. Over the past 50 years almost every researcher who has studied some aspect of erosion has offered reasons for yield decline as erosion occurs. But generally, data gathering has been fragmetary and soil system characteristics incomplete. We have selected research at Watkinsville, Georgia; Ibadan, Nigeria; and Waterloo, Ontario, as a basis for exploring this topic.

Watkinsville, Georgia (83° 24'W, 33° 54'N). The study was conducted on clayey, kaolinitic, thermic Typic Hapludult derived from granite, gneiss, and amphibolites, with considerable variation in depth to parent material and parent geology. Pedons in 40 farm fields on sites classified as slightly, moderately, and severely eroded were selected for study (70). A truncation of the pedon was shown to relate to erosion class. Although the Ap characteristics were modified in relation to erosion class (Table 5), the subsoil characteristics and dimensions also were modified (Table 6). Is it valid to attribute these differences to erosion? Crop yield apparently was affected the most by topsoil differences over the range of pedons sampled where depths to the C horizon were

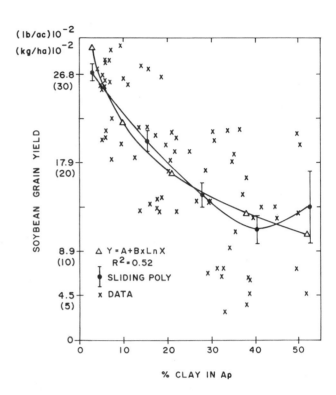

Figure 3. Soybean grain yield as a function of clay content of the Ap horizon on slightly, moderately, and severely eroded sites of Cecil-Pacolet soils.

always greater than 20 inches. A landscape position study of these pedons showed that most of the severely eroded sites were in a lower linear slope position that affords an opportunity for greater volumes of runoff water and possibly elevated velocities over the surface.

In this study the decrease in soybean yield with increasing clay content of the Ap horizon was highly significant and corresponded closely to the change from slightly to moderately to severely eroded class (Figure 3). This is a logical consequence of continued cultivation as erosion progresses on soils with clayey subsoils. Tillage and cropping practices, however, moderated the effect of soil loss from these sites on clay content in the Ap horizon. For example, tillage that includes moldboard plowing at a constant depth over time will be different than disk harrowing at a much shallower depth.

The researchers noted that there was a significant difference in soil water regime among erosion classes (Figure 4) and that only on the slightly eroded sites did rainfall during the critical fruiting period affect yield (Figure 5). They concluded that there was a significant reduction in rainfall infiltration associated with the increase in clay content, which created a less favorable soil water regime (18, 70). Further data analysis revealed that the C content in the top 4 inches related to yield on the severely eroded or high-clay-content sites but not on the slightly and moderately eroded sites. This suggests that the organic matter status of the surface soil in relation to the clay content is pivotal in assessing the influence of erosion on productivity, assuming adequate depth of solum

Table 5. Topsoil textures on slightly, moderately, and severely eroded sites on 40 farm fields on Cecil-Pacolet soils near Watkinsville, Georgia (70).

Erosion Level	Percent Sand		Percent Silt	Percent Clay	
	Mean	Range	Mean	Mean	Range
Slight	75	53-87	17	8	3-20
Moderate	63	47-78	15	22	10-36
Severe	49	30-63	14	37	26-53

Table 6. Mean depths of soil horizons on slightly, moderately, and severely eroded sites in 40 farm fields near Watkinsville, Georgia (70).

Horizon Description	Mean Depth (inches) for Erosion Level		
	Slight	Moderate	Severe
Ap depth	7.5	4.7	4.2
BE or B/A thickness	5.0 (37)*	3.5 (9)	3.7 (4)
Depth to bottom of Bt	35.5	25.8	21.8

*Number in parenthesis is the number of fields having these horizons. Not all pedons had an intermediate horizon between the Ap and Bt.

for water storage and accompanying root proliferation.

Ibadan, Nigeria (6° to 8° N and 3° to 6° E). Intensive soil erosion investigations began here in 1970 and are continuing (*47, 48, 49, 50*). Average rainfall is 43 to 59 inches, distributed bimodally such that there are two growing seasons. The soils are Oxic Paleustalfs derived from fine-grained biotite-gneiss and schist. Soils have a light to medium textured surface over a sandy clay to clay subsoil. Experiments were set up on newly cleared land and fertilized maize and cowpeas were grown on slopes from 1% to 15%. Although soil loss increased with percent slope and related significantly to yield decline, organic matter losses and degradation of physical properties influencing soil water regime influenced the yield decline on a given slope. As soil loss increased, Ca, K, Mg, P, N, pH, cation exchange capacity, infiltration rate, available water, and water retention at 0.1 bar decreased.

Waterloo County, Ontario (80° 35' W, 43° 25' N). Researchers here concluded that the effect of erosion on soil properties depended largely on the physical and chemical characteristics of a soil prior to erosion (*9*). One of the soils in the study was a Burford soil series, a coarse loamy over sandy skeletal, mixed, calcareous, mesic Typic Hapludult. It has an inherently medium- to coarse-textured surface underlain by very coarse, gravelly subsoils. This soil exhibited a considerably coarser surface soil texture in the eroded state. On the other hand, the Huron soils (clayey, illitic, calcareous, mesic Typic Hapludults) possessed a medium- to fine-textured A horizon underlain by fine- to very fine-textured B and C horizons. Erosion resulted in a finer textured surface soil. In general, eroded phases had lower organic matter contents, lower fertility, and poorer physical characteristics than the slightly eroded phase.

Table 7 shows the corn yield reductions attributed to severe

erosion and the major causes on the Waterloo county sites (*9*). Researchers attributed plant N deficiences to denitrification on the clayey soils and to rapid leaching on the coarse-textured soils during wet periods. In the drier season of 1983, N supply was the lesser problem and soil water the greater problem. Phosphorous and K did not appear yield-limiting. Erosion greatly reduced organic matter content in both fine- and coarse-textured soils (Table 8). Measurements of soil water at specific times related statistically to yield, indicating a significant reduction of rainfall infiltration on eroded sites.

The critical role of organic matter in plant nutrient and water supply emerges as a primary focus in any consideration of soil erosion-induced soil productivity. This conclusion is supported by global investigations. When the organic matter status of the surface soil is not maintained in the culture of a crop on a landscape site, then low infiltration occurs and low rainfall amounts are stored. In addition, the

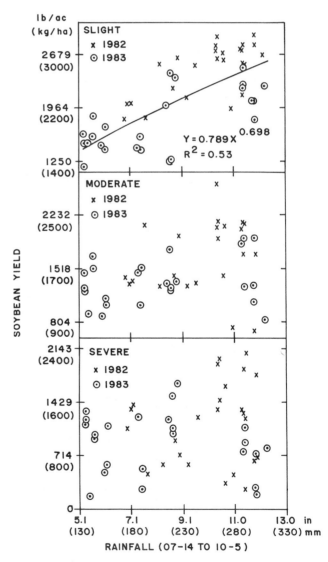

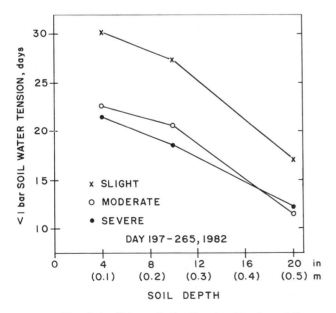

Figure 4. Number of days that soil water tension at three depths was greater than 1 bar during indicated interval in 1982 on slightly, moderately, and severely eroded sites of Cecil-Pacolet soils (*18*).

Figure 5. Soybean yields on slightly, moderately, and severely eroded sites on 40 fields in relation to rainfall amount on those sites between July 14 and October 5, 1982 and 1983 (*70*).

vulnerability of the surface to particle detachment and transport of soil material increases. Yield losses following periods of well-vegetated noncultivation are large in the initial stages of cultivation in runoff-prone situations because the higher organic matter soil volumes are transported first. There is then a tendency toward a yield plateau until erosion introduces another yield-reducing solum characteristic. This provides rationale for an exponential decrease of yield with cumulative soil loss. Nitrogen nutrition is involved inherently and contributes signficantly to crop performance.

Researchers frequently have cited available water capacity and soil water retention as a dominant effect of erosion on crop yield. Care needs to be exercised in arriving at such a conclusion. Organic C status influences both nutrient factors and factors associated with soil water supply, including infiltration and water movement. It is difficult, therefore, to isolate soil variable effects. Infiltration is dependent on a water-stable porous structure at the soil surface, which is a function of organic C supply and biological activity. Thus, it seems likely that the effect of soil erosion on infiltration is primary. If increases in available water capacity and water retention are to have significant impacts on yield, considerable depth of soil must be affected and the site must be in a climatic region in which crop performance relates closely to stored soil water and, thus, might benefit from additional storage capacity.

Yield recovery and maintenance

In general, if land users are to continue to use the soil asset in crop production, soil erosion must be kept within certain tolerance limits, and landowners must adopt cropping practices attuned to the potential of the soil and climate resource. This may include the use of engineering structures to manage runoff, such as grassed waterways and terraces, and adopt-

ing forest, grass, or other long-term rotations.

In many instances the landowner may require a crop culture aimed at yield recovery and yield maintenance. Based on the research cited previously, a crop culture that achieves yield recovery from historic erosion and maintains yield will (a) have a low risk of soil surface crusting that prevents achieving a satisfactory plant population, (b) provide nutritional requirements of the crop, and (c) achieve rainfall infiltration and water storage in the soil volume of major rooting that is feasible for the soil, climate, and landscape situation while maintaining the resource.

The quantity and quality of organic matter inputs into the culture, therefore, become paramount because both affect soil crusting and nutrient and water supply to the crop. Equally important is the management of the organic inputs and associated soil disturbance or tillage. In view of a recent review of restoration of crop productivity on eroded soils by Frye and associates (25), we will cite a few examples of research efforts that examine the influence of legumes in crop systems on soil characteristics, soil and water loss, and crop performance. Subsequently, we will discuss legume biomass inputs relative to the soil deficiencies created by soil erosion.

Soil organic matter and soil aggregation. Researchers have investigated the relationship between soil organic matter and soil aggregation for more than a half century (4, 30, 43). Aggregate stability relates to soil N in a similar manner as to soil organic matter. Clay and Fe_2O_3 content modify these relationships.

Bolton and Webber (14) measured the effect of cropping systems on the soil aggregation and organic matter content on a fine-loamy, mixed, mesic Typic Argiaquall in southern Ontario. At the 0- to 4-inch depth soil aggregation and organic matter content decreased in the order bluegrass > alfalfa-brome > oats > continuous corn. These effects were the same at the 4- to 8-inch depth but not at the 8- to 12-inch depth. The clipped 50-year-old bluegrass sod and 30 years of continuous corn provided a range in crop cultivation intensity. The second year of alfalfa-brome in the 5-year rotation of corn, corn, oats, alfalfa-brome, alfalfa-brome demonstrated a dramatic ameliorative effect. Adams and Dawson (1) reported that water-stable aggregates, organic C, and infiltration rates increased on a clayey, kaolinitic, thermic Typic Hapludult as the proportion of sod in the rotation increased. These data are representative of a body of research relative to the influence of grass-legume crops upon the accumulation of soil organic matter, increase in soil aggregation, and improved water infiltration (16, 17, 39). Accompanying the growth of these crops is a reduction in intensity or frequency of tillage.

Organic matter, soil water, and soil loss. Mills and associates (56) examined the effect of cropping and tillage on erosion probabilities at Watkinsville, Georgia, on Cecil-Pacolet soils (clayey, kaolinitic, thermic Typic Hapludults). They studied a series of summer and winter crops, including soybeans, grain sorghum, barley, wheat, and crimson clover, on a 6.7-acre watershed over a 10-year period. During the last 3 years when grain sorghum was no-till planted into crimson clover, the probability that soil loss would be more than 0.9 tons/acre was almost zero. For conventionally tilled soy-

Table 7. Corn grain yield reduction on severely eroded sites (9).

	1982		1983	
Site	Yield Reduction (%)	Major Causes(s)	Yield Reduction (%)	Major Causes(s)
2	38	N, H_2O	52	H_2O
3	43	H_2O	38	H_2O
5	32	H_2O	35	H_2O
6	66	H_2O, N	59	H_2O

Table 8. Surface soil physical properties of selected sites (9).

Site	Surface Texture	Erosion Phase	Organic Matter (%)	Dry Bulk Density (g/cm³)	Estimated Soil Profile Θva* (cm H_2O)
3	Fine sandy loam	Non-eroded	2.6	1.44	14.1
		Severely eroded	0.7	1.46	12.8
		Depositional	5.5	1.23	18.1
6	Gravelly sandy loam	Non-eroded	3.7	1.37	7.4
		Severely eroded	2.2	1.68	5.1
		Depositional	4.3	1.38	11.1

*Θva = estimated pre-season plant-available water in 39.4 inches of soil.

beans without a winter crop and grassed waterway, the probability that soil loss would be less than 5.0 tons/acre was less than 1%. Hargrove and associates (29) reported that average aboveground clover production on this same watershed was 2.29 tons/acre, which was similar to or slightly greater than residue amounts from small grain cover crops. Sorghum grain yields averaged 1.63 tons/acre without fertilizer-N additions. Yields were similar to yields following barley when 80 pounds/acre fertilizer-N was applied. They attributed increased rainfall capture and reduced risk of soil loss to the improved soil surface characteristics provided by the surface residues.

El Swaify and associates (24) intercropped rose clover in corn on an Oxic Aridisol in Hawaii under irrigation and fertilization. They found a 31% reduction in runoff and a 71% reduction in soil loss relative to fallow. Corn yield was not affected or was slightly lower than conventionally grown corn.

At Watkinsville researchers investigating yield recovery imposed treatments on each of three classes of erosion on Cecil-Pacolet soils (clayey, kaolinitic, thermic Typic Hapludults). They compared the effect on soil water regime and organic C of no-till planting grain sorghum into a winter crop of crimson clover with grain sorghum planted into a conventionally prepared seedbed with no winter crop. In addition to biomass production, soil water was measured at five depths to 59 inches three times/week during the growing season for grain sorghum over the duration of the experiment. Other aspects of the study were discussed by Langdale and associates (52). From the soil water data for 1982, 1983, and 1984 the researchers determined the soil water regime for the grain sorghum in terms of the days with less than 1 bar soil water tension at each depth. The treatments did not affect soil water regimes at 39 and 59 inches significantly. At the 4-, 10-, and 20-inch depths the soil water regimes across all erosion levels for each year were significantly better when grain sorghum was no-till planted into crimson clover than when it was planted into a conventionally prepared seedbed without a winter crop. Organic C content in the 0- to 0.4-inch depth in 1985 was higher under the clover treatment than the no clover treatment (52). Additional seasons of data will be required to document the renovation process more fully. However, the immediate and persistent annual effect of the clover treatment upon soil water regime encourages its use in yield recovery.

Legume biomass and N accumulation. The symbiotic fixation of N by many legume species and the range in species adaptation has made legumes important in cropping systems worldwide. As biomass contributors to cropping systems they achieve a special role as N accumulators as well. Because N frequently is deficient in eroded soil situations, legumes seem to have a role in renovation.

Temperature and rainfall regimes are critical in the design of cropping systems that effectively use the biomass and N inputs of the selected legume species. In the udic, thermic region Hoyt and Hargrove (33) have evaluated and summarized the use of winter legumes. These legumes have been used mainly as cover crops or mulches and N sources for corn or grain sorghum. Evaluations of species as modifiers of soil surface characteristics are needed for the range in

eroded soil conditions that are experienced.

Peele and associates (58) showed the effectiveness of winter legumes as mulches. Soil losses during the summer were 1.05 tons/acre with a rye-vetch cover crop during the previous cool season. On the other hand, soil losses were 8.4 tons/acre with no winter cover. They also observed that residues decay more slowly on heavier soils (58). Beale and associates (11) reported that vetch and rye mulches increased soil organic matter in the Ap horizon from the initial level of 1.5% to 2.1% over a 10-year period; crimson clover mulch increased soil organic matter from 1.6% to 2.0%. In check plots without the mulch organic matter fell to 1.2% during the period. Batchelder and Jones (8) demonstrated that surface straw mulches coupled with increased fertilization increased productivity of exposed subsoils of a clayey, mixed, mesic Typic Hapludult. They attributed the benefit of mulches to reduced water loss and increased rainfall capture. Researchers and farmers alike have accepted the beneficial aspects of mulches as far as rainfall capture and retention is concerned. But a critical question remains as to the effectiveness of surface residues (mulches) in improving soil properties.

There are concerns about losses of soil organic N and C by leaving legume residues on the soil surface. Albrecht (3) compared soil N accumulation with 2.5 tons/acre red clover applied on the soil surface and not incorporated until the next application of mulch 1 year later with the same amount of red clover incorporated into the surface 7 inches of soil immediately after application. Annual applications were made over 15 years. The plots were on level land, surrounded by borders, and covered with a screen to prevent soil loss by water or wind erosion. The plots were kept fallow, and volunteer plants were not allowed to become established in the plots. After 15 years, there was 3% more residual N in the soil on plots on which surface applications were made compared to where the red clover was incorporated to 7 inches. These results indicate clearly that surface-applied material did not increase net N losses and, by inference, that surface applications of leguminous materials did not increase N volatilization losses. There could have been compensating processes, such as greater denitrification in the surface-applied plots and higher leaching losses of N in the incorporated plots. The net effect was surprising in that accumulation of total soil N was similar with residues exposed at the soil surface or immediately incorporated to 7 inches.

Wilson and Hargrove (73) evaluated the release of N from crimson clover residues placed on the soil surface as in no-till or buried at plow-layer depth as in conventional tillage. They concluded that the initial rate of N disappearance from the residue was more rapid in the buried residue than in that applied at the soil surface. The percentage of initial residue N remaining after either placement was similar after 16 weeks. They concluded that the release of N from either placement of residues was sufficiently rapid to be of significant benefit to the summer crop.

Several researchers have conducted ¹⁵N trials where legume residues were incorporated into soil and ¹⁵N recoveries determined (6, 44, 45, 46). Heichel (31) in summarizing ¹⁵N-labeled legume research, indicated that 15% to 25% of symbiotic N in legumes is recovered in the first harvest of the subsequent crop, with perhaps 4% recovery in the second year. Such recoveries are 8% to 22% of those estimated by agronomic studies of fertilizer equivalency. Available information suggests that the bulk of the legume ^{15}N resides in the soil organic fraction. These ^{15}N studies were conducted with the labeled residues mixed with the soil. We are not aware of ^{15}N-labeled legume studies evaluating recovery of N from surface mulches under conservation tillage or no-till situations.

Hoyt and Hargrove (33) concluded that crimson clover, hairy vetch, and Austrian winter peas provide sufficient winter biomass to retard erosion, accumulate more than 134 pounds/acre N in organic N, and supply up to 89 pounds/acre N for the subsequent summer crop. Data from 48 vetch crop-location-year yields in North Carolina showed that mean forage dry matter production was 2,369 pounds/acre, with a standard deviation of 1,203 pounds/acre (40). Meyer and associates (55) indicated that 893 pounds/acre of mulch could retard erosion potential. The presence of roots and stubble were not included in the estimates, although each should aid materially in erosion control and increased water infiltration.

Studies by Burton (19) of the amount of N accumulated by cool-season annual legumes indicated that from 12% to 20% of the N in cool-season annual legumes occurred in the roots. Duggar (22) determined that about 20% of N in hairy vetch was in the roots and stubble at prebloom stage. The proportions of N in the root systems of ladino clover and alfalfa are slightly higher, according to Giddens (26). The majority of legume root biomass is in the upper 6 to 8 inches of soil (12). The importance of legume roots in providing channels for water infiltration and percolation to lower depths has not been fully evaluated.

Examples of summer legumes include *Glycine* species; *Arachis* species; sericea lespedeza; and *Medicago* species, including alfalfa. Soybeans harvested for grain generally contribute inadequate amounts of N for a following nonlegume crop when residues are turned under in the fall and the next crop planted the following spring (26). Small grain crops no-till planted or seeded directly in soybean residues may realize a gain in available N in the range of 18 to 27 pounds/acre N. Estimates of net N returned to the soil in soybean residues after seed harvest range from 12 to 110 pounds/acre in the southeastern Coastal Plain (34). Available information suggests that production of good forage yields following soybeans requires additional N fertilization (71).

Sericea lespedeza is a recommended legume for roadside stabilization and soil conservation purposes. It is tolerant of acid soils and persists with minimum management inputs. Studies by Adams and associates (2) and by Giddens (26) indicate that corn crops following sericea lespedeza require N fertilization similar to corn crops not preceded by legumes. Apparently, sericea does not supply much N to a succeeding crop. Sericea leaves from a 10-year-old stand contained 1.5% N, and stems contained 0.74% N (26). Dry matter accumulation in sericea appears to be mostly in stems. Apparent immobilization of N in organic forms is high. Unpublished research indicates that cereal grains planted in undisturbed sericea lespedeza sods become N-deficient and require N fertilization for good forage production (personal communication, C. S. Hoveland, Agronomy Department, University

of Georgia, Athens).

There is interest in the potential of no-till intercropping nonlegumes; for example, corn in living perennial summer legumes, such as alfalfa, crownvetch, and birdsfoot trefoil. This procedure has been tried in Illinois and Pennsylvania by using herbicides to control legume competition (23, 28). Researchers have intercropped no-till corn in grass sods successfully (20). However, new cropping and farm system designs and operations may yet use perennial summer legumes in particular soil and climatic situations.

Tillage consequences. Yield recovery of mismanaged soils clearly depends upon tillage practices. Soil mismanagement usually includes excessive tillage, but use of legumes in yield recovery implies reduced tillage. Wells (69) summarized the effects of no-till practices on soil behavior as follows: (a) 15% to 25% more water available during the growing season because of better rainfall infiltration and reduced evapotranspiration; (b) slower temperature changes with season and less diurnal temperature fluctuation; (c) greater organic matter, root activity, and microbial activity in the surface 2 inches of soil, whereas it is mixed somewhat uniformly to plow depth in conventional tillage; (d) residual soil N increases, probably in a slowly labile organic N pool; (e) bulk density changes usually have been small and not different from those of conventional tillage; and (f) soil acidity at the soil surface has increased because fertilizer N inputs are not incorporated and because organic acids are released during residue decomposition.

Wells (69) pointed out that vegetation plays a more influential role in relation to on-going soil formation factors with continuous no-till production. He also suggested that there is a greater build-up of immobilized N under no-till, and that the poorer yield response to N at low rates of N may be due to immobilization of N in the concentrated organic matter layer near the soil surface. Cumulative research findings with corn indicate a greater build-up of total N with continuous no-till production. An important question is how long does this N build-up occur and will it reach a new equilibrium? Recent results from a long-term experiment on a Maury silt loam (Typic Paleudalf) comparing corn yield responses to N fertilization under no-till and conventional tillage suggest that the N unavailability observed during the first 9 years at low N rates under no-till was a transient effect (60). In this study Rice and associates used rye as a winter cover crop and mulch for no-till. The availability of soil N with no-till approached that of conventional tillage after 10 years. While it is likely that the same immobilization phenomenon would occur with legumes as cover crops and mulches, the time period for comparable N availability in conservation tillage likely would be less. Wells (69) also concluded that no-till corn uses fertilizer-N more efficiently than conventionally grown corn. Planting corn in killed legume sods has the potential to reduce or compensate for the lower mineralization of soil organic N possibly inherent with no-till crop culture (69).

Power and Doran (59) summarized the effect of tillage practices on water content, organic matter, microbial biomass, and potentially mineralizable N (Tables 9 and 10).

There appears to be a great potential for using legumes as mulches in conservation tillage to rapidly build up a biologically active, carbonaceous surface layer that will have a potentially larger labile N pool than that in predominantly grass no-till systems. The microbial biomass, microbial by-products, and partially humified materials accumulated at the soil surface also would increase the pool of very slowly labile N. Legume residues have low C:N ratios. Thus, they should provide a more rapidly mineralizable N pool than grass residues. More rapid decomposition of legume residues provides a potential advantage for enhanced N use by the following nonlegume. This is supported by the generally positive effect of legumes on yields in conservation tillage at lower N inputs (33, 73).

Barnett and associates (7) showed that inputs of organic matter from grasses and legumes in rotation increased soil organic matter on a class IV Cecil sandy loam. In the same 10-year period soil organic matter was low and relatively constant with continuous cotton. When the plots that received organic matter from legumes and grasses in rotation were returned to continuous cotton, the level of soil organic matter returned to about the same level as those which had been in continuous cotton over the same interval. Allison (5) and Volk and Loeppert (67) pointed out that bonding between clay particles and organic matter is well recognized and that the retention of organic carbon was enhanced by bentonite clays (2:1 type clays) compared with kaolinitic clays (1:1 type lattice clays). Kamprath and associates (40) observed the greatest effect of winter legumes on subsequent crop yields on clay soils.

Many studies indicate that cultivation decreases soil organic matter (26, 37, 68). There are indications that enhanced soil organic matter levels will increase availability of nutrients other than N for plant growth through retention and reduced loss rates and potentially through the reduction of Al toxicity (41). Jenny (38) established the basic equation for soil N availability: $dn/dt = -k_1 N + k_2$, where N is the content of soil organic N, $-k_1$ is the proportion of N mineralized, and k_2 is the annual N addition. Changes in soil N are determined by rates of N loss and accumulation. Blevins and associates (13) showed that N fertilization in no-till agroecosystems can enhance levels of both organic C and N. Jenny (37) established that stable soil N levels are higher under good soil management systems than under poor systems. There is overwhelming evidence that management as well as environmental factors influence soil organic matter.

Legume residues with low C:N ratios may result in rapid net mineralization. Decomposition of a legume with a low C:N ratio is characterized by a rapid phase (few weeks), followed by a slower release rate that ultimately approaches the humus state of organic matter (42). The decomposition rate of the humus type of organic matter is about 1% to 2% per year. Research has shown the importance of a large pool of soil organic C and N for a stable productive ecosystem. This level of organic matter and its steady state value is determined by soil water, temperature, cultivation, levels of nutrient losses to crop uptake, leaching, and volatilization, as well as the amount and distribution of organic C and N inputs in specific agroecosystems.

Microbial soil biomass is "the eye of the needle through which all natural organic material must pass, often more than once, as they are degraded to inorganic compounds" (36).

Table 9. Effect of tillage practices on changes in water and organic matter contents, level of microbial biomass, and potentially mineralizable nitrogen [after Doran, 1978, 1981; taken from Power and Doran (59)].

Soil Depth (cm)	Relative difference*			
	Water Content	Organic Matter	Microbial Biomass	Potentially Mineralizable N
	%			
0 - 7.5	+28	+27	+32	+37
7.5-15.0	+ 1	- 4	- 5	- 2

*Calculated by [(no tillage/plow) - 1] X 100.

Table 10. Effect of degree of soil disturbance on the levels of microbial biomass, water content, and organic matter of surface soil (0-7.5 cm) [J. W. Doran, unpublished data; taken from Power and Doran (59)].

Management Practice	Degree of Disturbance	Microbial Biomass (pounds C/acre)	Volumetric Water Content (%)	Organic Matter (%)
Sod control	None	955	17.7	4.49
No-till	Minimum	790	14.3	3.80
Subtillage	Moderate	739	12.2	3.28
Plow	Maximum	587	10.6	2.42

This concept, credited to Jenkinson, is crucial in understanding the effect of erosion on soil productivity losses associated with soil organic matter losses. Conversely, restoration of productivity requires restoring soil organic matter to the level that is optimum for that agroecosystem. No-till systems have many similarities to natural systems in terms of crop culture. Natural systems use soil biota to control residue decomposition without the aid of tillage, while in conventional agroecosystems man uses tillage to enhance decomposition.

Crop residue placement is a major regulator of N cycling in agroecosystems. Soil tillage accelerates the rate of plant residue decomposition. No-till practices immobilize N concentrated in soil organic matter at the surface of soil. This immobilized N represents both conserved N and N unavailable for crop use. The large pool of immobilized N enhances storage capacity and promotes long-term retention. The accelerated decomposition from tillage when inadequate residues are returned to the soil may stress the system over the long term. One major benefit of no-till cropping may be the reorganization of N and C storage and delayed consequent movement throughout the soil system (32). Nitrogen mineralized at rates greater than that used by the crop represents a labile pool of N subject to loss by denitrification and leaching. Research has established that high N losses from leaching can occur from mineralized legume residues (61). The same work showed that the presence of live plants effectively reduced the loss of NO_3.

Woods and Schuman (74) postulated that there is a critical organic matter concentration essential to the formation of soil nutrient cycles. Under rangeland conditions the level required to maintain a sustainable soil N cycle ranged from 0.1% to 0.7% organic C. Jefferies and associates (35) pointed out that successful restoration of derelict land from coal mine spoils required an accumulation of organic N and a functional N cycle. Residues remaining at the soil surface may have a favorable effect in establishing this sustainable level of organic matter early, with minimum inputs, to support the microbial biomass and sustain functional soil mineral cycling. This sustainable cycling is essential to restoring and maintaining soil productivity as well as using fertilizer effectively. Legumes, with their lack of dependence on soil N, represent a promising source of organic N coupled with a source of residues for building sufficient labile and nonlabile pools of organic matter to support functional mineral cycles in eroded soil.

Conclusions

Because the influence of soil erosion on productivity is now a global concern, researchers have increased efforts to develop crop cultures that are economically productive while maintaining the soil resource. We conclude that erosion deteriorates the soil surface and progressively truncates the soil volume only to the extent that it may limit rooting and supplies of water and nutrients. The rate at which this deterioration occurs depends upon soil and climatic conditions, particularly rainfall amounts and intensity and crop cultural history. Yield decline and soil loss probably are exponential with time. Although water deficits most often limit crop yields on eroded sites, the quantity and status of organic C in the surface 0.4 to 0.8 inch seems to be the critical deficiency. Physical and chemical conditions of the soil surface can be modified by additions of organic biomass, which in turn affects surface sealing, infiltration, and N availability. A conceptual framework is emerging that permits development and application of alternative site management, based on principles of soil, crop, and biological science in the context of climate, hydrology, and geology and supported by considerable research data.

Evidence suggests that selected legumes, alone or in combination with nonlegumes, provide protection from soil erosion; increased organic C in the surface soil, water infiltration, and aggregate stability; and improved nutrient status. The effectiveness of legume species in modifying surface soil conditions and N supplies under a range of eroded soil conditions needs additional research. Research is needed to determine the quantity of biomass of a particular C:N ratio that is required to sustain an optimum physical condition at the soil surface and an optimum organic C and mineral cycling for effective climatic utilization by the crop.

REFERENCES

1. Adams, W. E., and R. N. Dawson. 1964. *Cropping system studies on Cecil soils, Watkinsville, Ga., 1943-62.* USDA-ARS-41-83. Agr. Res. Serv., U.S. Dept. Agr., Watkinsville, Ga.
2. Adams, W. E., H. D. Morris, Joel Giddens, R. N. Dawson, and G. W. Langdale. 1973. *Tillage and fertilization of corn grown on lespedeza sod.* Agron. J. 65: 653-655.
3. Albrecht, W. A. 1936. *Methods of incorporating organic matter with the soil in relation to nitrogen accumulation.* Res. Bull. 249. Mo. Agr. Exp. Sta., Columbia.
4. Allison, F. E. 1968. *Soil aggregation-some facts and fallacies as seen by a microbiologist.* Soil Sci. 106: 136-143.
5. Allison, F. E., M. S. Sherman, and L. A. Pinck. 1949. *Maintenance of soil organic matter. I. Inorganic soil colloid as a factor in retention of carbon during formation of humus.*

Soil Sci. 68: 463-478.
6. Azam, F., K. A. Malik, and M. I. Sajjad. 1985. *Transformations in soil and availability to plants of ^{15}N applied as inorganic fertilizers and legume residues.* Plant and Soil 86: 3-13.
7. Barnett, A. P., J. S. Rogers, W. E. Adams, and L. F. Welch. 1961. *Cropping systems, organic matter and nitrogen.* Ga. Agr. Res. 3(1): 10-11.
8. Batchelder, A. R., and J. N. Jones, Jr. 1972. *Soil management factors and growth of Zea Mays L. on top soil and exposed subsoil.* Agron. J. 64: 648-652.
9. Battiston, L. A., R. A. McBride, M. H. Miller, and M. J. Brklacich. 1985. *Soil erosion-productivity research in southern Ontario.* In *Erosion and Soil Productivity.* Pub. 8-85. Am. Soc. Agr. Eng., St. Joseph, Mich. p. 28-38.
10. Baver, L. D. 1950. *How serious is soil erosion.* Soil Sci. Soc. Proc. 15: 1-5.
11. Beale, O. W., G. B. Nutt, and T. C. Peele. 1955. *The effects of mulch tillage on run-off, erosion, soil properties, and crop yields.* Soil Sci. Soc. Am. Proc. 19: 244-247.
12. Bennett, O. L., and B. D. Doss. 1960. *Effect of soil moisture level on root distribution of cool season forage species.* Agron. J. 52: 204-207.
13. Blevins, R. L., G. W. Thomas, M. S. Smith, W. W. Frye, and P. L. Cornelius. 1983. *Changes in soil properties after 10 years continuous non-tilled and conventionally tilled corn.* Soil and Tillage Res. 3: 135-146.
14. Bolton, E. F., and L. R. Webber. 1952. *Effect of cropping systems on the aggregation of a Brookston clay soil at three depths.* Sci. Agr. 32: 555-558.
15. Borst, H. L., A. G. McCall, and F. G. Bell. 1945. *Investigations in erosion control and reclamation of eroded land at the Northwest Appalachian Conservation Experiment Station, Zanesville, Ohio, 1934-42.* Tech. Bull. No. 888. U.S. Dept. Agr., Washington, D.C.
16. Bruce, R. R. 1947. *The effect of cultural practices on some of the physical properties of a clay soil.* Thesis. Ont. Agr. Col. Guelph, Ont.
17. Bruce, R. R. 1955. *An instrument for the determination of soil compactibility.* Soil Sci. Soc. Am. Proc. 19: 253-257.
18. Bruce, R. R., A. W. White, H. F. Perkins, A. W. Thomas, and G. W. Langdale. 1986. *Soil erosion modifications of crop root zone in layered soils.* Trans., XIII Cong. Int. Soc. Soil Sci. IV: 1,573-1,574.
19. Burton, G. W. 1976. *Legume nitrogen versus fertilizer nitrogen for warm season grasses.* In C. S. Hoveland [ed.] *Biological N Fixation in Forage-Livestock Systems.* Spec. Pub. No. 28. Am. Soc. Agron., Madison, Wisc. pp. 54-72.
20. Carreker, John. R., S. R. Wilkinson, A. P. Barnett, and J. E. Box, Jr. 1977. *Soil and water management systems for sloping land.* ARS-S-160. Agr. Res. Serv., U.S. Dept. Agr., 76 pp.
21. Daniels, R. B., J. W. Gilliam, D. K. Cassel, and L. A. Nelson. 1985. *Soil erosion class and landscape position in the North Carolina Piedmont.* Soil Sci. Am. J. 49: 991-995.
22. Duggar, J. F. 1899. *Winter pasturage, hay and fertility afforded by hairy vetch.* Bull 105. Ala. Agr. Exp. Sta., Auburn. pp. 129-160.
23. Elkins, Donald, Duane Frederking, Reza Marashi, and Byron McVay. 1983. *Living mulch for no-till corn and soybeans.* J. Soil and Water Cons. 38: 431-433.
24. El-Swaify, S. A., A. Lo, R. Joy, L. Shinshiro, and R. S. Yost. 1986. *Benefits of legume intercropping to crop yields and soil conservation in the tropics.* Trans., XIII Cong. Soc. Soil Sci. IV: 1,580-1,581.
25. Frye, W. W., O. L. Bennett, and G. J. Buntley. 1985. *Restoration of crop productivity on eroded or degraded soils.* In R. F. Follett, and B. A. Stewart [eds.] *Soil Erosion and Crop Productivity.* Am. Soc. Agron., Madison, Wisc. pp. 335-356.
26. Giddens, Joel. 1985. *Nitrogen cycling in Georgia soils.* Res.

Bull. 327. Univ. Ga., Athens. 36 pp.

27. Hall, G. F., R. B. Daniels, and J. E. Foss. 1982. *Rate of soil formation and renewal in the U.S.A.* In *Determinants of Soil Loss Tolerance.* Pub. No. 45. Am. Soc. Agron., Madison, Wisc. pp. 23-39.

28. Hall, J. K., N. L. Hartwig, and L. D. Hoffman. 1984. *Cyanazine losses in run off from no-tillage corn in "living" and dead mulches vs. unmulched conventional tillage.* J. Environ. Quality 13: 105-110.

29. Hargrove, W. L., G. W. Langdale, and A. W. Thomas. 1984. *Role of legume cover crops in conservation tillage production systems.* Paper No. 84-2038. Am. Soc. Agr. Eng., St. Joseph, Mich.

30. Harris, R. F., G. Chesters, and O. N. Allen. 1966. *Dynamics of soil aggregation.* Adv. Agron. 18: 107-109.

31. Heichel, G. H. 1985. *Nitrogen recovery by crops that follow legumes.* In Robert F. Barnes, P. Roger Ball, Raymond W. Brougham, Gordon C. Marten, and Dennis J. Minson [eds.] *Forage Legumes for Energy Efficient Animal Production.* Proc., Trilateral workshop. Nat. Tech. Info. Serv., Springfield, Va. p. 183-190.

32. House, Garfield, J., Benjamin R. Stinner, D. A. Crossley, Jr., Eugene P. Odum, and George W. Langdale. 1984. *Nitrogen cycling in conventional and no-tillage agroecosystems in the Southern Piedmont.* J. Soil and Water Cons. 39: 194-200.

33. Hoyt, Greg. D., and William L. Hargrove. 1986. *Legume cover crops for improving crop and soil management in the southern United States.* Hortsci. 21(3): 397-492.

34. Hunt, P. G., T. A. Matheny, and A. G. Wollum II. 1985. Rhizobium japonicum *nodular occupancy, nitrogen accumulation, and yield for determinate soybean under conservation and conventional tillage.* Agron. J. 77: 579-584.

35. Jefferies, R. A., K. Willson, and A. D. Bradshaw. 1981. *The potential of legumes as a nitrogen source for the reclamation of derelict land.* Plant and Soil 59: 173-177.

36. Jenkinson, D. S. 1977. *The soil biomass.* N.Z. Soil News 25(6): 213-218.

37. Jenny, Hans. 1933. *Soil fertility losses under Missouri conditions.* Bull. 324. Univ. of Mo., Columbia.

38. Jenny, H. 1941. *Factors of soil formation.* McGraw-Hill, New York, N.Y. p. 381.

39. Johnston, J. R., G. M. Browning, and M. B. Russell. 1942. *The effect of cropping practices on aggregation, organic matter content and loss of soil water in the Marshall silt loam.* Soil Sci. Soc. Am. Proc. 7: 105-107.

40. Kamprath, Eugene J., W. V. Chandler, B. A. Krantz. 1958. *Winter cover crops: Their effects on corn yields and soil properties.* Tech. Bull. No. 129. N. Car. Agr. Exp. Sta., Raleigh. 47 pp.

41. Kapland, D. I., and G. O. Estes. 1985. *Organic matter relationship to soil nutrient status and aluminum toxicity in alfalfa.* Agron. J. 77: 735-738.

42. Keeney, Dennis. 1985. *Mineralization of nitrogen from legume residues.* In Robert F. Barnes, P. Roger Ball, Raymond W. Brougham, Gordon C. Marten, and Dennis J. Minson [eds.] *Forage Legumes for Energy Efficient Animal Production.* Proc., Trilateral workshop. Nat. Tech. Info. Serv., Springfield, Va. pp. 177-182.

43. Kemper, W. D., and E. J. Koch. 1966. *Aggregate stability of soils from western United States and Canada.* Tech. Bull. No. 1355. Agr. Res. Serv., U.S. Dept. Agr., Washington, D.C.

44. Ladd, J. N., J. M. Oades, M. Amato. 1981. *The distribution and recovery of nitrogen from legume residues decomposing in soils sown to wheat in the field.* Soil Biol. Biochem. 13: 251-256.

45. Ladd, J. N., and M. Amato. 1986. *The fate of nitrogen from legume and fertilizer sources in soil successively cropped with wheat under field conditions.* Soil Biol. Biochem. 18: 417-425.

46. Ladd, J. N., M. Amato, R. B. Jackson, and J. H. A. Butler. 1983. *Utilization by wheat crops of nitrogen from legume residues decomposing in soils in the field.* Soil Biol. Biochem. 15: 231-238.

47. Lal, R. 1976. *Soil erosion on Alfisols in western Nigeria. I. Effects of slope, crop rotation, and residue management.* Geoderma 16: 363-375.

48. Lal, R. 1976. *Soil erosion on Alfisols in western Nigeria. IV. Nutrient element losses in runoff and eroded sediments.* Geoderma 16: 403-417.

49. Lal, R. 1976. *Soil erosion on Alfisols in western Nigeria. V. The changes in physical properties and the response of crops.* Geoderma 16: 419-431.

50. Lal, R. 1981. *Soil erosion problems on Alfisols in western Nigeria. VI. Effects of erosion on experimental plots.* Geoderma 25: 215-230.

51. Langdale, G. W., J. E. Box, Jr., R. A. Leonard, A. P. Barnett, and W. G. Fleming. 1979. *Corn yield reduction on eroded Southern Piedmont soils.* J. Soil and Water Cons. 34: 226-228.

52. Langdale, G. W., R. R. Bruce, and A. W. Thomas. 1987. *Restoration of eroded Southern Piedmont land in conservation tillage systems.* In *The Role of Legumes in Conservation Tillage Systems.* Soil Cons. Soc. Am., Ankeny, Iowa. pp. 142-144.

53. Langdale, G. W., and W. D. Shrader. 1982. *Soil erosion effects on soil productivity of cultivated cropland.* In *Determinants of Soil Loss Tolerance.* Pub. No. 45. Am. Soc. Agron., Madison, Wisc. pp. 41-51.

54. McDaniel, T. A., and B. J. Hajek. 1985. *Soil erosion effects on crop productivity and soil properties in Alabama.* In *Erosion and Soil Productivity.* Pub. 8-85. Am. Soc. Agr. Eng., St. Joseph, Mich. pp. 48-58.

55. Meyer, L. D., W. H. Wischmeier, and G. R. Foster. 1970. *Mulch rates required for erosion control on steep slopes.* Soil Sci. Soc. Am. Proc. 34: 928-931.

56. Mills, W. C., A. W. Thomas, and G. W. Langdale. 1986. *Estimating soil loss probabilities for Southern Piedmont cropping tillage systems.* Trans., ASAE 29: 948-955.

57. Papendick, R. I., D. L. Young, D. K. McCool, and H. A. Krauss. 1985. *Regional effects of soil erosion on crop productivity. The Palouse area of the Pacific Northwest.* In R. F. Follett and B. A. Stewart [eds.] *Soil Erosion and Crop Productivity.* Am. Soc. Agron., Madison, Wisc. pp. 305-320.

58. Peele, T. C., G. B. Nutt, and O. W. Beale. 1946. *Utilization of plant residues as mulches in the production of corn and oats.* Soil Sci. Soc. Am. Proc. 11: 356-360.

59. Power, J. F., and J. W. Doran. 1984. *Nitrogen use in organic farming.* In Roland D. Hauck [ed.] *Nitrogen in Crop Production.* Am. Soc. Agron., Madison, Wisc. pp. 585-598.

60. Rice, C. W., M. S. Smith, and R. L. Blevins. 1986. *Soil nitrogen availability after long-term continuous no-tillage and conventional tillage corn production.* Soil Sci. Soc. Am. J. 50: 1,206-1,210.

61. Roberts, George. 1937. *Legumes in cropping systems.* Bull. 374. Ky. Agr. Exp. Sta., Lexington. pp. 121-153.

62. Rogowski, A. S. 1985. *Evaluation of potential topsoil productivity.* Env. Geoch. and Health. 7: 87-97.

63. Stocking, Michael. 1984. *Erosion and soil productivity: A review.* Food and Agr. Org., United Nations, Rome, Italy. 102 pp.

64. Stocking, M., and L. Peake. 1985. *Erosion-induced loss in soil productivity: Trends in research and international cooperation.* Food and Agr. Org., United Nations, Rome, Italy. 52 pp.

65. Stone, J. R., J. W. Gilliam, D. K. Cassel, R. B. Daniels, and L. A. Nelson, and H. J. Kleiss. 1985. *Effect of erosion and landscape position on the productivity of Piedmont soils.* Soil Sci. Soc. Am. J. 49: 987-990.

66. U.S. Soil Conservation Service. 1975. *Soil taxonomy. A basic*

system of soil classification for making and interpreting soil surveys. Agr. Handbk. 436. U.S. Dept. Agr., Washington, D.C.

67. Volk, Bob G., and Richard H. Loeppert. 1982. *Soil organic matter.* In Victor J. Kilmer [ed.] *Handbook of Soils and Climate in Agriculture.* CRC Press, Inc. Boca Raton, FL. pp. 211-268.

68. Welch, L. F. 1976. *The Morrow plots-hundred years of research.* Ann. Agron. 27: 881-890.

69. Wells, K. L. 1984. *Nitrogen management in the no-till system.* In Roland D. Hauck [ed.] *Nitrogen in Crop Production.* Am. Soc. Agron., Madison, Wisc. pp. 535-550.

70. White, A. W., Jr., R. R. Bruce, A. W. Thomas, G. W. Langdale, and H. F. Perkins. 1985. *Characterizing productivity of eroded soils in the Southern Piedmont.* In *Erosion and Soil Productivity.* Pub. 8-85. Am. Soc. Agr. Eng., St. Joseph, Mich. pp. 83-95.

71. Wilkinson, S. R., and J. A. Stuedemann. 1983. *Increased use of Southern Piedmont land and climate resources by interseeding small grains in dormant coastal bermudagrass.* In J. Allen Smith and Virgil W. Harp, [eds.] Proc. XIV Int. Grassland Congress. Westview Press, Boulder, Colo. pp. 568-571.

72. Williams, J. R. (Chm); National Soil Erosion-Soil Productivity Research Planning Committee. 1981. *Soil erosion effects on soil productivity: A research perspective.* J. Soil and Water Cons. 36: 82-90.

73. Wilson, D. O., and W. L. Hargrove. 1986. *Release of nitrogen from crimson clover residue under two tillage systems.* Soil Sci. Soc. Am. J. 50: 1,251-1,254.

74. Woods, L. E., and G. E. Schuman. 1986. *Influence of soil organic matter concentrations on carbon and nitrogen activity.* Soil Sci. Soc. Am. J. 50: 1,241-1,245.

Maize yields and soil losss with conservation and conventional tillage practices on a tropical Aridisol

K. S. Fahrney, S. A. El-Swaify, A. K. F. Lo, and R. J. Joy

Conservation tillage systems are characterized by a minimum of soil disturbance and the maintenance of crop residues on soil surfaces. Crop residues absorb the energy of raindrop impact, thus preventing detachment and entrainment of soil particles. Soil losses decrease dramatically with increasing mulch rates (6).

Cover crops, when included in cropping systems, shield soils from raindrop impact while decreasing leaching losses of soil nutrients. If the cover crop is a legume, it will often enhance soil fertility (2, 4).

In the tropics high temperatures may accelerate the decomposition of residues, while feed and fuel demands often preclude the use of crop residues as protective mulches (3). In arid and semiarid regions of the tropics, lack of soil moisture may limit the regrowth of natural vegetative cover, leaving soil surfaces bare and unprotected at the onset of violent monsoonal storms.

Sunn Hemp (*Crotalaira juncea* L.) is a drought-tolerant, fast-growing herbaceous legume native to India. It is grown throughout the tropics primarily as a green manure. *Crotalaria juncea* (cv. Tropic Sun) has been estimated to contribute 134 to 147 pounds of N to the soil in 60 days (5), offering farmers an important incentive to keep fallow soils covered. Our study examined the immediate and residual effects of various tillage treatments and residue from a Crotalaria cover crop on maize yields and soil and water losses.

Study methods

Two crops of maize (*Zea mays* L.)—tropical hybrid Pioneer X304C—were grown on runoff plots (4% to 6% slope) at the Soil Conservation Service's Hawaii Plant Materials Center on the island of Molokai. Annual precipitation is 15 to 20 inches. The soil is an Aridisol (Ustollic Camborthid) and is moderately to highly erodible (*1*).

We used a split-plot, randomized complete block design with three replications. Mainplot treatments were tillage by moldboard or chisel plowing and no-till. We retained a set of fallow plots for comparison. Subplot treatments were the presence or absence of residue from a Crotalaria cover crop. In the latter treatments, Crotalaria was broadcast seeded, irrigated, grown for 5 weeks, and then spray-killed with glyphosate. We then imposed tillage treatments perpendicular

K. S. Fahrney is a graduate student, S. A. El-Swaify is a professor of soil science and chairman, and A. K. F. Lo is an assistant soil scientist, Department of Agronomy and Soil Science, University of Hawaii, Honolulu, 96822. R. J. Joy is a plant materials specialist, Hawaii Plant Materials Center, Soil Conservation Service, U.S. Department of Agriculture, Hoolehua, Molokai, 96729.

to the slope. Fallow and plowed subplots were shallow roto-tilled to form a smooth, uniform seedbed. The maize was planted in rows running downslope.

We equipped each treatment with moisture sensors. Subplots had independent water delivery systems to control irrigation. We applied optimal levels of fertilizer except for N, which we applied as urea at about half the recommended rate.

To investigate residual treatment effects tillage and cover crop treatments were not reimposed prior to the second maize crop. Photographic slides were taken periodically from above each subplot to estimate canopy and residue cover. We measured maize height at tasseling and harvested the maize at the black layer stage.

Results

Crotalaria reached an average height of 4 feet and provided 100% soil coverage in 5 weeks. Tissue analysis indicated an aboveground N content of 88 pounds/acre. One week after herbicide spraying, cover on the no-till plots decreased to 83%. Total percent cover remained fairly constant, peaking at 89% for these plots 33 days after maize planting and decreasing to 84% by 76 days. As maize canopy increased, percent residue cover decreased inversely.

Five significant rainfall events occurred during the first maize crop (August-January 1985), with totals ranging from 1.1 to 3.1 inches. No storms occurred during the second crop (April-August 1986). The first storm occurred at 64 days after planting. By this time the maize canopy was about 70% of total cover. Runoff exceeded the capacity of collection barrels during the first, second, and fourth storms, falsely equalizing treatments. Figure 1 shows mean values from the third and fifth storms.

For all tillage treatments, Crotalaria residue cover decreased runoff significantly—by 20%. As expected, runoff was substantially higher on the fallow plots compared to planted treatments, emphasizing the importance of canopy cover. Runoff from no-till, chisel, and moldboard treatments was 44%, 30%, and 38% of runoff from fallow, respectively.

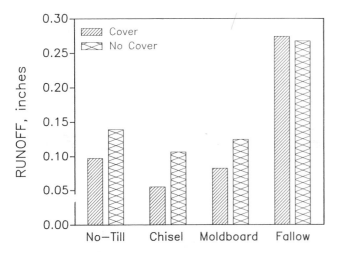

Figure 1. Runoff for third and fifth storms.

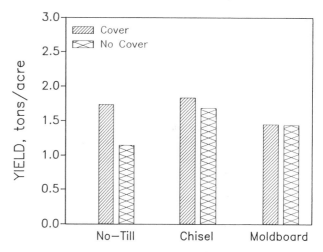

Figure 3. Maize grain yield for the first crop.

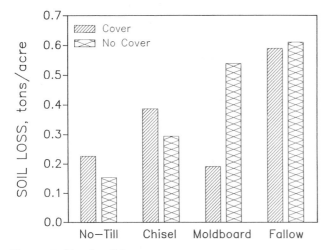

Figure 2. Total soil loss.

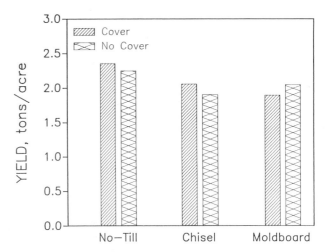

Figure 4. Maize grain yield for the second crop.

Figure 2 shows total soil loss from the five storms. Over all tillage treatments, Crotalaria residue cover reduced soil loss by 13%, but this was not statistically significant. Soil loss was greater for fallow treatments than for planted treatments. Soil loss from no-till, chisel, and moldboard treatments was 32%, 57%, and 61% of fallow, respectively.

Figures 3 and 4 show maize grain yields for the first and second crops. First crop grain yields were 18% higher on subplots with a prior Crotalaria cover crop. There was no residual fertility benefit from the Crotalaria cover crop for the second crop. We noted no definite trends in comparing the effect of tillage treatments on grain yields between the two maize crops.

Conclusions

Crotalaria as a cover crop provides full canopy protection to the soil within 4 to 5 weeks. As a surface-applied green manure, Crotalaria can significantly increase maize yields when N is limiting. Effectiveness of the residue mulch in controling soil and water losses is explained in view of the timing of rainfall events relative to the crop's growth.

REFERENCES
1. Dangler, E. W., S. A. El-Swaify, L. A. Ahuga, and A. P. Barnett. 1976. *Erodibility of selected Hawaii soils by rainfall simulation.* ARS W-35. Agr. Res. Serv., U.S. Dept. Agr., Washington, D.C.
2. Ebelhar, S. A., W. W. Frye, and R. L. Blevins. 1984. *Nitrogen from legume cover crops for no-tillage corn.* Agron. J. 76(1): 51-55.
3. El-Swaify, S. A., E. W. Dangler, and C. L. Armstrong. 1982. *Soil erosion by water in the tropics.* Hawaii Inst. Tropical Agr. and Human Resources, Univ. Hawaii, Honolulu.
4. Frye, W. W., W. G. Smith, and R. J. Williams. 1985. *Economics of winter cover crops as a source of nitrogen for no-till corn.* J. Soil and Water Cons. 40(2): 246-249.
5. Rotar, P. P., and R. J. Joy. 1983. *'Tropic Sun' sunn hemp,* Crotalaria juncea *L.* Res. Ext. Series 36. Hawaii Inst. Tropical Agr. and Human Resources, Univ. Hawaii, Honolulu.
6. Wischmeier, W. H., and D. D. Smith. 1978. *Predicting rainfall-erosion losses - a guide to conservation planning.* Agr. Handbk. no. 537. U.S. Dept. Agr., Washington, D.C.

Effect of tillage on runoff, erosion, and first-year alfalfa yields

S. J. Sturgul and T. C. Daniel

In the dairy state of Wisconsin alfalfa is the most important and abundant legume crop. An established alfalfa stand offers excellent erosion protection. But the 5- to 6-week period required for stand establishment is also a time of maximum erosion potential, especially when the alfalfa is sown directly into a well-prepared, fine seedbed—a practice (direct seeding) that is becoming increasingly popular statewide.

The need for less erosive farming alternatives while maintaining crop yield and quality has precipitated research into establishing alfalfa using various conservation tillage methods. In the spring of 1986 the University of Wisconsin-Madison Soil Science Department intitiated a project to examine the interaction of tillage and alfalfa on the Arlington Experimental Farm in southern Wisconsin. The tillage methods analyzed relative to direct seeding were direct seeding with an oat nurse crop, chisel plow, and no-till. Stand establishment, growth and yield, and soil loss were evaluated as a function of tillage.

Study methods

We established 90-foot by 30-foot plots in a randomized block design on a Plano silt loam soil (Typic Argiudoll) with a uniform slope of about 5%. The previous crop was corn, and ample surface residue was present. Tillage methods included moldboard plowing, disking, and harrowing in the direct seeding and nurse crop treatments; one pass with a chisel plow, disking, and harrowing in the chisel treatments; and no tillage.

We used a broadcast seeder to plant the alfalfa on the direct seeding and nurse crop treatments and a no-till grain drill for the chisel and no-till treatments. All plots were seeded at a uniform rate of 15 pounds/acre.

In June and August of 1986 we applied simulated rainfall on the plots to generate runoff and subsequent erosion. Simulated rainfall was applied for 1 hour with a modified version of the Purdue sprinkling infiltrometer at an intensity equal to a storm with a 100-year return period in Wisconsin (2.75 inches). We recorded runoff volumes from the individual plots and took samples to determine sediment content and particle size distribution. We conducted the first rainfall simulation 7 weeks after planting, at which time the alfalfa seedling canopy was 30% to 40% for all treatments with the exception of the nurse crop, on which the canopy was 80% to 90% due to the oats. The second simulated rainfall occurred 17 weeks after planting; alfalfa canopies were removed just prior to rainfall.

S. J. Sturgul is a research assistant and T. C. Daniel is a professor, Soil Science Department, University of Wisconsin, Madison, 53706.

We harvested the plots three times during the growing season to determine yields.

Results and discussion

Table 1 shows the runoff volume, sediment concentration, and soil loss data collected from the two erosive rainfalls. Figure 1 shows the significant differences in soil loss with respect to tillage methods. Erosion was highest on the direct-seeded plots in both events compared to the other three tillage practices. In the June run direct seeding was nearly twice as erosive as the chisel plow treatments, nearly four times as erosive as the no-till treatments, and nearly seven times that of the nurse crop treatments. The same trends in soil loss occurred in August, with the exception of the nurse crop treatment, which was only slightly more erosive than the no-till treatment because the oat canopy was not present to absorb the force of the raindrops in this latter rainfall.

Greater soil loss occurred during the second rainfall for all treatments. The runoff volume data showed that every treatment experienced more runoff during the second rainfall than the first (Table 1); this increased the energy for transport in the erosion process. The second rainfall generated more runoff due to reduced infiltration capacity of the soil caused by the surface being sealed and crusted as a result of the first rainfall. Also, the lack of a crop canopy during the second rainfall event exposed the soil to the full erosive force of the falling raindrops, providing energy for soil particle detachment. The combined larger forces of overland flow and raindrop impact caused the sediment concentration of the runoff to increase for every treatment. Interrill erosion of the plots increased and led to greater soil losses during the second rainfall.

In addition to the four tillage treatments, we simulated rainfall on no-till plots with the surface residue removed. The objective of this treatment was to examine the effect of surface residue under no-till soil surface conditions. The effect of surface residue was clearly evident (Table 1 and Figure 1). Greater soil loss occurred without the residue. Lacking residue to absorb the overland flow and impact energy of

Table 1. Average runoff volume, sediment concentration, and soil loss produced by two erosive rainfalls on first-year alfalfa.

Tillage Method and Rainfall Event	Runoff Volume (gallons)	Sediment Concentration (ppm)	Soil Loss (tons/acre)
First rainfall			
Direct seeding	4.14	2,380	0.22
Nurse crop	4.13	260	0.03
Chisel	2.93	1,610	0.11
No-till	1.71	1,220	0.05
No-till, no residue	5.80	2,840	0.46
Second rainfall			
Direct seeding	7.03	2,630	0.42
Nurse crop	4.49	1,370	0.13
Chisel	4.50	2,150	0.24
No-till	2.84	1,450	0.12
No-till, no residue	8.39	3,160	0.61

an erosive storm, the energies for detachment and transport of soil particles increased. These factors, combined with the reduced infiltration capacity of a no-till soil surface, led to greater soil loss.

In addition to soil loss data, we analyzed the sediment in the runoff for aggregate and particle size distribution. This aspect of the study is incomplete; thus, no data are presented.

Figure 2 shows alfalfa yield data from three cuttings in the 1986 growing season. There was no significant difference in the yield data for the direct-seeded, chisel, and no-till treatments. Alfalfa yields on the nurse crop treatment were significantly lower due to oat competition, which reduced yield in the first cutting. The 1986 data were similar to data collected at the site in 1984 and 1985. Data from these years showed no significant yield differences in the yields of unreplicated, direct-seeded, chisel, and no-till alfalfa demonstration plots.

Conclusions

The first year of this study has shown that under the appropriate conditions farmers can use conservation tillage methods to establish alfalfa and protect the soil from erosion

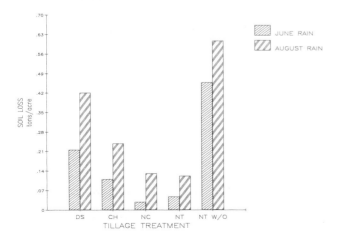

Figure 1. The effect of four tillage treatments on soil loss produced by two erosive rainfalls on first-year alfalfa.

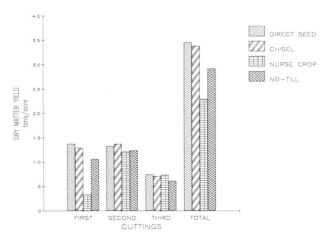

Figure 2. The effect of four tillage treatments on first-year alfalfa yields.

without reducing yields. In this study no-till methods offered the best soil protection without lowering yields. Chisel plowing offers less soil protection than no-till, but significantly more than direct seeding.

Using no-till methods, as well as other forms of conservation tillage, to establish alfalfa is a relatively new idea. As such, advances in the technology associated with no-till are needed before its broad-scale implementation. One of the more pressing advances required is the development and labeling of herbicides that can safely function in the no-till environment. In addition, mechanical and economical advances in the machinery associated with these methods are required. The more compacted and residue-covered soil associated with conservation tillage requires specialized machinery to insure good seed-soil contact. A limited number of drills are available, and those that function well are expensive. Along with herbicide and mechanical advances, growers will have to increase their planning and management of the system. For example, carryover of herbicides may be more of a problem under a no-till system than under traditional methods.

Restoration of eroded Southern Piedmont land in conservation tillage systems

G. W. Langdale, R. R. Bruce, and
A. W. Thomas

Soil water stress is a critical plant growth controlling factor on eroded Southern Piedmont lands (7). Researchers used perennial legumes to control runoff and accompanying soil losses for decades on these landscapes (1, 6). Most of these legumes were associated with long-term cropping systems of low economic intensity. Research recently has demonstrated the value of cool-season annual legumes to provide mulches and biologically fixed N in conservation production systems (2, 3). Hydrologic importance of the latter production system also has been documented (5). Research suggests that cool-season legumes have the capacity to biologically fix N for summer annual nonleguminous crops and return large quantities of crop residue to the soil. Rainfall retention also increases as organic C and N levels are significantly improved by managing cool-season legumes on eroded soils in regions where rainfall far exceeds evapotranspiration.

Study methods

We established a plot study in 1982 on a Cecil soil (clayey, kaolinitic, thermic, Typic Halpudults) representing three levels of soil erosion. These soil erosion levels were severe (3-inch Ap with 21% clay), moderate (5-inch Ap with 11% clay), and slight (7-inch Ap with 66% clay), located in a nonrandomized spatial manner. Randomized and replicated tillage-cropping treatments on each soil erosion level included a double crop of crimson clover (*Trifolium incarnatum* L.-cv. Tibbee) followed by no-tilled grain sorghum (*Sorghum bicolor* L. Moench-cv. Dekalb BR64) and monocropped, conventionally tilled grain sorghum. The irrigation treatments were rainfed and irrigated when soil water tension at the 8-inch depth reached 0.3 bar. We applied N fertilizer variably among treatments in an attempt to eliminate N stress in the grain sorghum (Table 1). Lime, P, and K were differentially applied with respect to soil erosion level to eliminate soil test deficiencies. On the severely eroded site, dolomitic limestone requirements were threefold (3 tons/acre) and P fertilizer requirements were twofold (113 pounds/acre) those of other soil erosion sites.

Results and discussion

Planting grain sorghum into crimson clover residues increased grain sorghum yields on the moderately eroded site at each soil water level. Grain yields on all irrigation treatments

G. W. Langdale and R. R. Bruce are soil scientists and A. W. Thomas is an agricultural engineer, Southern Piedmont Conservation Research Center, Agricultural Research Service, U.S. Department of Agriculture, Watkinsville, Georgia 30677.

Table 1. Effect of soil erosion, nitrogen supply, irrigation, and tillage on grain sorghum yields, 3-year average.

Winter Cover	Nitrogen Fertilizer (pounds/acre)	Water	Tillage	Soil Erosion Level			
				Slight	Moderate	Severe	Mean
				tons/acre			
Fallow	80	Rainfed	Conventional tillage	1.65	1.74	2.24	1.96
Crimson clover	0	Rainfed	No-till	1.76	2.55	1.76	2.04
Fallow	130	Irrigated	Conventional tillage	2.60	2.55	2.91	2.69
Crimson clover	50	Irrigated	No-till	2.46	2.69	2.76	2.63
Mean				2.13	2.38	2.41	
Standard error of mean				0.21	0.20	0.21	
Grain sorghum stover				2.64	2.91	2.94	
Crimson clover stover				2.52	3.09	2.60	

Table 2. Effect of soil erosion, nitrogen supply, irrigation, and tillage on percent organic carbon in the 0- to 0.4-inch soil depth, 1985.

Winter Cover	Nitrogen Fertilizer (pounds/acre)	Water	Tillage	Soil Erosion Level		
				Slight	Moderate	Severe
				%		
Fallow	80	Rainfed	Conventional tillage	0.60	1.10	1.07
Crimson clover	0	Rainfed	No-till	1.71	2.31	3.55
Fallow	130	Irrigated	Conventional tillage	0.65	1.18	1.29
Crimson clover	50	Irrigated	No-till	1.92	2.90	3.73
Standard error of mean				0.22	0.39	0.47
Original organic carbon, 0- to 3-inch depth				0.60	1.11	1.21
Percent clay, 0- to 3-inch depth				6.0	10.7	20.81

were more than 2.5 tons/acre at all levels of soil erosion. Lower levels of applied N fertilizer and biologically fixed N appeared to be inadequate on slight and severely eroded sites. All restoration treatment responses were more consistent on the moderately eroded site. Kamprath and associates (4) demonstrated that growth of cool-season legumes was greater on heavier textured soils. This is also reflected in stover yields of both grain sorghum and crimson clover (Table 1).

Sorghum yields on the severely eroded site suggested that crimson clover could not supply quantities of fixed N equal to 80 pounds/acre of applied N fertilizer during the first year on the slightly eroded site and the first 2 years on the severely eroded site. This is reflected in the 3-year yield means (Table 1). During the third year, grain sorghum response to tillage-cropping and N was not significant on the severely eroded site. Grain sorghum yields during the third year were 2.3 and 2.8 tons/acre for the rainfed and irrigated treatments, respectively, on the severely eroded site. These yields were nearly equal to or greater than yields on the moderately or slightly eroded site. Improved grain yields during the third year on the severely eroded site may be explained partially by accompanying increases in organic C values in the 0- to 0.4-inch soil depth (Table 2). These C values reflected the effect of crimson clover and grain sorghum residues associated with the treatment applied on each soil erosion level.

Conclusion

Although additional quantities of P fertilizer and dolomitic lime are required occasionally to restore productivity on eroded soils, the use of cool-season legumes in conservation tillage modes can improve rainfall infiltration, increase soil C, and restore crop yields. The time frame required to restore soil productivity on eroded Southern Piedmont soils may be a function of soil erosion severity and favorable climatic cycles.

REFERENCES

1. Carreker, J. R., S. R. Wilkinson, A. P. Barnett, and J. E. Box, Jr. 1977. *Soil and water management systems for sloping lands.* ARS-S-160. Agr. Res. Serv., U.S. Dept. Agr., Washington, D.C.
2. Hargrove, W. L. 1986. *Winter legumes as a nitrogen source for no-till grain sorghum.* Agron. J. 78: 70-74.
3. Hargrove, W. L., G. W. Langdale, and A. W. Thomas. 1984. *Role of legume cover crops in conservation tillage production systems.* Paper 84-2038. Am. Soc. Agr. Eng., St. Joseph, Mich.
4. Kamprath, E. J., W. V. Chandler, and B. A. Krantz. 1958. *Winter cover crops - Their effects on corn yields and soil properties.* Tech. Bul. No. 129. N. Car. Agr. Exp. Sta., Raleigh.
5. Mills, W. C., A. W. Thomas, and G. W. Langdale. 1986. *Estimating soil loss probabilities for Southern Piedmont cropping-tillage systems.* Trans., ASAE 29(4): 948-955.
6. Pieters, H. J., P. R. Henson, W. E. Adams, and A. P. Barnett. 1950. *Sericea and other perennial lespedeza for forage and soil conservation.* Circ. No. 863. U.S. Dept. Agr., Washington, D.C.
7. White, Jr., A. W., R. R. Bruce, A. W. Thomas, G. W. Langdale, and H. F. Perkins. 1985. *Characterizing productivity of eroded soils in the southern Piedmont.* In C. K. Mutchler [ed.] *Erosion and Soil Productivity.* Publ. 8-85. Am. Soc. Agr. Eng., St. Joseph, Mich. pp. 83-95.

ECONOMICS

Economics of using legumes as a nitrogen source in conservation tillage systems

John R. Allison and Stephen L. Ott

Using legumes to provide N for other crops is not a new idea. Ancient Greeks and Chinese used legumes to improve subsequent crop yields (13). In the United States the use of legumes as a green manure gained in popularity during the early 1900s. After World War II, production of synthetic N from natural gas led to relatively inexpensive N fertilizer. Farmers switched from using legumes to applying N fertilizer. A renewed interest in using legumes for N occurred after the 1974 energy crisis. Nitrogen fertilizer prices more than doubled from 1973 to 1975 and failed to return to their pre-1974 level. During this period of increased N fertilizer cost, farmers also have been adopting conservation tillage practices. Thus, it is only natural that the interest in using legumes in conservation tillage systems should increase.

The future role of legume N in conservation tillage systems depends upon its realized economic benefits and costs. To focus on the different factors affecting profitability, we have selected agronomic data that provide insights on the economic desirability of using legumes as a N source.

Use of legumes as a N source can be divided into two categories. A legume can be a full-season crop in a rotation or used as a winter cover crop. When used as a full-season crop, a legume can provide livestock feed as well as soil N. Its seed may also be harvested. The primary functions of legumes used as cover crops are N production and soil erosion control.

Legumes as cover crops

With any discussion of the economics of legumes as cover crops, one must also look at the economics of having any type of cover crop. Cover crops can improve grain crop yields by reducing soil erosion, conserving water, recycling nutrients, and improving soil physical characteristics (13). Unfortunately, the research results concerning these benefits are mixed. Grass cover crops resulted in lower yields of corn in Georgia[1], North Carolina (17), and Tennessee (1); sorghum in south Georgia[1]; and cotton in Tennessee (1). Corn yield in Kentucky (6) and sorghum yield in north Georgia[1] were greater following a grass cover crop than following no cover crop. Thus, the benefits of using a grass cover crop in improving yields are not always certain. Unfortunately, cover crop establishment costs occur every year. If the value of the increased crop yield does not cover the establishment cost, then using a grass cover crop reduces profits.

The high cost of establishing grass cover crops and increasing N prices combined to generate research interest in using legume cover crops. Legume cover crops provide the same soil benefits of grass cover, plus they provide N. Some researchers hypothesized that the N produced would pay for the legume establishment cost (13).

Economics of using legume cover crops. The economics of using legume cover crops depends upon their relative profitability to grass cover crops and to no cover crop systems. Profitability is determined by two factors: costs and revenues.

Two obvious costs to compare are legume establishment and N fertilizer expenses. Legume establishment costs include seed, any fertilizer for the cover crop, labor, fuel, and machinery. With a seeding rate of 22 pounds/acre and a price of 98 cents/pound, crimson clover seed costs $21.56/acre. Planting costs add at least another $5/acre. Interest charges increase these establishment costs because legume planting

---

John R. Allison is a professor and Stephen L. Ott is an assistant professor, Department of Agricultural Economics, Georgia Experiment Station, University of Georgia, Experiment, 30212.

[1]Unpublished data, William Hargrove, Department of Agronomy, University of Georgia, Experiment.

occurs 4 to 6 months before N application. Thus, legume establishment costs would purchase about 125 pounds of N fertilizer based on a fall 1986 price of 23 cents/pound. This quantity is greater than the quantity of N supplied by legume cover crops to succeeding grain crops in Kentucky (7) and Georgia (8). However, in North Carolina (17) legumes have replaced up to 180 pounds/acre N fertilizer in corn production. Summer legume cover crops in Florida (15) have supplied N in excess of 200 pounds/acre for use by winter grass crops. Thus, in comparing legume cover crop establishment costs against the value of N fertilizer saved, neither N source dominates the other. In addition, it appears that the economic desirability of using legume cover crops depends upon geographic location.

There can be other costs associated with using legume cover crops—opportunity costs. An opportunity cost is the value of something that is sacrificed to produce the desired product. One possible opportunity cost for legume cover crops is reduced grain crop yield if spring planting is delayed. Delayed planting occurs when a producer allows the legume more time to fix N. In Pennsylvania Cramer (2) found that allowing a legume cover crop an additional 3 weeks growth resulted in more N produced for corn. This additional N did not, however, compensate for lower grain yields due to the delay in corn planting date. As a consequence, profits were lower for the late-planted corn. Thus, there is a trade-off between producing N from legumes and grain crop yield.

The time required to plant the legume cover crop is another opportunity cost. To allow the winter legume cover crop adequate growing time before cold temperatures set in, farmers may have to harvest the preceding grain crop earlier or speed up that harvest. Speeding up harvest incurs higher fixed charges, as combine size or number increase. Earlier harvesting may reduce crop yields. Also, fall planting of legume cover crops may take labor away from other farm activities, such as harvesting other crops. Delaying the harvest of other crops can reduce yields, especially during a wet fall. All these opportunity costs work against using legume cover crops.

On the benefit side, the advantages of using a legume cover crop include potential grain yield increases from improved soil tilth, decreased soil erosion, and reduced fossil energy consumption. Corn yields in Alabama's sand mountain region (16), North Carolina (17), and Tennessee (1) were greater following a legume cover crop than following no cover crop for a given N rate. Yield benefits are not guaranteed, however. Corn yields in Alabama's coastal plain region (16) and cotton yields in Tennessee (1) were lower for a given N rate when planted into a legume cover crop compared to planting into no cover crop.

A legume cover crop protects the soil until there is enough crop canopy from the grain crop. This reduced soil erosion benefit may be overrated. Researchers in Georgia and Washington found that on a per-acre basis each ton of soil saved from erosion had a discounted value of less than $1 (3). The benefits of reducing soil erosion usually are not immediate but accrue sometime in the future. Discounting to the present reduces this benefit to its low value.

The quantity of fossil energy required to produce N fertilizers is more than that required to produce legume cover crop seed. As a result, using a legume cover crop reduces the quantity of fossil energy needed to produce a bushel of grain crop (9, 14). Like reduced soil erosion, this value may be overstated. Because the value of reduced N fertilizer already has been included in the analysis, any economic benefit of reduced energy savings is based on externalities. Externalities are any costs not incorporated in the market price and thus have little impact on an individual decision-maker.

Gross revenue minus costs determines profit. The value of using a legume cover crop is in the enhanced profit it provides. To correctly determine the value of using legume cover crops, all cover crop systems need evaluation at their maximum profit. Determining maximum profit is a complex process, however. With profit maximization, resource use expands until its benefit no longer exceeds its cost. Each system is then at maximum profit when all inputs are at their optimal use. Every cover crop can have its own grain crop yield response to N fertilizer and its own optimal quantity of N fertilizer use. Optimal quantities of N fertilizer are determined by using N-grain yield response functions and prices of grain and N. For some legume cover crops, the grain yield response function may not be significant. Herein, we use the average grain yield with no N fertilizer. Unfortunately, many published reports do not include such response or production functions. Instead they include results for a few specific N levels. This forces analysts to choose a level that may not be optimum.

To properly evaluate maximum profits requires knowing optimal legume seeding rates in addition to optimal N fertilizer rates. There is little research to date on the grain crop yield effects of various legume cover seeding rates. Until such data are available, the single seeding rate used in experiments will have to suffice.

What has research shown on the relative profitability of using legume cover crops? There have been few economic studies. Frye and associates (7) found that using crimson clover and hairy vetch as cover crops increased monocropped corn yields compared to yields with no cover crop at each tested N fertilizer rate. The increase in corn yield was more than enough to pay for the hairy vetch but not enough to pay for the crimson clover. They also found that using hairy vetch as a cover crop increased the corn yield from the previous year. There was no such increase for the other cover crops. However, they did not attempt to calculate the value of this benefit.

The limitation of this study was the low N rates that were analyzed. The maximum N rate was 90 pounds/acre. Economically optimum N rates were probably greater than 90 pounds/acre. Thus, the no-cover crop production system was disadvantaged. Frye and associates, however, did project the profitability of using 135 pounds/acre N and found that hairy vetch still was the best option over no cover crop. Thus, the economics of using hairy vetch as a legume cover are promising in Kentucky while those of crimson clover are not.

We have analyzed the profitability of using crimson clover as a legume cover crop for monocropped grain sorghum production in both northern and southern Georgia. We have developed production functions for legume, grass (wheat or rye), and no-cover crop systems. Using optimal N fertilizer rates, grain sorghum yields were lower with the legume cover crop in both areas of the state. Consequently, crimson clover was less profitable than either the no-cover crop or grass cover crop production systems. Thus, in Georgia the profit prospects of using crimson clover as a legume cover crop in grain sorghum production are not encouraging.

Doering (5) completed a whole-farm economic analysis of using a legume cover crop. Simulating use of hairy vetch as a winter cover crop on a corn-soybean farm in Indiana, he sought to find more energy efficient methods of producing corn and soybeans. Using hairy vetch as a winter legume cover crop reduced the quantity of N fertilizer required to produce corn. Therefore, energy use was lower than when a legume cover crop was used. However, using a legume cover crop was not as profitable as using N fertilizer in a standard corn-soybean rotation.

Profit risks associated with legume cover crops. Comparing average relative profits is not the only way to evaluate different cover crop systems. Profit riskiness can affect the desirability of using a particular cover crop. A risk-adverse producer will accept increased risks only if compensated by increases in expected (average) profits. However, as risk increases, the additional expected profit necessary to compensate for this risk increases at an increasing rate. Thus, if legume cover crop systems are riskier than the other cover crop systems, then expected profits must be substantially greater.

The risks associated with legume cover crops are many. Legume cover crops reduce soil moisture in the spring (6). If it is dry during spring planting, there may not be enough moisture to adequately germinate the planted grain crop. The grower may have to replant the grain crop at an additional expense to avoid reduced plant stands, which lower the yield potential. Dabney and associates (4) reported such problems in Louisiana.

Another risk of using legume cover crops is winter kill. Low winter temperatures can kill legume cover crops. Crimson clover winter kills at temperatures of 5°F or lower. At Experiment, Georgia, temperatures have fallen below 5°F in one out of six winters during the past 20 years. If legumes winter kill, growers must apply a N fertilizer the next spring at an additional cost.

Yield variability is another risk that farmers face. Based on sorghum research plot data in Georgia, grain yield variance in any one year was greater following a legume cover crop than following no cover crop. However, over a 5-year period, sorghum yields did not vary significantly following the two types of cover crops. This suggests that a farmer may experience more yield variations within and among individual fields when using a legume cover crop than when using no cover crop. Over time, however, the variance in total farm grain yield would be independent of cover crop used.

Legume cover crops also can reduce yield risk. After the grain crop starts growing, a legume cover crop may minimize the impact of a dry growing season. Frye and associates (7) found this to be true in Kentucky.

Using simulation and stochastic dominance theory, Ott and Hargrove analyzed the impact of risk using grain sorghum

data from northern Georgia[2]. They found that without any risks, N fertilizer price must increase by 30% for the legume cover crop to be as profitable as no cover crop. The impact of a greater yield variance from using a legume cover crop was to raise the amount by which N prices must increase by an additional 4 percentage points. The risk of winter kill raised the break-even N price even further, to an increase of 65%.

Thus, the economic desirability of using legume cover crops is not great and varies from location to location. At current N prices the savings from using legumes is not enough to warrant their use. Legume cover crops are profitable only when they provide a significant increase in grain crop yield. To enhance the profitability of using legume cover crops, researchers need to focus on reducing establishment costs and improving the legume cover crop N yield.

Full-season legumes in crop rotation

The economic return of legumes in a crop rotation is a function of the yield interactions between the legume and the succeeding crops in the rotation, the costs incurred for the legume crop, and the price of N. Complicating any such evaluation are variances in the relationships caused by climatic, insect, and disease factors. The benefits of using a legume derive from the value of harvested forage and/or seed plus reduced per-unit cost of output of the following crops realized from the presence of the legume in the rotation. The potential reduction in per-unit output costs occurs from reduced input use and/or from greater production per acre. Legumes in a rotation can replace fertilizer N with biological N. Legumes also may stimulate yields of succeeding crops by increasing soil organic matter and improving soil tilth and/or structure.

The cost of establishing a legume is a straight-forward cost, and although it is very important it may not be the most critical cost. Unless the legume replaces fallow in the rotation, a farmer will lose income from the crop replaced by the legume. This lost income or opportunity cost is critical in determining the profitability of legumes in a crop rotation. It is a function of the price and yield of the crop replaced and the price and yield of the legume crop.

The value of the legume forage may or may not be easy to determine. If the grower produces forage for sale, then the market price is the appropriate value. If the forage is intended for feed, then its value is the lower of the price of farm-delivered equivalent forage or the market value of the feed ingredients replaced by the forage.

Influences on succeeding crops in the rotation also may be hard to estimate. If all the benefits occur in the following production season, then a farmer can document and quantify changes in output. However, changes in soil tilth, delayed release of plant nutrients, and/or reduced soil loss may affect crop yields only marginally in any one year. Thus, a farmer may not be able to observe and quantify output modifications in a single season.

[2]S. L. Ott and W. L. Hargrove. 1987. "Profits and Risks Associated with Using Crimson Clover as a Nitrogen Fertilizer Substitute in Grain Sorghum Production." Paper presented at Southern Agricultural Economics Association annual meeting, Nashville, Tennessee.

The benefits of using legumes in rotation also can be a function of the environment. Single-year observations of benefits, which are a function of the environment, can overstate or understate the benefits. If responses vary because of environmental factors or biological reactions, then measures of variation become important and should be incorporated in any economic analysis.

Nitrogen response relationships need to be estimated so the farmer can compare the legume rotation and nonlegume rotation at economic optimum. This is particularly critical if the crop following the legume can use more N than that supplied by the legume.

If the legume crop itself has a value, the economic optimal rotation composition is a function of the relative price and yield of the legume crop, the price and yield of the crop replaced by the legume, and the reduction in per-unit output cost of succeeding crops in the rotation.

There is some published research on the economics of legumes in crop rotations, but only a few of these studies dealt with conservation tillage practices. Fortunately, many agronomic aspects of legume crops in rotation are similar under conventional and conservation tillage. The major differences occur in the grain stand response to planting in conventionally prepared seed beds after legumes versus planting in a legume mulch or stubble of legumes killed by herbicides. Therefore, for this analysis we felt that any potential error caused by the difference in stand response was not great enough to exclude research that used conventional tillage.

As in many areas of research, researchers have reached different conclusions on the role of legumes in crop rotation. The reasons behind these different conclusions are as important or more important than the actual conclusions of many studies. In Pennsylvania Lazarus and associates (12) compared legume and nonlegume crop rotations under conventional tillage and no-till. They used linear programming to optimize farm organization on productive soils. Their analysis assumed a non-N yield response of 14 bushels/acre for corn following 2 or more years of alfalfa. With this assumption, the alfalfa-corn rotation was more profitable than continuous corn when corn occupied 40% of the rotation. When they reduced the alfalfa-corn rotation to 33% corn, the rotation was less profitable than continuous corn. Feeding the hay and grain to dairy cattle and comparing differences in feed costs did not change this relationship. Results for no-till systems were similar to results to those for conventional tillage. Doering (5) compared one grain-alfalfa rotation with three continuous grain rotations. The alfalfa grain rotation was optimum only under very high alfalfa prices.

Hesterman and associates (11) compared the economics of alfalfa-corn and continuous grain rotations using conventional tillage practices. A major advantage of their study was that it was based on yields obtained from agronomic research (10) that measured yields under several N regimes for continuous corn and alfalfa-corn rotations. They found that the alfalfa-corn rotation was much more profitable than continuous grain at all locations (11). They concluded that the forage value of the alfalfa crop influenced profitability of the alfalfa-corn rotation more than the value of the N contribution. They performed price sensitivity analysis by using a fixed price relationship between corn and alfalfa and parameterizing the price of N. They used a value of alfalfa that was a composite price of a per-unit value of crude protein and per-unit value of total digestible nitrogen. The composite alfalfa price used was relatively high compared to the corn price. Using their alfalfa price, a continuous alfalfa rotation would have been more profitable than the alfalfa-corn and continuous corn rotations.

Break-even analysis of using legumes in crop rotations. Two studies by Hesterman and associates (*10, 11*) are important in terms of the economics of using legumes in rotation in conservation tillage. Their excellent N response functions and cost data permit developing an estimate of the corn, alfalfa, and N price relationships for determining under which price ratio alfalfa-corn and corn-corn rotations are optimum. From these relationships, break-even prices of alfalfa, corn, and N can be estimated and expressed as a function of the other two prices. We performed this break-even analysis for alfalfa-corn and corn-corn rotations using Hesterman and associates data from two geographic locations, Becker, and Lamberton, Minnesota. We used an average yield of the two alfalfa varieties.

We used the corn N response functions and cash operating costs from Becker to formulate gross margin functions for each rotation. We expressed the gross margin functions in terms of alfalfa, corn, and N prices. We then determined equilibrium price functions for the break-even relationships between the rotations by equating the gross margin equations and solving for the corn price as a function of alfalfa and N prices.

Rather than continue with three dimensional analyses, we used two prices of N—\$.23/pound N, representing December 1986 prices, and \$.35/pound N, a 50% increase. Figure 1 shows the plots of the corn and alfalfa break-even prices for the corn-corn and alfalfa-corn rotations by the two N prices.

Alfalfa prices in the range of the current market prices are high enough to make the alfalfa-corn rotation profitable, using the December 1986 market price of corn in Minnesota (\$1.39/bushel). Adding the government deficiency payments of \$1.21/bushel, the corn-corn rotation is more profitable. The 50% increase in N price had relatively little influence,

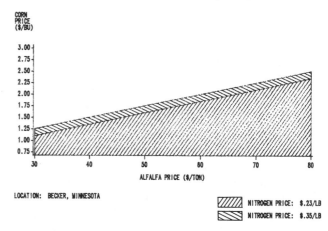

Figure 1. Price map of alfalfa in rotation to supply nitrogen, Becker, Minnesota.

which is consistent with Hesterman's analysis.

We used the Lamberton data to determine relationships between legumes and succeeding crops on less N-responsive soils and to demonstrate a variation in response. First-year corn had a yield response only up to 50 pounds N, and yields were comparable to the Becker area. The growing season during the second season was dry; thus, there was no significant corn yield response to N for either corn-corn or corn-alfalfa. The average corn yield over all N levels for all rotations was 34% lower than the first-year corn yield.

For the yield variation analysis, we made assumptions about N response in a normal year. We assumed that there would not be a N yield response for corn following alfalfa in a normal year and that the yield would equal that of the corn in the first year of the study with 50 pounds N. With these assumptions, we used the same procedure as for the Becker area to develop break-even corn and alfalfa prices for two good years (Figure 2) and for one good year followed by a poor year (Figure 3).

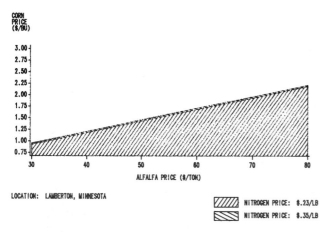

Figure 2. Price map of alfalfa in rotation to supply nitrogen in two good years, Lamberton, Minnesota.

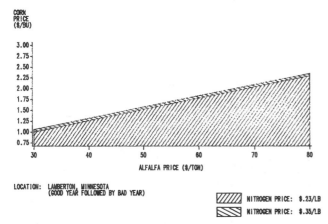

Figure 3. Price map of alfalfa in rotation to supply nitrogen in a good year followed by bad year, Lamberton, Minnesota.

These relationships were similar to the Becker relationship (Figure 1). The difference: the alfalfa-corn rotation was slightly less profitable and the influence of N prices was almost negligible in a good-yield year and only slightly more influential in a poor-yield year. Thus, with current N prices, the value of the forage crop is more critical than the price of N.

Relative riskiness is important in determining the desirability of using legumes in crop rotations. However, unless the legume in the rotation is less sensitive to poor growing conditions than is the grain crop, adding a legume to the rotation may not reduce risks. Several years of research data are needed to develop risk measures of a legume-grain crop rotation.

No studies, including our analysis, have gone beyond merely mentioning potential long-term benefits of legumes in grain rotations. Researchers need to conduct long-term studies to demonstrate if such benefits are large enough to influence economic decisions about legumes in grain rotations. Researchers also may need to determine grain crop yield effects of seeding into legume sod versus grain crop stubble. Farmers also have to know if conservation tillage systems can provide suitable environments—soil pH, weed control, etc.—for legume crop establishment in a crop rotation.

Conclusions

Agronomically, there is no question that legumes in either rotation or as a cover crop can improve crop yields and/or reduce the quantity of N fertilizer required. However, is it economical to substitute legumes for N fertilizer? The answer to that question is not clear at this point. The economics of such a substitution is a function of a few key parameters. As a cover crop, the most important parameters are the increase in succeeding crop yield and legume establishment cost. Legume cover crops are profitable if they increase the yield of the succeeding crop sufficiently over that of using no cover crop. The economic desirability of legumes decreases when they are valued primarily as a N substitute.

When legumes are part of a crop rotation, the most important parameters are the prices of the legume forage and competing crops and any yield increases attributed to legumes in rotation. When forage prices are high relative to other crop prices, legumes are profitable.

Changes in the price of N fertilizer have only marginal effects on the profitability of legumes used as either cover crops or in rotations. This is because N fertilizer is very cost-effective. Nitrogen fertilizer prices would have to increase substantially for legume cover crops to become cost-effective N sources. The N price that makes biological N cost-effective in a rotation is a function of the legume crop and grain crop values. Therefore, legumes as a N source will have limited application until large N fertilizer price increases occur or the cost efficiency of growing legumes improves.

REFERENCES

1. Buntley, G. J. 1986. *Tennessee no-till update*. In Proc., S. Region No-Tillage Conf. Ky. Agr. Exp. Sta., Lexington.

2. Cramer, Graig. 1986. *Plow legumes early.* The New Farm (May/June): 26-27.
3. Christensen, L. A., and D. E. McElyea. 1986. *Economic implications of soil characteristics and productivity in the southern piedmont, with comparisons to the pacific northwest.* In *Erosion and Soil Productivity.* Am. Soc. Agr. Eng., St. Joseph, Mich. pp. 233-242.
4. Dabney, S. M., D. J. Boethel, D. J. Boquet, J. L. Griffith, W. B. Hallmark, R. L. Hutchinson, L. F. Mason, and J. L. Rabb. 1986. *Update of No-tillage in Louisiana.* In Proc. S. Region No-Tillage Conf. Ky. Agr. Exp. Sta., Lexington.
5. Doering, Otto C. III. 1977. *An energy based analysis of alternative production methods and cropping systems in the corn belt.* Nat. Sci. Found., Res. Applications Rpt. 770125. Ind. Agr. Exp. Sta., Purdue Univ., West Lafayette. 44 pp.
6. Frye, W. W. 1986. *Kentucky no-tillage update.* In Proc., S. Region No-Tillage Conf., Ky. Agr. Exp. Sta., Lexington.
7. Frye, W. W., W. G. Smith, and R. J. Williams. 1985. *Economics of winter cover crops as a source of nitrogen for no-till corn.* J. Soil and Water Cons. 40(2): 248-249.
8. Hargrove, W. L. 1986. *Winter legumes as a nitrogen source for no-till grain sorghum.* Agron. J. 78(1): 70-74.
9. Hargrove, W. L., and G. W. Langdale. 1986. *No-tillage research in Georgia.* In Proc., S. Region No-Tillage Conf., Ky. Agr. Exp. Sta., Lexington.
10. Hesterman, O. B., C. C. Sheaffer, D. K. Barnes, W. E. Lueschen, and J. H. Ford. 1986. *Alfalfa dry matter and nitrogen production, and fertilizer nitrogen response in legume-corn rotation.* Agron. J. 78(1): 19-23.
11. Hesterman, O. B., C. C. Sheaffer, and E. I. Fuller. 1986. *Economic comparisons of crop rotations including alfalfa, soybean, and corn.* Agron. J. 78(1): 24-28.
12. Lazarus, W. F., L. D. Hoffman, E. J. Partenheimer. 1980. *Economic comparisons of selected cropping systems on Pennsylvania cash crop and dairy farms with highly productive land.* Bull. 828, Pa. Agr. Exp. Sta., Pa. St. Univ., University Park. 19 pp.
13. Martin, G. W., and J. T. Touchton. 1983. *Legumes as cover crop and source of nitrogen.* J. of Soil and Water Cons. 38(3): 214-216.
14. Ott, S. L. 1986. *Reducing energy inputs in crop production by supplying nitrogen from legume cover crops: An economic case study.* In Proc., Sixth Ann. Solar, Biomass and Wind Energy Wrkshp. Office of Energy, U.S. Dept. Agr., Washington, D.C.
15. Reddy, K. C., A. R. Soffes, and G. M. Prine. 1986. *Tropical legumes for green manure. I. Nitrogen production and the effects on succeeding crop yields.* Agron. J. 78(1): 1-4.
16. Reeves, D. W., D. H. Rickerl, C. B. Elkins, and J. T. Touchton. 1986. *No-tillage update report - Alabama.* In Proc., S. Region No-Tillage Conf. Ky. Agr. Exp. Sta., Lexington.
17. Worsham, A. D. 1986. *No-tillage research update - North Carolina.* In Proc., S. Region No-Tillage Conf., Ky. Agr. Exp. Sta., Lexington.

An economic and energy analysis of crimson clover as a nitrogen fertilizer substitute in grain sorghum production

Stephen L. Ott

It is possible to produce no-till sorghum using a winter legume cover crop as the only source of N (2). To test the economics of substituting a legume cover crop for N fertilizer, I compared the net returns of three types of winter cover crops at two different locations and for various prices of sorghum and N.

Study methods

Cover crops analyzed included crimson clover (a legume), rye or wheat (grasses), and stubble (no planted cover crop). The locations, the Piedmont and the Upper Coastal Plain of Georgia, represent two major production areas in the Southeast. (The Piedmont and Upper Coastal Plain will be referred to as north and south Georgia, respectively.)

I used two sorghum and two N fertilizer prices to test the influence of prices. This provided four different price scenarios. The federal government's effective loan rate of $1.74/bushel and support price of $2.88/bushel were the two sorghum prices. The two N prices selected were the current price of $.26/pound and a 50% increase to $.39/pound.

In addition to testing the economic efficiency of using legume cover crops, I also calculated the energy efficiency. For each combination of cover crop, location, and price, I determined the direct energy used in producing sorghum, expressing it on a per bushel basis.

I used 3 years of yield data from north Georgia and 2 years of data from south Georgia to develop quadratic N response functions for each cover crop. An indicator variable accounted for any differences in productivity between the two locations. The response function was not significant for crimson clover. I then used the average yield for crimson clover for each location. For the other two cover crops, I calculated economic optimum quantities of N for each price scenario. Using a farm unit as the model, I determined total operating costs and net return to fixed resources with a budget generator.

Results and conclusions

Table 1 shows results of my analysis. Sorghum yields in southern Georgia for all cover crops were greater than those in north Georgia. In north Georgia sorghum yields using a grass cover crop were higher than yields using stubble. But this extra yield did not pay the expense of planting a grass cover crop. In south Georgia sorghum yields with a grass cover crop were lower than yields with stubble. Using a

Stephen L. Ott is an assistant professor, Department of Agricultural Economics, Georgia Experiment Station, University of Georgia, Experiment, 30212.

Table 1. Grain sorghum yields, gross revenue, operating costs, net returns, and direct energy for three types of no-till cover crops and four price scenarios in Georgia.

Cover Crop and Location in Georgia	Yield* (bushels/acre)	Gross Revenue ($/acre)	Operating Costs ($/acre)†	Net Returns ($/acre)‡	Direct Energy (btu/bushels)§
Price scenario I: sorghum, $1.74/bushel; N, $.26/pound					
Stubble					
North	64	111.36	86.85	24.51	131,000
South	89	154.86	99.53	55.33	100,400
Crimson clover					
North	62	107.88	90.46	17.42	116,300
South	82	142.68	103.14	39.54	88,000
Grass‖					
North	66	114.84	98.76	16.08	145,600
South	87	151.38	111.44	39.94	116,800
Price scenario II: sorghum, $1.74/bushel; N, $.39/pound					
Stubble					
North	60	104.40	87.53	16.87	123,100
South	85	147.90	100.21	47.69	93,400
Crimson clover					
North	62	107.88	90.46	17.42	116,300
South	82	142.68	103.14	39.54	88,000
Grass‖					
North	63	109.62	103.79	5.83	143,600
South	84	146.16	116.47	29.69	114,300
Price scenario III: sorghum, $2.88/bushel; N, $.26/pound					
Stubble					
North	66	190.08	92.02	98.06	139,500
South	91	262.08	104.70	157.38	107,300
Crimson clover					
North	62	178.56	90.46	88.10	116,300
South	82	236.16	103.14	133.02	88,000
Grass‖					
North	67	192.96	101.48	91.48	149,900
South	88	253.44	114.16	139.28	120,400
Price scenario IV: sorghum, $2.88/bushel; N, $.39/pound					
Stubble					
North	64	184.32	98.95	85.37	134,400
South	90	259.20	111.63	147.57	101,700
Crimson clover					
North	62	178.56	90.46	88.10	116,300
South	82	236.16	103.14	133.02	88,000
Grass‖					
North	66	190.08	109.91	80.17	146,900
South	87	250.56	122.59	127.97	117,800

*Economic optimum for each price scenario.
†Variable inputs including labor and interest on operating capital.
‡Return to fixed resources of land and machinery.
§Energy invested in variable inputs.
‖Rye in north Georgia, wheat in south Georgia.

cover crop is justified based on reduced soil erosion. However, 1 ton of soil in the Georgia Piedmont has a value of less than $1.00 (*1*). Thus, to justify planting crimson clover in the Piedmont, it must reduce soil losses by at least 7 additional tons of soil/acre over stubble.

REFERENCES
1. Christensen, Lee A., and David E. McElyea. 1985. *Economic implications of soil characteristics and productivity in the southern piedmont, with comparisons to the pacific northwest.* In *Erosion and Soil Productivity.* Am. Soc. Agr. Eng., St. Joseph, Mich. pp. 233-242.
2. Hargrove, W. L. 1986. *Winter legumes as a nitrogen source for no-till grain sorghum.* Agron. J. 78(1): 70-74.

crimson clover cover crop at both locations generally resulted in lower yields than the two other cover crops. At current N prices net returns for crimson clover were lower than those of stubble and grass. The 50% increase in N price was not enough to make crimson clover net returns higher than net returns for stubble. But the N price increase did make crimson clover more profitable than grass. Overall, crimson clover was more energy efficient than stubble, which was more energy efficient than grass.

At the present time it appears that planting a cover crop does not pay. For crimson clover to be profitable, seeding cost must decline by one-half or yields must increase 6% in north Georgia and 11% in south Georgia. Sometimes a

Economics of legume cover crops in corn production

W. Donald Shurley

The Food Security Act of 1985 contains provisions to reduce soil erosion on cropland. The conservation compliance provision requires farmers to develop approved conservation plans if they wish to produce crops on highly erodible land. Failure to do so will disqualify the farmer from certain U.S. Department of Agriculture programs.

This provision could increase interest in no-till production methods. Producers using conventional methods may need only to switch to no-till methods to meet conservation compliance limits on tolerance erosion.

The results from an experiment on the effects of winter cover crops on no-till corn yields may offer these farmers some alternatives. The experiment was conducted at the University of Kentucky Research and Education Center farm at Princeton between 1980-1986. One objective of the study was to determine if growing a legume cover crop during the winter using no-till practices could supply a portion of the N needed by the corn crop, thereby increasing profits through higher corn yields and/or lower production costs.

Study methods

The experiment was set up in 1979 on a moderately well-drained Zanesville silt loam soil. Winter cover crops included hairy vetch, bigflower vetch, and rye. Corn residues as a winter cover served as the check treatment. Cover crops were overseeded into the corn crop in mid-September. Corn was then planted directly into the cover crop in mid- to late-May. Before or during planting, a mixture of Paraquat, Bladex, and Lasso was used to kill the cover crop and control weeds.

From 1980 to 1983 each treatment received three levels of ammonium nitrate—0, 45, and 90 pounds/acre N. From 1984 to 1986 the rates were 0, 75, and 150 pounds/acre, respectively. Dry conditions existed in 1980, 1983, and 1986.

To evaluate relative profitability, I calculated per-acre net returns above operating costs for each treatment. The net return above operating costs is gross income (price times yield) minus direct cash expenses, including seed, fertilizer, chemicals, fuel and repairs, custom hire, equipment rental, and drying. I also included an interest charge on these expenses. Corn prices were the average prices received by Kentucky farmers during the period (1). I used the loan rate of $1.84/bushel for 1986, however. I estimated operating costs using University of Kentucky budgets (2, 3) and the USDA index of prices paid series (4).

Corn yields

Average corn yields were highest on the bigflower vetch and hairy vetch plots with added N. Between 1980 and 1983

W. Donald Shurley is an associate extension professor, Department of Agricultural Economics, University of Kentucky, Lexington, 40546.

yields on plots with hairy vetch and 90 pounds/acre N averaged 11 bushels/acre higher than yields on plots with 90 pounds/acre N and no legume cover. Yields on plots with bigflower vetch and 90 pounds/N increased an average of nearly 12 bushels/acre compared with corn residue and 90 pounds/acre N (Table 1).

During this same period, bigflower vetch and hairy vetch supplied the equivalent of over 50 pounds/acre N/year. Yields on plots with corn residue and 45 pounds/acre N averaged 33 bushels/acre. Plots with hairy vetch and no supplemental N yielded 20 bushels more (53.3 bushels/acre).

Between 1984 and 1986, plots with bigflower vetch and hairy vetch with no supplemental N yielded only 10 to 15 bushels/acre less than those with corn residue and 75 pounds/acre N. Highest yields during the 1984-1986 period occurred on plots of hairy vetch and bigflower vetch with 150 pounds/acre N. Yields averaged 7 to 10 bushels/acre higher than yields on plots with corn residue. Over the entire study period, the hairy vetch and bigflower vetch treatments with the higher rates of supplemental N averaged

Table 1. Average corn yields by nitrogen treatment, Princeton, Kentucky, 1980-1986.

Years and N Rates	Corn Yields (bushels/acre) by Cover Crop			
	Corn Residue	Rye	Bigflower Vetch	Hairy Vetch
1980-1983				
No N	16.3	9.8	41.5	53.3
45 pounds/acre N	33.0	21.3	62.5	60.0
90 pounds/acre N	72.5	62.3	84.3	83.5
1984-1986				
No N	18.7	19.0	46.7	42.3
75 pounds/acre N	56.7	55.0	81.7	78.3
150 pounds/acre N	96.7	94.0	116.7	113.3
7-year average				
No N	17.3	13.7	43.7	48.6
Mid-range N	43.1	35.7	70.7	67.7
High-range N	82.7	75.7	99.0	96.3

Table 2. Per-acre net returns above operating costs by nitrogen treatment, Princeton, Kentucky, 1980-1986.

Years and N Rates	Net Returns ($/acre) by Cover Crop			
	Corn Residue	Rye	Bigflower Vetch	Hairy Vetch
1980-1983				
No N	−59	−100	−38	+7
45 pounds/acre	−34	−82	+3	+9
90 pounds/acre N	+61	+12	+56	+55
1984-1986				
No N	−69	−96	−60	−51
75 pounds/acre N	−4	−36	+1	+8
150 pounds/acre N	+65	+31	+65	+66
7-year average				
No N	−63	−98	−47	−18
Mid-range N	−17	−62	+2	+9
High-range N	+63	+20	+60	+60

about 15 bushels/acre more corn yield than corn residue plots with the same amount of N applied.

Net returns

Net returns above operating costs were the highest for the corn residue treatment with 90 pounds/acre N in the first 3-year period. Net returns on bigflower vetch and hairy vetch plots with 90 pounds/acre were $5/acre lower. From 1984 to 1986 bigflower vetch and hairy vetch plots with 150 pounds/acre N yielded net returns equivalent to those on the corn residue treatment (Table 2). For the 7-year period, net returns on the bigflower vetch and hairy vetch plots with the higher rates of N were slightly less than net returns on the residue plots.

However, during 1982 and 1985—when growing conditions were favorable—average net returns on the bigflower vetch and hairy vetch plots with the higher N rates averaged $22/acre/year higher than net returns on the corn residue treatments.

Conclusions

The results of this study suggests that legume cover crops can provide an economical source of N for no-till corn production in Kentucky. Bigflower vetch and hairy vetch can provide as much as 50 pounds/acre N. A combination of bigflower vetch or hairy vetch along with 150 pounds/acre N produced corn yields 7 to 10 bushels/acre higher than corn residue with the same amount of N.

Net returns using bigflower vetch and hairy vetch at the higher rates of N were about the same as the residue treatment even though 3 of the 7 years during the study were dry years. Even under these dry conditions, additional corn yields from the bigflower vetch and hairy vetch treatments over the 7-year period were sufficient to recover seeding costs.

A dollar value was not placed on soil savings due to the cover crops. Given that net returns were essentially the same over the same period, soil savings would result in added benefits in the long term.

REFERENCES
1. Kentucky Crop and Livestock Reporting Service. 1986. *Kentucky agricultural statistics, 1986-1886.* Frankfort, Ky.
2. Shurley, W. Donald, Larry D. Jones, David L. Debertin, and Charles L. Moore, Sr. 1983. *Estimated costs and returns for production of various crops and livestock in Kentucky during 1983.* Agr. Econ. Ext. No. 16. Univ. Ky., Lexington.
3. Shurley, W. Donald, and Richard L. Trimble. 1986. *Crop and forage budget estimates for Kentucky for 1986.* Agr. Econ. Ext. No. 55. Univ. Ky., Lexington.
4. United States Department of Agriculture. 1983. *Agricultural prices: Annual summary 1982.* Washington, D.C. pp. 100, 132-134.